The Complete Book of

MEAT

The Complete Book of
MEAT

Edited by Frank Gerrard and F. J. Mallion

Assistant Editor: Mabel Quin

VIRTUE: *London & Coulsdon*

British Library Cataloguing in Publication Data

The complete book of meat. – 2nd ed.
1. Meat
I. Gerrard, Frank   II. Mallion, F J
III. Quin, Mabel
641.36   TX373

ISBN 0–900778–17–2

Copyright © 1977 Virtue & Company Limited
25 Breakfield, Coulsdon, Surrey

This edition first published 1977
Second edition 1980

Printed in Great Britain by
Ebenezer Baylis & Son Ltd
The Trinity Press, Worcester

PUBLISHER'S PREFACE

Publication of THE COMPLETE BOOK OF MEAT represents a milestone in the meat industry. For many years the "farm gate" was a tangible barrier, restricting the producers' interest in the livestock, on the one hand, and the butchers' concern with the final produce—the carcase—on the other. It is now appreciated that efficient meat production, in terms of quality and economics, should represent an unbroken chain reaching from conception to consumption. The Common Agricultural Policy, the new and important slaughtering techniques and efficient refrigeration are all links in the chain and these are fully dealt with.

Frank Gerrard and F. J. Mallion are two names well-known throughout the meat trade. The former was previously Head of Department Smithfield College and is a Past President of The Institute of Meat. The latter is Head of Department of Food Commodities, College for the Distributive Trades, London and a Past President of The Institute of Meat. Both contribute to the Meat Trades Journal and are authors of standard works for the Meat Trade.

These two experts have combined the wealth of their knowledge to produce a unique volume. Under their editorship 20 different authors have contributed chapters on their own special subject, thus making this book a complete work on meat.

The farmer's interest in the joint should be complemented by the butcher's awareness of the complexities of animal production, and we are confident that THE COMPLETE BOOK OF MEAT will contribute to this better understanding within the trade. It will be an invaluable source of information to the farmer, the wholesaler, the retail butcher and to the student who aspires to a successful career in the meat trade.

Throughout this book metric and imperial measures are shown side by side where possible. It has not always been practical to translate them accurately but Conversion Tables will be found at the end of the book. Where costs are shown in the book they are only for the purpose of example. In these times of continual inflation it is obviously not possible to produce accurate or lasting costs.

We wish to thank Dr H. K. Baker and Mr G. Harrington of the Meat and Livestock Commission and Mr A. V. Pettit of The Institute of Meat for very considerable help and for the loan of many photographs and illustrations.

BIOGRAPHIES

Frank Gerrard, MBE, F Inst M, M Inst R, MRSH, Gold Medallist Worshipful Company of Butchers

Medallion Ecole Professionnelle de la Boucherie (Paris). Diploma Polish Government. Formerly Head of Department Smithfield College, London. Founder Member and Past President of the Institute of Meat. Consultant to various Government bodies throughout the world. Technical Editor *Meat Trades Journal*. Author of a number of standard books on Meat Technology and Processed Meats.

F. J. Mallion, FRSH, F Inst R, F Inst M, Gold Medallist Worshipful Company of Butchers

Head of Department of Food Commodities, College for the Distributive Trades, London. Past President of the Institute of Meat and Council Member of the Royal Smithfield Club.

A. R. P. Bardsley, BSc(Agric)

Technical Officer in charge of Pedigree cattle on-farm Recording Programme at Meat and Livestock Commission (MLC). Graduate of the University of Aberdeen. Previously Planting and Advisory work in Nigeria and in the field for Pig Industry Development Authority (PIDA).

Michael B. Barnes, Dip M, A Inst M, Cert Ed

Lecturer in Management and Accounts in the Smithfield Department of Food Technology of the College for the Distributive Trades, London. Assistant Examiner in Principles of Accounts 'O' level, University of London.

L. Croft, BSc(Agric)

Senior Pig Specialist Officer, Eastern Region at MLC. Graduate of Leeds University. Previously PIDA Field Officer, Lincolnshire.

D. Croston, BSc(Hons)

Technical Officer, Sheep. Development and operations of Sheep Pedigree Recording Scheme at MLC. Graduate of Newcastle University.

Alastair Cuthbertson, BSc, Dip Agric(Cantab)

Currently Head of Carcase and Meat Quality Section, MLC. Formerly Carcase Evaluation Officer, PIDA. Contributor to many papers published in scientific journals and in the farming and meat trade press.

John E. Downer, MEHA, Ronald Williams Silver Medallist

Lecturer at the Department of Food Commodities, College for the Distributive Trades, London; formerly Principal Environmental Health Officer (Food) London Borough of Islington and Divisional Health Officer (Food) Borough of Southend-on-Sea. Examiner for the Royal Society of Health's Certificate in the Hygiene of Food Retailing and Catering.

David Geary, BSc, F Inst R, F Inst M, MRSH, Gold Medallist Worshipful Company of Butchers

Practising Master Butcher. Former Editor of *Meat Marketing* and Managing Director of Messrs Payne and Sons, Butchers. Past President of Institute of Meat. Trading Standards Consultant to many local authorities.

B. W. Holt, M Inst M

Head of Meat Plant Advisory Service, MLC. Formerly Chief Engineer at IWEL Engineering Ltd. (meat plant construction and engineering). Member of the Royal Society of Health.

J. B. Kilkenny, BSc(Hons) (Dip Stat)

Systems and Production Officer in charge of Systems Development for beef cattle, sheep and pigs at MLC. Graduate of Leeds University. Previously with Beef Recording Association.

S. V. Lambert

After training with famous shopfitting firm became Works Manager. Adviser and designer for Shopfitting Division of Messrs Rushbrooke of Smithfield, London. Specialised in all spheres of the Retail Butchery Trade throughout the country.

A. P. Landon

Technical Officer Pigs, in charge of on-farm testing at MLC. Previously PIDA Field Officer.

W. H. E. Lewis, BSc(M Agric Sci)

Technical Officer in charge of Performance Testing Programme for beef bulls at MLC. Graduate of Reading University. Previously at Beef Recording Association.

R. Malton, M Inst R, Lightfoot Medallist

Senior Scientific Officer at the Agricultural Research Council's Meat Research Institute working on commercial meat refrigeration. Formerly on field work for the Torry Research Station. Author of many papers on commercial meat refrigeration.

H. F. Marks, MSc(Econ)

Chief Economist, MLC, concerned with providing market information for the Meat and Livestock industry. Formerly Assistant Economist, Commonwealth Economic Committee; Economist, Ministry of Agriculture, Fisheries and Food; and Head of Economics Department PIDA.

D. C. Rixson, M Inst M, ARSH, AAMI, MMGI

Principal Lecturer for the Department of Food Commodities, College for the Distributive Trades, London. Formerly fully engaged in the family business F. R. Rixson and Company Limited, retail butchers. Regular contributor to the *Meat Trades Journal* and has published books on the Pricing and on the Cutting of Meat, and a report on animal bones from archaeological sites in London.

T. A. Roberts, B. Pharm, Ph D, MA(Cantab)

Head of Microbiology Section and of the Food Quality and Control Division of the Agricultural Research Council's Meat Research Institute. Formerly Lecturer at the School of Pharmacy, London University; joined the Low Temperature Research Station, Cambridge, to work on the microbiology of food irradiation, interaction of the many physical and chemical factors combining to control or prevent growth of food-poisoning bacteria in foods. Published about 100 original scientific papers and contributed to others.

B. G. Shaw, B Tech, Ph D

Senior Scientific Officer, Microbiology Section of the Agricultural Research Council's Meat Research Institute studying the microbiology of the spoilage of meat. Graduate of Bradford University. Carried out research on farm waste disposal at Rowett Research Institute, Aberdeen and awarded his doctorate by Aberdeen University. Published several papers on the microbiology of farm waste disposal, and on meat and fish spoilage.

Clifford Simmons, BSc

Senior Lecturer at the Department of Food Commodities, College for the Distributive Trades, London. Author of many articles on food technology. Consultant in the compositional and bacteriological aspects of food science.

J. Southgate, BSc(Hons) (Agric)

Technical Officer in charge of Progeny testing and Commercial Breed Comparisons at the MLC. Secretary of British Society of Animal Production. Graduate of Wye College, University of London. Formerly with the Beef Recording Association.

Donald Sproat

Poultry and Game Wholesaler, Smithfield, London.

Miss R. Stollard, BSc(Hons) (Agric Sci)

Technical Officer in charge of Commercial Recording Scheme for Cattle at the MLC. Graduate of Nottingham University.

CONTENTS

ix

CONTENTS
continued

CONTENTS
continued

CONTENTS
continued

CONTENTS
continued

COLOUR PLATES

BLACK AND WHITE PLATES

xv

The World Meat Situation

Beef and veal output has increased in most producing countries during the past decade. The United States of America is the world's principal producer, as well as the largest importer. Production in 1977, at 11·8 million tonnes, was over 1·8 million tonnes higher than the average for 1969–73, and almost double the output of any other single producer. The United States has one of the highest per head consumption rates in the world amounting to 58·9 kg in 1977. As a result, most of the country's production is retained for domestic use and relatively little is exported.

BEEF AND VEAL

United States

The United States is, however, the world's largest importer, although beef and veal imports, which are used mainly for manufacturing purposes, accounted for only 7% of the total consumed in 1977. Imports of beef are controlled by a system of import quotas but during the past few years have in fact been regulated by "voluntary agreements". The legislative powers, however, remain. Imports in 1978 amounted to 675,000 tonnes; in addition over one million live cattle were imported. Australia has, in recent years, grown in importance as an exporter to the United States and in 1978 accounted for 53% of the United States' imports. Most of the imports of live cattle are of Mexican and Canadian origin.

Soviet Union

The Soviet Union is the world's second largest producer of beef and veal, with an output in 1977 of 6·2 million tonnes. The USSR has shown an increased interest in importing beef in recent years but normally only enters the market at times of world surplus when prices are low. In 1975 the USSR purchased considerable quantities of "surplus" EEC beef and contracts have also been concluded with Australia and Argentina. Imports in 1978 are estimated at only 120,000 tonnes.

Argentina

Argentina is the third most important producer of beef and veal. Production in 1977 amounted to 2·9 million tonnes. Until 1971, when Australia increased in importance, Argentina was the world's largest exporter of beef and veal, but the quantities sent abroad remain substantial. Exports to the EEC have been relatively low in recent years due to restrictions and high import duties. Exports in 1978 amounted to 328,000 tonnes, of which 30% was sent to the EEC.

Australia

Australian production of beef and veal has shown a considerable recovery in recent years and in 1977 amounted to 2·1 million tonnes. Australia normally exports between 30 and 40% of its production. However, exports to the EEC have also been affected recently by the virtual ban on imports from non-EEC countries. The United States is by far the most important single market, and has for many years

2

accounted for over half of Australia's beef and veal exports. However, Japan has become more important as an outlet for Australian beef and, in 1977/78, was the second largest market and is the largest single market for Australian chilled beef.

In the original six member states of the European Community, beef production relies heavily on the dairy herd. This has meant that there has been a conflict between the policies of trying to restrict milk production while at the same time increasing the output of beef. France produced 1·65 million tonnes of beef and veal in 1977. In Western Germany 1·38 million tonnes were produced in 1977. In the enlarged Community, the consumption per head is highest in France, Italy and Belgium/Luxembourg, and with the exception of Denmark where consumption has tended to decline in recent years, the level of consumption is also relatively high in the other member states. Within the EEC as a whole, there is a significant amount of intra-Community trade in beef and veal and substantial quantities are also normally imported from third countries. However, there was a virtual ban on imports from third countries between 21 July 1974 and 31 March 1977, and imports continue to be restricted by high import duties. In 1978 the United Kingdom imported 273,000 tonnes of beef and veal of which 142,000 were from the Irish Republic.

EEC

Italy has in recent years become the world's largest importer of live cattle and calves reaching a peak level of 2·6 million head in 1973. Since then there has been some decline. Most of the cattle and calves are of French and West German origin, but there are also substantial imports from Eastern Europe. The United Kingdom also imports a large number of Irish store cattle.

Unlike the production and trade in beef and veal, which is carried on by a large number of countries, the production and trade in mutton and lamb is more concentrated—only a small number of countries account for a relatively large share of both world production and international trade.

MUTTON AND LAMB

3

Soviet Union

The Soviet Union has traditionally been the world's largest producer of mutton and lamb and production in 1977 amounted to approximately 850,000 tonnes. Imports of mutton and lamb are normally very small but the Soviet Union is one of the world's largest importers of live sheep. Approximately 800,000 sheep were imported in 1977.

Australia

Sheepmeat production in Australia has fallen sharply since the beginning of the 1970's. Apart from seasonal factors the trend in mutton and lamb output is to a large extent influenced by changes in the price of wool. In 1970–72 the fall in wool prices was one of the factors which led to a reduction in sheep numbers and a consequent increase in sheepmeat production, which reached 948,000 tonnes in the twelve months ending June 1972. In the 1972–73 season, however, wool prices recovered and as a result many more lambs were retained for rearing with the result that slaughterings declined. Mutton and lamb production in 1977 as a whole is estimated at 544,000 tonnes. The per capita consumption of mutton and lamb is very high in Australia, although it has fallen in recent years and only amounted to 17·4 kg per head in 1978. Because of the high production levels, however, a substantial quantity is available for export, 187,000 tonnes in 1977/78, and Australia is the world's second largest exporter of mutton and lamb. The bulk of Australia's sheepmeat exports consists of mutton; Japan, where imports of mutton and lamb are not normally subject to quantitative controls, is normally the largest single importer of Australian mutton. The United Kingdom, the United States, Canada and Greece are the other major markets. There is also an important export market for live sheep in the Middle East.

New Zealand

New Zealand mutton and lamb production has been at a relatively low level in recent years and in 1977 amounted to 487,000 tonnes, 77,000 tonnes less than the 1969–73 average. However, the sheep flock is again currently expanding. New Zealand normally exports about 75% of its production, 391,000 tonnes in 1978. Most of the sheep meat exports consist of lamb and although there has been increased diversification the bulk is still shipped to the United

4

Kingdom (180,000 tonnes in 1977/78). New Zealand is also sending increasing quantities of mutton and lamb to Japan, South Korea and the Middle East. Exports to the United States have also increased and there are occasional large shipments of mutton to the USSR. New Zealand also has a per capita consumption rate which is very high relative to other parts of the world. Consumption in New Zealand amounted to 30·5 kg per head in 1978, 75% more than in Australia and ten times the EEC average.

In the United States sheep numbers and mutton and lamb production have been declining gradually during the past decade. Production only amounted to 180,000 tonnes in 1978. Imports are small and consumption of sheepmeat is also very low and was only 0·7 kg per head in 1978.

United States

The United Kingdom is the largest sheepmeat producer in the EEC and in 1977 produced 223,000 tonnes of mutton and lamb. Indeed, it is only in the United Kingdom, France, the Irish Republic and Italy that the production of sheepmeat in the Community takes place on any substantial scale. The United Kingdom and the Irish Republic are also the only substantial consumers of mutton and lamb in the Community. However, while the per capita consumption level in France remains low, it has been increasing in recent years. The United Kingdom remains the world's largest mutton and lamb importer and has traditionally imported, largely from New Zealand, a quantity in excess of her own production. Total United Kingdom imports in 1978 amounted to 221,000 tonnes.

United Kingdom

In Western Europe as a whole, Greece and in particular Spain, stand out as fairly substantial producers of sheepmeat; Greece is one of Australia's major export markets for mutton and lamb.

The trade in live sheep takes place on a large scale in a number of countries; as in the case of cattle, Italy is the world's largest importer of live sheep. The other Community countries which participate in the trade do so at more modest levels. Some Middle Eastern countries also import significant numbers of live sheep.

5

PIGMEAT

The production of pigmeat takes place on a substantial scale in many countries, and in the past decade, because of superior feed conversion ratios, world production of pigmeat has tended to increase at a faster rate than that of other meats. However, while production is widespread, it is in Europe that most of the international trade in pigmeat takes place.

United States

The United States produces more pigmeat than any other country in the world, 6·0 million tonnes in 1977, and has a relatively high level of per capita consumption. The United States also imports canned pigmeat from Denmark, the Netherlands and Poland—this in fact is the only substantial inter-continental trade which takes place. Exports of pigmeat, although relatively unimportant, have increased in recent years and Japan is now the United States' principal customer. It is interesting to note that although Japan is not considered a particular important meat producer, her production of pigmeat has increased considerably in recent years and in 1977 reached an estimated 1·2 million tonnes—a quantity which was in excess of the production of such countries as Denmark, the Netherlands and the United Kingdom.

Soviet Union

The Soviet Union produces considerably less pigmeat than the United States but production amounted to about 4·30 million tonnes in 1978; the USSR is, therefore, the world's second largest producer. Pigmeat consumption in the USSR has increased rapidly in recent years and in 1978 is estimated at 21 kg per head.

EEC

There is a substantial volume of trade in pigmeat and live pigs within Europe and a particularly large quantity is traded within the EEC which tends, as a whole, to be slightly more than self-sufficient in pigmeat. Western Germany is the largest producer in the Community, and the third largest in the world; in 1977 production totalled over 2·53 million tonnes. Western Germany also has by far the highest per head rate of consumption in the world—55·7 kg in 1978. Most of the pigmeat is consumed in processed or manufactured form. Although Western Germany is a major producer of pigmeat,

6

imports—principally from the Netherlands and Belgium—have increased rapidly in recent years; there has been a similar trend in live pig imports, most of which also come from the Netherlands and Belgium. Denmark is the world's leading bacon and ham exporter, and traditionally most of her exports are sent to the United Kingdom. Since joining the EEC, however, Denmark has also been able to increase its exports of pigmeat to the other EEC countries. France is the Community's second largest producer of pigmeat, and the per capita level of consumption has increased in recent years. As in the case of Western Germany, France exports little pigmeat, but imports a considerable quantity, principally from the Netherlands and Belgium. In 1978 the United Kingdom produced 631,000 tonnes of pork and 214,000 tonnes of bacon and ham. The United Kingdom is the world's largest importer of bacon, most of which is of Danish origin, but imports only small quantities of pork from abroad. Imports of canned pigmeat, however, are also substantial. The Netherlands and Belgium have both rapidly increased their production and export levels during the past decade, and both are now important suppliers for Western Germany, France and Italy. The Irish Republic also exports bacon and ham to the United Kingdom, although the quantities involved are small in comparison with Denmark. Pigmeat production and consumption in Italy remains relatively unimportant although this situation is also beginning to change.

Eastern Europe

Pigmeat production takes place on a substantial scale in some Eastern European countries—Poland and East Germany in particular, but significant amounts are also produced in Czechoslovakia, Hungary and Yugoslavia. Poland is also a major exporter of bacon and in 1978 supplied about 3% of the United Kingdom's requirements. Pigmeat has also become important as an export for Hungary.

Much of the world trade in live pigs takes place within the European Community—many of the pigs being imported by Western Germany and France. Italy imports a considerable number of weaners for fattening from Western Germany and other EEC countries. Hong Kong, however, is the world's largest live pig importer and in 1977 imported about 2·9 million pigs, a large percentage of them from China. It is probable that China is a major producer of pig-meat, although no statistics are available to confirm this.

TABLE 1

Estimated Production of Meat in Various Countries* (Source: USDA)

	Average 1969–73	1974	1975	1976	1977	1978†
BEEF AND VEAL (000 tonnes)						
EEC						
Belgium/Luxembourg	269	318	309	296	285	275
Denmark	187	238	235	242	238	240
France	1529	1792	1746	1799	1648	1700
Irish Republic	212	342	420	328	375	370
Italy	1062	1076	965	1018	1050	1090
Netherlands	296	363	373	376	368	307
United Kingdom	911	1073	1216	1074	997	985
Western Germany	1288	1394	1336	1402	1380	1445
Total EEC‡	5755	6595	6600	6533	6341	6412
Argentina	2370	2163	2439	2811	2909	2955
Australia	1176	1268	1697	1870	2093	1900
Canada	878	942	1049	1139	1181	1125
Japan	275	292	353	298	360	385
New Zealand	404	405	508	612	535	486
South Africa	498	524	454	495	550	575
Uruguay	319	320	345	405	314	400
United States	10076	10716	11271	12166	11844	11198
USSR	5225	5937	6020	5992	6231	6500
MUTTON AND LAMB§ (000 tonnes)						
EEC						
Belgium/Luxembourg	4	3	3	3	3	3
Denmark	2	1	1	1	1	1
France	127	138	139	155	153	156
Irish Republic	44	45	46	40	41	35
Italy	52	46	49	50	51	52
Netherlands	10	15	16	16	17	17
United Kingdom	224	252	259	248	223	246
Western Germany	11	16	21	22	24	26
Total EEC‡	473	515	534	534	512	534
Argentina	161	112	123	127	134	130
Australia	769	468	548	594	544	495
Canada	9	8	8	8	6	6
New Zealand	564	498	491	513	487	516
Greece	90	112	115	115	118	120
Spain	137	155	148	146	143	140
United States	246	211	186	168	159	153
USSR	920	925	926	855	850	870

8

	Average 1969–73	1974	1975	1976	1977	1978‡
PIGMEAT (000 tonnes)						
EEC						
Belgium/Luxembourg	479	638	610	612	625	628
Denmark	741	743	734	716	740	760
France	1314	1428	1470	1499	1533	1550
Irish Republic	149	131	99	121	129	126
Italy	593	709	735	781	860	915
Netherlands	706	828	843	889	916	966
United Kingdom	985	1006	842	889	957	850
Western Germany	2273	2380	2415	2467	2530	2630
Total EEC‡	7240	7862	7747	7974	8290	8425
Argentina	227	241	255	258	238	250
Australia	195	187	172	179	191	189
Canada	614	611	521	512	553	603
Japan	804	958	1039	1056	1165	1200
New Zealand	38	34	34	33	44	44
Sweden	255	280	279	290	302	309
United States	6586	6501	5343	5755	6009	6106
USSR	3436	3877	4042	3023	3585	4000

★ Carcase weight
† Forecast
‡ Due to rounding figures do not necessarily add up to totals shown
§ Including goatmeat

TABLE 2

Estimated Consumption of Meat in Various Countries (Source: ZMP Bonn)

	1970	1974	1975	1976	1977	1978
BEEF AND VEAL (kg per head)						
EEC						
Belgium/Luxembourg	26·7	30·7	30·5	29·0	29·0	28·0
Denmark	19·7	14·5	15·6	16·0	15·1	17·1
France	30·0	29·6	29·6	31·2	31·3	32·1
Irish Republic	19·0	23·0	29·1	24·8	23·8	28·1
Italy	25·5	25·0	23·1	23·5	24·4	24·9
Netherlands	21·8	22·3	22·3	22·9	22·6	21·6
United Kingdom	23·5	24·0	25·2	23·5	23·6	24·5
Western Germany	24·6	23·2	23·0	24·1	23·7	24·2
Total EEC★	25·4	25·1	25·1	25·3	25·4	26·0
Argentina	84·0	75·1	86·6	88·9	88·6	94·3
Australia	38·8	41·1	64·3	68·6	69·7	69·0
Greece	na	15·0	17·8	20·5	21·7	21·8
Portugal	na	14·5	12·7	13·6	14·1	14·3
Spain	11·9	12·0	14·0	13·7	13·3	12·5
United States	52·9	54·0	56·4	60·5	58·9	55·8

	1970	1974	1975	1976	1977	1978
MUTTON AND LAMB (kg per head) **EEC**						
Belgium/Luxembourg	1·0	1·2	1·3	1·6	1·8	1·8
Denmark	0·4	0·4	0·4	0·4	0·5	0·6
France	3·0	3·5	3·6	3·7	3·7	3·8
Irish Republic	10·9	11·0	10·9	10·4	10·3	9·0
Italy	1·2	0·9	1·1	1·1	1·1	1·2
Netherlands	0·2	0·2	0·2	0·2	0·3	0·4
United Kingdom	9·9	7·9	8·5	7·8	7·1	7·1
Western Germany	0·2	0·4	0·6	0·7	0·7	0·8
Total EEC★	3·3	2·9	3·1	3·0	2·9	3·0
Australia	38·5	23·7	26·7	24·2	17·7	17·4
Greece	na	13·4	14·0	14·0	13·3	13·4
New Zealand	na	37·7	37·1	33·0	30·0	30·5
Portugal	na	2·9	2·5	2·4	2·5	2·5
Spain	4·1	4·5	4·2	4·2	4·0	4·0
United States	1·5	1·0	0·9	0·9	0·8	0·7
PIGMEAT (kg per head) **EEC**						
Belgium/Luxembourg	33·6	39·0	36·3	36·6	36·9	39·8
Denmark	29·4	34·9	39·3	39·6	41·7	44·3
France	30·6	33·0	34·1	34·8	35·6	37·2
Irish Republic	30·9	31·7	27·2	29·1	27·9	29·5
Italy	12·7	16·9	17·8	18·8	20·1	21·4
Netherlands	28·6	34·2	35·5	35·5	35·3	38·8
United Kingdom	26·9	26·1	23·0	23·7	24·9	25·3
Western Germany	45·3	50·2	50·7	51·6	52·6	55·7
Total EEC★	29·5	32·5	32·3	33·0	34·0	35·9
Greece	na	11·8	11·5	11·6	13·8	13·8
Portugal	na	15·0	15·5	15·1	16·1	15·7
Spain	13·7	20·2	18·6	19·6	20·5	23·0
United States	30·1	31·3	25·4	27·0	27·9	27·9

na Not available

★ Due to rounding figures do not necessarily add up to totals shown.

TABLE 3

Imports of Meat and Livestock into Various Countries (Source: ZMP Bonn)

	1970	1974	1975	1976	1977	1978
BEEF AND VEAL (000 tonnes) **EEC**						
Belgium/Luxembourg	19	20	29	33	35	50
Denmark	2	1	1	1	1	3

	1970	1974	1975	1976	1977	1978
France	72	117	160	157	206	261
Irish Republic	..	1	2	1	2	.
Italy	290	297	320	293	323	320
Netherlands	44	48	45	59	67	71
United Kingdom	265	249	196	214	258	273
Western Germany	185	176	196	210	213	200
Total EEC★	877	909	947	968	1105	1177
Greece	68	24	37	79	90	100
Japan	23	54	45	92	84	100
Spain	99	14	27	44	50	69
United States	525	489	549	587	565	675

LIVE CATTLE (000)
EEC

	1970	1974	1975	1976	1977	1978
Belgium/Luxembourg	141	95	156	159	195	185
Denmark	1	..	..	..	..	..
France	44	25	74	148	258	299
Irish Republic	163	93	86	89	142	119
Italy	2100	1786	2304	2355	1895	2145
Netherlands	66	32	70	82	126	122
United Kingdom	524	384	504	230	304	254
Western Germany	210	167	207	186	208	210
Total EEC★	3250	2581	3402	3250	3127	3434
Greece	12	34	16	28	26	na
Spain	9	37	13	7	5	11

MUTTON AND LAMB
(000 tonnes)
EEC

	1970	1974	1975	1976	1977	1978
Belgium/Luxembourg	5	8	10	12	14	15
Denmark	1	1	2	2	2	3
France	31	44	52	42	46	47
Irish Republic	—	—	—	—	—	—
Italy	7	6	12	13	12	17
Netherlands	2	1	1	2	2	2
United Kingdom	331	213	244	226	219	221
Western Germany	5	10	20	25	29	29
Total EEC★	382	283	339	320	323	334
Greece	40	8	12	13	8	10
Japan	111	90	131	136	148	140
United States	38	10	12	16	10	18

LIVE SHEEP (000)
EEC

	1970	1974	1975	1976	1977	1978
Belgium/Luxembourg	158	118	275	262	299	401
Denmark	—	—	—	—	—	—

	1970	1974	1975	1976	1977	1978
France	182	320	450	441	449	504
Irish Republic	146	139	187	186	221	352
Italy	1468	1163	1311	1012	933	832
Netherlands	14	14	50	43	39	35
United Kingdom	71	67	78	80	62	86
Western Germany	4	114	257	386	359	513
Total EEC★	2043	1934	2608	2410	2363	2722
Greece	482	—	—	—	—	—
Iran	na	na	1518	1800	2500†	na
Kuwait	na	na	410	417	600†	na
Saudi Arabia	na	na	1305	1296	1300†	na
Spain	15	30	2	1	1	—
USSR	na	na	1014	860	800†	na

PIGMEAT
(000 tonnes)

EEC

	1970	1974	1975	1976	1977	1978
Belgium/Luxembourg	12	9	14	13	10	9
Denmark	1	1	1	—	—	2
France	183	191	193	198	207	220
Irish Republic	—	1	2	2	2	2
Italy	101	218	234	255	240	262
Netherlands	2	7	7	7	11	11
United Kingdom—Pork	11	7	16	12	16	39
Bacon and Ham	385	299	287	269	279	312
Western Germany	91	249	242	241	259	292
Total EEC★	786	982	997	998	1022	1148
Japan	17	42	125	149	110	110
Spain	1	8	44	53	7	38
United States	139	153	141	133	127	147

LIVE PIGS (000)

EEC

	1970	1974	1975	1976	1977	1978
Belgium/Luxembourg	70	77	355	354	377	501
Denmark	—	—	—	—	—	—
France	1128	909	1172	1329	1376	1663
Irish Republic	17	85	18	9	2	3
Italy	445	675	570	686	497	479
Netherlands	88	85	35	91	43	,56
United Kingdom	1	5	62	58	76	72
Western Germany	372	1039	1050	805	901	988
Total EEC★	2120	2875	3253	2734	3271	3761
Austria	3	158	56	13	20	124

★ Due to rounding figures do not necessarily add up to totals shown
† Estimate
na Not available
.. Negligible

TABLE 4

Exports of Meat and Livestock from Various Countries (Source: ZMP Bonn)

	1970	1974	1975	1976	1977	1978
BEEF AND VEAL (000 tonnes) EEC						
Belgium/Luxembourg	17	29	32	38	35	45
Denmark	70	105	129	107	137	154
France	115	252	293	275	216	184
Irish Republic	140	199	270	180	262	262
Italy	1	—	3	1	2	5
Netherlands	115	134	137	133	153	169
United Kingdom	10	60	115	101	88	95
Western Germany	55	117	138	142	185	209
Total EEC★	523	896	1116	976	1079	1121
Argentina	348	106	78	224	278	328
Australia	340	340	515	581	714	773
New Zealand	184	181	202	249	245	234
Uruguay	131	100	79	143	107	105
Yugoslavia	48	35	37	57	59	100
LIVE CATTLE (000) EEC						
Belgium/Luxembourg	98	67	127	140	145	120
Denmark	116	33	30	21	18	15
France	692	937	1358	1377	1256	1320
Irish Republic	529	447	695	370	453	549
Italy	—	1	1	1	—	—
Netherlands	80	124	238	248	211	208
United Kingdom	232	87	211	297	511	522
Western Germany	597	466	572	532	407	440
Total EEC★	2345	2163	3233	2985	3000	3173
Austria	117	122	117	103	101	103
Hungary	220	165	198	161	168	155
Yugoslavia	77	8	50	35	23	na
MUTTON AND LAMB (000 tonnes) EEC						
Belgium/Luxembourg	3	—	—	—	—	—
Denmark	—	—	—	—	—	—
France	—	—	—	—	1	1
Irish Republic	8	11	12	7	6	15
Italy	—	—	—	—	—	—
Netherlands	9	13	15	14	14	19
United Kingdom	11	27	34	33	45	41
Western Germany	2	2	6	8	8	8
Total EEC★	33	53	66	64	75	78
Argentina	35	18	21	26	28	24
Australia	168	83	135	183	206	173
New Zealand	422	372	380	399	405	391

13

	1970	1974	1975	1976	1977	1978
LIVE SHEEP (000)						
EEC						
Belgium/Luxembourg	1	79	222	231	250	357
Denmark	—	—	8	9	7	5
France	2	6	7	3	3	4
Irish Republic	67	120	208	104	116	102
Italy	1	11	43	40	21	27
Netherlands	2	37	75	70	88	86
United Kingdom	268	81	235	322	356	587
Western Germany	106	184	228	292	315	301
Total EEC★	447	518	1025	1071	1156	1468
Australia	na	na	1809	3347	4437	4930
Hungary	1008	856	879	720	822	na
Spain	34	17	17	23	51	na
Yugoslavia	132	31	118	55	60	na
PIGMEAT						
(000 tonnes)						
EEC						
Belgium/Luxembourg	118	198	175	172	170	185
Denmark	345	337	359	333	347	380
of which bacon and ham	290	245	236	218	218	222
France	9	18	17	19	15	21
Irish Republic	45	28	13	27	32	27
of which bacon and ham	28	19	10	15	21	20
Italy	3	4	6	8	8	7
Netherlands	206	287	305	346	382	446
of which bacon and ham	9	13	23	27	32	43
United Kingdom	17	21	7	13	19	15
Western Germany	13	5	3	7	16	14
Total EEC★	755	896	885	927	989	1095
LIVE PIGS (000)						
EEC						
Belgium/Luxembourg	1097	1082	1039	1097	1222	1164
Denmark	138	98	74	82	66	57
France	99	114	182	318	307	221
Irish Republic	1	4	62	43	76	71
Italy	—	—	1	—	—	—
Netherlands	362	991	1069	978	1282	1686
United Kingdom	66	99	31	31	29	34
Western Germany	480	181	131	216	133	217
Total EEC★	2243	2569	2587	2766	3115	3450

★ Due to rounding figures do not necessarily add up to totals shown
na Not available

The Home Market

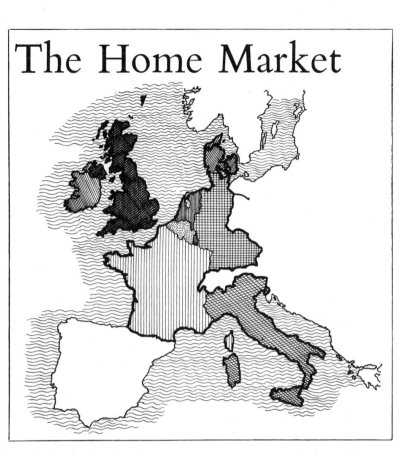

The aim of this chapter is to give a brief account of the meat production and supply situation in the United Kingdom. For simplicity's sake the chapter has been divided into three separate sections: beef and veal, mutton and lamb, and pigmeat.

The value of output of fat cattle and calves at the farm gate has been increasing rapidly in recent years and in 1978 (forecast) amounted to £1310 million. Fat cattle production accounted for 47% of total livestock output and 19% of total agricultural output (see Table 1, p. 16). Beef production continues to be very much a

BEEF AND VEAL
Home Production

TABLE 1

The Value of Agricultural Production (£ million) (Source: MAFF)

	1976	1977	1978*
LIVESTOCK			
Fat Cattle and Calves	995	1063	1310
Fat Sheep and Lambs	240	267	322
Fat Pigs	556	641	691
Poultry	344	421	422
Other	24	31	32
	2159	2423	2778
LIVESTOCK PRODUCTS			
Milk and Milk Products	1294	1485	1591
Eggs	342	391	373
Clip Wool	24	30	32
Other	10	10	10
	1670	1917	2007
Farm Crops	1447	1365	1528
Horticulture	593	712	709
Sundry Output	32	38	42
TOTAL OUTPUT†	5900	6455	7064

* Forecast

† Due to rounding, figures do not necessarily add up to totals shown

by-product of the dairy herd but only a small proportion of the calves are finished on dairy farms. Most of them are sold to fatteners and in fact the cattle may change hands several times—being sold as calves, then as stores and finally reaching the farm where they are finished. The proportion of "beef type" animals has, however, increased rapidly until 1975 but there has since been a reversal of the trend and by 1978 the proportion of cows and heifers intended mainly for rearing beef calves in the United Kingdom had again fallen to only 32·5%. In England and Wales the proportion was 22%, but in Scotland the proportion was very much higher—59%. There are many systems of beef production and producers very often use more than one system. The most intensive system—intensive cereal beef—is based on a high cost concentrate ration. Because the animals are housed, virtually no land is required. The less intensive grass/cereal systems include several methods of utilising dairy-bred calves, which can be born at almost any time of the year. Many producers also finish suckled calves and stores under differing conditions. It is estimated that during the past few years approximately

5% of the beef was produced intensively in under twelve months, 52% was produced by semi-intensive methods and was yard finished and 43% was grass finished. In the case of the latter two systems the animals are slaughtered at about 18 months and 2 years respectively. Beef can be produced from many breeds—it is estimated that approximately 39% are of Friesian and Friesian cross origin and 28% from beef type breeds such as Aberdeen Angus, Hereford, Charolais and South Devon.

In 1978 approximately 1·03 m. tonnes of beef and veal were produced in the United Kingdom. Of this 68% was produced from home-bred steers and heifers, 9% from Irish store cattle and 23% from culled cow and bull production.

TABLE 2

Cattle Slaughterings and Beef and Veal Production (Source: MAFF)

Year	Steers and Heifers	Cows and Bulls	Calves	Total	Beef and Veal Production
	(000)	(000)	(000)	(000)	(000 tonnes)
1970	2896	791	356	4043	947
1973	2547	745	141	3433	878
1974	3134	1045	416	4595	1073
1975	3611	1217	530	5359	1217
1976★	3196	988	294	4478	1074
1977	2901	947	264	4112	1002
1978	3003	879	157	4039	1027

★ Fifty-three week year

Table 2 shows the slaughterings of various types of cattle and beef and veal production since 1970. Cattle slaughterings rose to a peak in 1975 but have since then fallen. In 1978 4·0 million cattle were slaughtered—of these 3·0 million were steers and heifers, 879,000 cows and bulls and 157,000 calves. Beef and veal production in 1978 amounted to 1·03 million tonnes. A small reduction is expected in 1979.

The United Kingdom adopted the Common Agricultural Policy for beef and veal in 1973. This provides for the support of domestic prices by means of intervention (and in the case of the United Kingdom variable premia which are paid to producers) as well as protection against imports from third countries. See page 29 for a description of the common market beef regime. Beef producers are also assisted by means of hill livestock compensatory allowances.

Trade

Imports of beef and veal, which had fallen to the very low level of 196,000 tonnes in 1975, have since then again increased and by 1978 amounted to 273,000 tonnes. Of special interest is the change in the source of United Kingdom beef imports which became particularly marked when the EEC restrictions on imports of beef from third countries were introduced in mid 1974. Although these restrictions were removed in March 1977 the continuing high level of import duties has meant a low volume of imports from third countries. In 1970 non-Community countries accounted for about 55% of United Kingdom beef imports. Argentina supplied 22% and Australia 12%. In 1978 third countries only accounted for 13% of United Kingdom imports. Imports from other EEC countries into the United Kingdom have been encouraged by the high level of monetary compensatory amounts paid on exports from many of the member states. The Irish Republic's exports of beef, in particular, have steadily increased in importance. In 1966 only 14% of total United Kingdom imports were from this source, but by 1978 the Irish Republic, at 52%, was the largest supplier. Moreover, in recent years, there has been a change in the type of beef imported—the amount of frozen boneless beef, large quantities of which were imported from South America and Australia, before the imposition of restrictions, has declined.

A substantial import trade in Irish store cattle also takes place although this has tended to diminish in recent years. In 1978 354,000 live cattle were imported, virtually all from the Irish Republic.

The United Kingdom, although an importer, has also become an exporter of beef and live cattle. The figures are shown in Table 4. The exporting of live animals for slaughter, other than to the Irish Republic from Northern Ireland, was banned for welfare reasons from mid-1973 until the beginning of 1975, when it was felt that sufficient progress had been made in establishing international welfare safeguards. A substantial trade with the Continent did not

18

TABLE 3

Imports of Beef and Veal and Live Cattle
(000 tonnes)

	1971	1972	1973	1974	1975	1976	1977	1978
BEEF								
Boneless								
Fresh and chilled								
Irish Republic	5·9	5·8	4·3	8·9	3·4	2·7	2·6	3·9
Argentina	18·9	30·7	33·2	18·8	0·1	0·7	0·7	2·9
Botswana	(a)	0·1	0·1	1·3	4·2	6·8	8·2	0·6
Other countries	4·5	9·1	8·4	8·9	5·4	2·1	2·3	3·3
Total★	29·3	45·6	46·1	38·0	13·0	12·3	13·7	10·8
Frozen								
Australia	30·5	64·7	83·2	22·0	10·1	9·5	6·4	12·0
Botswana	6·0	6·7	17·2	2·1	6·5	9·8	9·4	3·0
New Zealand	12·1	10·8	11·5	7·2	5·7	9·4	7·9	2·9
Irish Republic	1·0	0·9	1·6	38·2	33·7	19·1	35·1	30·5
Argentina	19·9	33·1	26·3	9·3	2·3	7·6	6·2	10·5
France	12·7	7·3	6·1	10·1	20·0	23·4	22·7	19·1
Other countries	18·9	25·8	19·8	7·7	11·9	19·6	18·7	25·0
Total★	101·0	149·1	165·8	96·6	90·4	98·3	106·4	103·0
Total boneless★	130·3	194·7	211·9	134·6	103·4	110·7	120·1	113·8
Bone-in								
Fresh and chilled								
Irish Republic	98·1	72·5	42·6	53·3	55·9	55·3	82·5	106·5
France	11·3	0·6	0·7	3·3	2·0	1·2	0·4	0·2
Western Germany	1·0	0·1	3·1	22·2	14·4	16·0	17·8	19·7
Denmark	(a)	..	2·5	23·0	12·0	19·8	21·1	16·5
Netherlands	(a)	(a)	(a)	(a)	(a)	2·4	1·3	1·9
Other countries	6·3	1·4	1·9	4·1	1·5	0·1	0·1	..
Total★	116·7	74·7	50·8	105·8	85·8	94·7	123·3	144·8
Frozen								
Irish Republic	0·2	0·5	0·2	4·1	3·2	2·6	8·5	1·2
Australia	0·3	1·4	0·6	..	0·1	(a)	(a)	(a)
New Zealand	2·0	2·1	2·6	0·3	0·2	(a)	(a)	(a)
Other countries	2·3	1·6	0·9	1·2	0·5	2·7	1·7	8·7
Total★	4·9	5·6	4·2	5·6	4·0	5·3	10·2	9·9
Total bone-in★	121·6	80·4	55·0	111·5	89·9	100·0	133·5	154·6
Total beef★	251·9	275·1	266·9	246·1	193·2	210·7	253·6	268·4
VEAL	0·9	2·8	3·3	3·2	3·1	3·0	4·1	4·8
Total beef and veal★	252·8	277·8	270·2	249·2	196·3	213·7	257·7	273·2
LIVE CATTLE								
AND CALVES (000)	619·6	493·9	343·5	384·3	504·1	229·8	304·3	353·5

(a) Included, if any, in other countries
★ Because of individual rounding, figures do not necessarily add up to totals shown
.. Negligible

Source: 1971–1975 inclusive Commonwealth Secretariat Meat and Dairy Product Bulletin, 1976 to date United Kingdom customs returns

TABLE 4

Exports of Meat and Live Animals

	1971	1972	1973	1974	1975	1976	1977	1978
BEEF AND VEAL (000 tonnes)								
France	4·7	36·5	43·3	40·0	63·8	51·7	53·3	60·4
West Germany	1·6	4·0	3·5	6·3	20·3	24·7	19·4	18·6
Other countries	7·5	12·5	18·6	14·0	30·5	24·2	14·9	15·7
Total*	13·7	52·9	65·5	60·3	114·6	100·6	87·7	94·6
LIVE CATTLE AND CALVES (000)								
Irish Republic	96·7	99·8	79·9	76·9	64·4	39·4	133·5	114·9
Belgium/Luxembourg	7·5	35·4	39·6	—	46·0	60·5	70·4	39·7
France	8·8	10·9	7·6	2·0	44·4	114·4	207·7	249·4
Other countries	10·4	45·8	47·7	8·2	56·0	83·1	99·6	117·5
Total*	123·4	191·9	174·8	87·1	210·8	297·4	511·3	521·5
MUTTON AND LAMB (000 tonnes)								
France	7·9	17·0	22·5	21·0	25·9	15·6	19·3	10·2
Belgium/Luxembourg	2·8	2·5	2·1	1·7	2·4	6·7	10·3	13·7
West Germany	0·5	0·5	1·0	1·7	2·6	5·5	9·3	11·3
Other countries	4·3	3·1	1·7	2·2	2·7	4·9	5·7	6·2
Total*	15·4	23·1	27·4	26·6	33·6	32·7	44·6	41·4
LIVE SHEEP AND GOATS (000)								
Belgium/Luxembourg	43·6	121·6	26·7	..	88·1	118·3	116·9	206·5
Irish Republic	177·7	126·1	99·2	78·5	89·1	94·3	153·6	312·3
Total	222·9	251·3	129·9	81·1	235·4	322·4	356·1	587·2
PORK (000 tonnes)								
France	1·9	1·3	6·3	10·4	4·2	5·3	8·4	5·1
West Germany	0·4	0·5	2·3	5·8	1·3	3·4	6·9	5·9
Other countries	8·0	2·4	3·0	1·6	0·6	2·4	1·2	1·7
Total*	10·3	4·2	11·6	17·8	6·1	11·1	16·6	12·6
LIVE PIGS (000)								
Total	26·6	28·7	55·5	98·5	30·9	30·7	28·8	33·7

.. Negligible

* Due to individual rounding, figures do not necessarily add up to totals shown

Source: 1971–1975 inclusive Commonwealth Secretariat Meat and Dairy Product Bulletin, 1976 to date United Kingdom customs returns

resume until late in 1975. There can be little doubt that British live-stock producers and meat traders are becoming increasingly export conscious. In past years there has been a tendency for purchasers from abroad to come to this country to buy in order to augment their own supplies. Increasingly, British traders are now looking for export opportunities and are building up a regular trade.

TABLE 5

Estimated Per Head Meat Supplies Available for Consumption (kg per head) (Source: MAFF)

	1971	1972	1973	1974	1975	1976	1977	1978
Beef—bone in	18·6	16·6	14·8	19·7	21·4	18·7	18·7	19·5
bone out*	3·0	4·4	4·5	2·7	2·2	2·4	2·4	2·5
Total	(21·6)	(21·0)	(19·3)	(22·4)	(23·6)	(21·1)	(21·1)	(22·0)
Mutton and Lamb	10·0	9·2	8·3	7·7	8·3	7·6	7·0	6·9
Pork	11·8	12·3	12·1	12·0	10·3	10·2	11·5	11·7
Offal	3·5	3·6	3·4	3·4	4·0	3·7	3·9	4·0
Poultry	10·7	12·0	11·7	11·6	11·4	11·5	12·1‡	12·5‡
Bacon and Ham	11·9	11·2	10·1	9·5	8·7	8·5	8·9	9·2
Game and Rabbits	0·3	0·3	0·3	0·2	0·2	0·1	0·1	0·1
Imported meat products	3·4	3·6	3·5	3·1	3·2	3·9	3·9	3·9
Total (converted to edible weight)†	60·8	60·6	56·8	57·0	57·0	54·7	56·2‡	57·9‡

* Imported boneless beef converted to bone-in equivalent weight
† Using revised conversion factors from 1974 to date
‡ Not comparable with earlier years

Consumption

The United Kingdom has traditionally been regarded as a country with a high level of beef consumption. In the early seventies however the situation changed; per head beef supplies available for consumption fell continually between 1970 and 1973—consumption in the latter year amounted to only 20·6 kg (41·4 lb) per head. Consumption began to rise again in 1974, when it averaged 21·68 kg (47·8 lb) per head, and probably reached a peak in 1975—supply statistics indicate that consumption increased to about 22·68 kg (50 lb) per head. Since then consumption has fallen to 22·0 kg per head in 1978.

MUTTON AND LAMB

The structure of sheep production in the United Kingdom is complex and the general pattern of the industry is known as stratification. The poorer uplands carry hardy hill breeds which are adapted to difficult environments. They are not very productive but can survive on poor grazing. The ewes are generally moved to easier conditions after two or three lamb crops, where they then produce a more productive hybrid generation which is suited to lowland farming conditions. Lambs from the hills are sold as stores, or if they are required for breeding they are wintered away on better land and returned to the hills in the following spring. Cross-bred ewes are sold to farmers on the richer lowlands. The crossbred ewes are mated with Down rams which contribute essential qualities for fattening, such as improved liveweight gain, depth of fleshing and early maturity. The profitability of sheep production is highly sensitive to choice of breed or cross.

There are three main methods of producing lamb in the United Kingdom:

OUT OF SEASON PRODUCTION: The ewes lamb in December/January, the object being to produce fat lambs which are ready for sale when prices are highest in the early spring. This is a high feed cost system and only accounts for about 2% of total annual lamb slaughterings.

PRODUCTION OFF GRASS: The lambs are sold fat before the end of the grazing season. Marketing takes place from May to November and this system accounts for about 69% of lamb slaughterings.

PRODUCTION OFF GRASS AND FORAGE: This involves the retention of the lambs after the end of the grazing season when they will be finished on forage crops. This system accounts for about 29% of slaughterings. There is a considerable overlap between systems 2 and 3.

Lambs retained after the end of the year are known as hoggets. Hogget production accounts for approximately one-quarter of the annual lamb crop slaughtered.

Table 6 shows the slaughterings of sheep and lambs as well as the production of mutton and lamb since 1970.

Sheep producers continue to benefit from the government's policy of guaranteed prices and deficiency payments. This means that producers receive a deficiency payment amounting to the difference between the average market price and the guaranteed price applicable for the week concerned. This is converted into a scale of weekly standard prices which are highest in the spring and lowest in the autumn. However, as the market has been firm there have been no significant deficiency payments since the autumn of 1977. The

TABLE 6

Sheep Slaughterings and Mutton and Lamb Production (Source: MAFF)

	1970	1973	1974	1975	1976★	1977	1978
SLAUGHTERINGS (000 head)							
Ewes and Rams	1310	1304	1344	1533	1363	1195	1198
Sheep and Lambs	10147	10457	11601	11600	11308	10159	10372
Total	11457	11761	12945	13133	12671	11354	11570
PRODUCTION (000 tonnes)							
Mutton and Lamb	227	234	252	260	248	223	228

★ Fifty-three week year

TABLE 7

Imports of Mutton and Lamb into the United Kingdom
(000 tonnes)

	1971	1972	1973	1974	1975	1976	1977	1978
MUTTON								
Fresh, Chilled or Frozen								
Australia	13·7	15·4	9·6	3·1	4·9	10·6	3·1	3·9
New Zealand	19·7	12·7	8·5	3·9	5·5	13·1	5·5	15·1
Other countries	0·6	1·9	0·5	0·4	0·4	0·1	0·1	0·1
Total★	33·9	30·0	18·6	7·4	10·8	23·8	8·6	19·0
LAMB								
Fresh or Chilled(a)								
Total★	8·2	2·6	0·4	0·4	1·5	0·4	0·9	0·2
Frozen								
Australia	18·8	16·1	13·7	4·4	2·7	1·1	1·5	7·7
New Zealand	290·7	281·1	230·7	199·6	228·6	200·4	207·7	198·9
Other countries	1·8	1·5	2·2	0·9	0·4	0·1	0·1	0·2
Total★	311·2	298·6	246·7	204·9	231·6	201·6	209·4	206·7
TOTAL LAMB★	319·4	301·3	247·0	205·3	233·1	201·9	210·3	206·9
TOTAL MUTTON AND LAMB★	353·3	331·2	265·6	212·7	243·8	225·7	218·9	225·9

(a) 1971–1974 inclusive. "Fresh" only, "chilled" included with "frozen". Virtually all from Irish Republic
★ Due to individual rounding, figures do not necessarily add up to totals shown

Source: 1971–1975 inclusive Commonwealth Secretariat Meat and Dairy Produce Bulletin, 1976 to date United Kingdom customs returns

hill livestock compensatory allowances are also of considerable value to the industry. There is at present no EEC regulation for sheep meat but discussions are currently taking place and it is possible that there will be a Common Agricultural Policy for sheep meat by 1980.

Trade

Imports of mutton and lamb have fallen in recent years and in 1978 amounted to only 226,000 tonnes. Over 90% of the lamb is imported from New Zealand, although small quantities also come from the Irish Republic and Australia. Imports of lamb from Argentina are banned because of the danger of foot and mouth disease—Argentina had previously sent a small quantity. New Zealand sheep production has been tailored to suit the United Kingdom market.

Although the United Kingdom continues to import about 60% of its mutton and lamb requirements, it has also in recent years become an exporter of lamb and live sheep—although, except in the case of the Irish Republic, the latter were banned between November 1973 and January 1975. Exports of mutton and lamb in 1978 amounted to 41,000 tonnes and 587,000 live sheep were also exported. Although it may seem surprising that a major importing country should also be an exporter of lamb the trades are in many ways complementary. The United Kingdom imports frozen lamb from New Zealand and exports fresh lamb to France and to a lesser extent other EEC countries (see Table 4, p. 20).

Consumption

The per head consumption of mutton and lamb has fallen in recent years and in 1978 only amounted to 6·9 kg compared with 10·52 kg in 1968. The decline in consumption was mainly due to a reduction in New Zealand supplies. There are considerable regional differences in the consumption of the various types of meat. The differences are especially marked in the case of mutton and lamb. In 1977, for example, the average household consumption per head of population was about 6·2 kg in Scotland compared with 13·7 kg in London. In general it appears that the consumption of mutton and lamb declines as one moves north from London although it increases in the western part of the country.

24

The value of output of pigmeat has also been increasing in recent years and the farm gate value in 1978 was £691 million—24·9% of total livestock output (see Table 1, p. 16). Slaughtering fell to a low level in 1978 when approximately 13·77 million pigs were slaughtered in the United Kingdom—of these 2·1 million were used wholly for bacon, 4·0 million were used in part for bacon and 7·3 million were used for other purposes (fresh pork and manufactured products). The slaughtering statistics since 1970 are shown in Table 8. Of special interest is the increase in the proportion of pigs used in part for bacon. The bacon factories are no longer only curers of whole carcases of Wiltshire-type bacon from bacon pigs (59 kg–77 kg) liveweight—they also purchase other types and weight ranges of pigs and use the pigmeat for the manufacture of sausages, pies etc. as well as bacon and the sale of fresh pork. The fresh meat trade, in the main, buys porkers which are much lighter and are sold as whole carcases. The United Kingdom is the only country in the EEC which produces so many different types of pigs. The average slaughter weight for all types is approximately 64 kg deadweight—this compares with about 99 kg in Italy, 85 kg in Western Germany and between 80 kg and 85 kg in France, the Netherlands and Belgium. In 1978 the United Kingdom produced 631,000 tonnes of pork and 214,000 tonnes of bacon and ham.

The structure of pig production has changed rapidly in recent years and the industry has become one of Europe's most efficient. The

**PIGMEAT
Home Production**

TABLE 8

Slaughterings of Pigs and Pork, and Bacon and Ham Production (Source: MAFF)

	1970	1973	1974	1975	1976*	1977	1978
SLAUGHTERINGS (000)							
Sows and Boars	323	413	469	334	346	392	342
Other Pigs:							
Used wholly for bacon	3,076	2,877	2,726	2,260	2,487	2,450	2,125
Used in part for bacon	3,542	4,059	4,154	3,697	3,861	3,626	4,001
Others	7,450	7,742	7,925	6,477	6,816	7,727	7,650
Total	14,391	15,091	15,274	12,768	13,510	14,195	13,776
PRODUCTION (000 tonnes)							
Pork	624	682	690	572	597	650	632
Bacon and Ham	251	252	243	210	227	219	214

* Fifty-three week year

25

number of holdings with sows and gilts has been falling rapidly—
from 35,000 in England and Wales in 1973 to 22,000 in 1977. The
average number of sows and gilts per holding has, however,
increased from 23·7 in 1973 to 31·8 in 1977. This trend can be
expected to continue. The number of holdings with pigs of any
description has fallen from 47,000 in 1973 to 30,000 in 1977. The
average herd size has increased from 157 in 1973 to 222 pigs in 1977.
There have also been regional changes in the location of pig production. Pig farming has tended to move away from such areas as the
South West and Wales—areas where cereal prices tend to be
relatively high—to the grain growing areas of the East. A rapid
expansion has also occurred in Yorkshire and Lancashire. The
average herd size is highest in the East (367) and South East—about
332 pigs in 1977.

There have also been considerable changes in the marketing of
slaughter pigs. The proportion of pigs sold liveweight has fallen to a
very low level. In the first five months of 1977 11% were sold
liveweight and 89% deadweight. All types of pigs continue to be
sold through the auction markets but it is rare for a good quality
bacon pig to be marketed in this manner. About 90% of the bacon
pigs are sold on a long term contract to the bacon factories.

A further development which is of interest is the considerable
amount of vertical and horizontal integration which has developed
in the pig industry. The feedingstuffs companies have tried to secure
their markets: in some cases they have begun to produce pigs and
they have also become involved in "marketing groups". The buyers
and processors of pigmeat are also showing an increasing interest in
integrating their businesses with pig production. Some provide
breeding stock and others have gone into pig production on their
own account.

Whereas integrated organisations only provide a small percentage of
pigmeat supplies, a much more common arrangement is the marketing of pigs under a contract scheme. Such arrangements account for
about 45% of all pigmeat sales. The contracts differ in depth and
scope but the basic essential is that the pig fattener provides the
wholesaler with a supply of slaughter pigs (normally a guaranteed

number of pigs per month or per year) in return for a fixed or mutually agreed price. Similar arrangements exist in the production process, i.e. in the supply of weaners and feed. These arrangements by-pass the regular marketing channels and as a consequence assist in the short-term stability of the market.

The Fatstock Guarantee Scheme for pigs ended in mid-1975 and the United Kingdom has now adopted the Common Agricultural Policy for pigs.

Price Structure

The United Kingdom is virtually self-sufficient as regards its pork supplies—only small quantities are imported, mostly from Denmark, Finland and the Irish Republic (see Table 9).

Trade

The situation in the case of bacon and ham is, however, very different. The United Kingdom continues to import about 58% of

TABLE 9

Imports of Pigmeat into the United Kingdom
(000 tonnes)

	1971	1972	1973	1974	1975	1976	1977	1978
BACON								
Denmark	285·2	263·2	248·6	241·5	232·0	208·1	210·5	233·1
Netherlands	7·4	8·8	9·6	12·0	21·1	25·5	31·2	40·8
Poland	42·3	38·2	32·2	19·4	18·4	16·6	13·1	14·8
Irish Republic	27·4	28·1	20·7	19·3	9·4	14·0	21·0	20·6
Sweden	10·5	9·9	7·5	5·1	4·3	1·1	..	—
Other countries	2·6	2·4	1·6	1·2	2·2	3·9	2·9	2·4
Total★	375·5	350·6	320·1	298·4	287·4	269·3	278·8	311·7
PORK								
Fresh, Chilled and Frozen								
Irish Republic	18·8	26·4	6·6	1·1	0·4	1·1	1·0	0·5
Denmark	0·8	2·8	6·1	3·3	13·4	10·2	12·5	28·2
Other countries	8·4	17·0	5·8	0·8	2·7	0·9	2·7	9·9
Total★	28·1	46·3	18·5	5·2	16·5	12·2	16·2	38·6

★ Due to individual rounding, figures do not necessarily add up to totals shown
.. Negligible

Source: 1971-1975 inclusive Commonwealth Secretariat Meat and Dairy Produce Bulletin, 1976 to date United Kingdom customs returns

its bacon and ham supplies. Of the 312,000 tons imported in 1978, 75% was of Danish origin, 13% was imported from Poland and 5% from the Irish Republic. The United Kingdom only exports an insignificant quantity of pigmeat—13,000 tonnes of pork in 1978; 34,000 live pigs were also exported (see Table 4).

Consumption

The per head level of pigmeat consumption is shown in Table 5. Pork consumption, which had risen sharply between 1976 and 1977, in 1978 reached 11·7 kg per head. Bacon and ham consumption rose to 9·2 kg in 1978.

The Common Agricultural Policy

The aim of the Common Agricultural Policy is the establishment of a common or single market for agricultural products throughout the Community providing for free and unimpeded trade between the member states. The production in each country should be free to compete in the markets of the other members, and to achieve this, all subsidies on production or restraints which distort competition or affect trade flows between member states are not permitted.

BEEF AND VEAL

The United Kingdom adopted the Common Agricultural Policy for beef in February 1973 and two months later the fatstock guarantee

29

scheme for fat cattle which had been in existence since after the war and provided for guaranteed prices and deficiency payments was ended. The United Kingdom, which opposes complete dependence on intervention buying as a means of supporting market prices, has been allowed to adopt—not necessarily on a permanent basis—a beef support system based on target prices and variable premium payments in addition to the normal EEC support arrangements described below.

The following account therefore refers to those measures which are operative in normal market conditions.

The commodities covered by the regulations include adult cattle (other than pedigree breeding stock) and calves; edible meat and offal whether fresh, chilled, frozen, salted, in brine, dried or smoked; and certain preparations containing beef or offal as well as fat from cattle or calves whether rendered or unrendered.

The Guide Price

The Community system is based upon the guide prices for adult cattle and calves. These do not vary seasonally and normally apply throughout an April/March marketing year. They are sometimes adjusted during the marketing year.

The 1979/80 EEC guide price is 154·58 ECU per 100 kg lw (89·85p per kg lw) (converted to sterling at the rate of 1·72039 ECU = £1).

Theoretically, there are four main methods of ensuring that producer returns will not fall below the desired level:—

(a) A system of protection against imports from third countries by means of a combination of customs duties and variable levies.
(b) A system of restitution payments or export subsidies.
(c) Support buying and private storage.
(d) Member states have been permitted to operate premium schemes since August 1974.

Ad valorem duties and variable levies are both applied to a range of products. The customs duties as prescribed in the regulations are as follows:—

	Per cent ad valorem charge
Live animals (other than pure-bred)	16
Meat: fresh, chilled and frozen	20
salted or in brine, dried or smoked	24
Fresh, chilled/frozen beef offals: liver	11
other	7
Other preparations of meat and offal: uncooked, more than 40% meat	20
other	26
Fats, rendered or unrendered	7

These apply to imports into the EEC as do variable levies.

The variable levy affecting trade in live animals and fresh, chilled, salted, in brine, dried and smoked beef and veal is calculated weekly.

The levy for adult cattle as at July 1979 consists of the difference between the guide price and the duty-paid import price, but the actual rate of levy applied is dependent upon the relationship between the guide price and the EEC reference price (market price) as follows:—

Reference Price as % of Guide Price	Rate of Levy %
Less than 90	114
90 or over but not more than 96	110
96 or over but not more than 98	105
98 or over but not more than 100	100
Over 100 but not more than 102	75★
Over 102 but not more than 104	50
Over 104 but not more than 106	25
Over 106	Nil

★ The levy for frozen beef may be fixed between 75 and 100 per cent of the basic levy

Variable levies for the different cuts of fresh and chilled beef and veal are derived from the live cattle levy by a system of coefficients. Levies on frozen beef are calculated separately.

Special arrangements, however, exist for the importing of beef under preferential levy arrangements.

1. The GATT quota whereby, under certain conditions, a specified amount of frozen beef may be imported levy-free into the various EEC member states from any third country. There is also a quota under which defined categories of breeding animals can be imported on preferential terms.

2. Favourable levy arrangements also exist for the importing of a specified quantity of frozen beef for manufacturing purposes; the size of the quota is determined by a balance sheet assessment of Community requirements of manufacturing grade beef and available supplies.

3. Under the Lomé convention a certain quantity of beef can also be imported free of duty from four African countries—Botswana, Kenya, Madagascar and Swaziland.

Export Restitution Payments

This allows EEC traders to participate on a competitive basis in the world trade for beef and veal. The ability to do this is a further aid to the maintenance of desired price levels.

Support Buying

Support buying measures normally apply to cattle and beef only—calves and veal are excluded. Direct purchases of specified commodities may be made by the Intervention Authority in the member states.★

The "permanent" intervention system is the effective support measure although there is provision for other support buying measures. The Intervention Authorities buy fresh or chilled beef of defined qualities originating in the Community and from animals dressed according to the Community arrangements (which require

★ In the United Kingdom the Intervention Board for Agricultural Produce is the Intervention Authority. The Meat and Livestock Commission, however, act as their agents for livestock and meat.

the removal of kidney knob and channel fat). The intervention price is currently set at 90% of the guide price. The actual buying-in prices are fixed at the beginning of each marketing year using a system of coefficients. During the course of the year, offers may be made to the Intervention Authorities at the fixed buying-in prices.

Under a regulation introduced in May 1978, support buying may be suspended for any particular category of beef in a certain country under certain market conditions. These conditions are based on the relationship between the country's reference price for that category and its upper buying-in price:—

(a) If the reference price for that category remains between 100 and 102% of its upper buying-in price for three weeks, the EEC Commission *may*, after consulting the Beef Management Committee, suspend support-buying of that particular category in the country concerned.

(b) If the reference price remains above 102% of the upper buying-in price for three weeks, the Commission may suspend support-buying for that category without having to consult the Management Committee.

(c) If support buying for a category is suspended and the reference price remains at or below 100% of its upper buying-in price for two weeks then the Commission *shall* automatically re-introduce support buying for that category in the country concerned.

(d) In the Irish Republic, support buying under these arrangements may not take place unless the Irish reference price is more than 85% of the EEC guide price.

Private Storage

EEC private storage aids may be paid to traders who agree to store specified minimum quantities of beef for a certain period of time. The rates of aid may be fixed in advance or by tender and may either relate to all types of beef (e.g. cows, steers, heifers) or to certain specified types. The schemes can apply separately to carcases,

33

forequarters and hindquarters and generally storage has been for a period of five or six months.

Monetary Compensatory Amounts

The aim of the system of monetary compensatory amounts is to compensate for the exchange rate movements between different Community currencies and between them and the European currency unit. While the £ sterling has devalued in relation to its reference rate, the United Kingdom MCA is applied as a charge on exports to all countries and as a refund on imports from all other countries.

PIGMEAT

The market price level, as in the case of beef and veal, is supported by protection against imports, export restitution payments and support buying. The details, however, are very different.

Protection Against Imports From Third Countries

Levies

Basic levies, imposed at the common external frontier, are one of the means of protection afforded to pig producers. The levy on the "basic product" i.e., carcase or sides, consists of two elements:—

1. An element equal to the difference in cost within the Community and on world markets of the quantity of feed grain required to produce a unit of carcase meat.

2. An element which is equal to 7% of the average of the sluice gate prices for carcase pigmeat applicable during the four quarters preceding 1 May of each year; this component is fixed for the twelve months beginning 1 August.

The levies on cuts of pigmeat and other pigmeat are derived from the levy on carcases or half carcases by a system of coefficients.

An additional protective element is also charged on certain processed pigmeat products.

34

In addition to the levies a system of sluice gate prices safeguards the Community market from low priced imports. The sluice gate price is fixed at a level which reflects the cost of pigmeat production in third countries, taking into account the cost of cereals and other feedingstuffs on the world market as well as other production and marketing costs. The basic sluice gate price is determined for pigmeat carcases and the sluice gate prices for a number of other commodities are derived by applying the coefficients used in the levy arrangements. These products, together with pigmeat carcases, are known as "pilot products".

Where products are imported into the Community below the sluice gate price, they may be subject to supplementary levies.

The levels of basic levies and sluice gate prices are subject to review on 1 May, 1 August, 1 November and 1 February. Any changes in the Community levels of cereal prices are reflected in adjustments to the levies at the August review.

Export restitution payments are made to enable exporters in the Community to compete on equal terms with suppliers in third countries.

The pigmeat regulations include provision for two types of intervention. Direct purchases may be made by the Intervention Authority in the member state or aid can be provided to enable private meat traders to put pigmeat into store.

The EEC Commission is empowered to decide the date at which the intervention measures are to be applied, their duration and the means by which the products held by the intervention agencies are to be released on the market.

The operation of the support buying measures is dependent upon the relationship between the Community BASIC PRICE and the Community REFERENCE PRICE for standard quality carcase pigmeat. Normally the basic price is fixed annually to apply

35

throughout a 12-month period commencing on 1 November, at a level designed to stabilise prices without leading to surplus production in the Community. The basic price in the EEC in July 1979 was 148·222 ECU per 100 kg dw (220 lb) (86·16p per kg). The basic price applies to carcases of Commercial Grade II, as defined in the EEC system of carcase classification.

When the reference price (the average price of carcases of reference quality) falls to 103% of the basic price and is likely to remain at that level, support buying measures *may* come into operation. In these circumstances direct support buying may be undertaken by Intervention Authorities at prices between 78 and 92% of the basic price.

Support buying may apply to pigmeat carcases, bellies and backfat. The support buying price for pigmeat carcases is normally specified for the standard quality, but the Community system provides for support prices to be specified for other qualities. The same price support level applies throughout the EEC and there is no provision for separate national or regional intervention arrangements.

Private Storage

In practice, no support buying has taken place since 1971 and private storage schemes have been operated when considered necessary. These may be introduced when the EEC reference price is below, and likely to remain below, 103% of the basic price. Fixed amounts are paid to traders who store minimum quantities of specific types of pigmeat (e.g. carcases, loins, bellies etc.) for specified periods of time.

Monetary Compensatory Amounts

Monetary Compensatory Amounts operate in the same way as for the beef sector.

Classification Scheme

The Community has adopted a common grading scale for pigmeat carcases in order to facilitate price collection and intervention arrangements on a common basis, as well as the development of trade within the Community. The original system was based on a combination of backfat measurements in relation to weight and the subjective assessment of muscular development. It was framed to

36

meet the particular trade requirements in the original six member states. However, the system has been adapted to meet the special circumstances of the trade in the United Kingdom, Denmark and the Irish Republic.

At present, the EEC has no common agricultural policy for sheepmeat and only charges common external ad valorem duties laid down under GATT on imports from third countries as follows:—

	% duty
Live sheep	15
Mutton and lamb, fresh, chilled or frozen	20
Sheep offals, fresh, chilled or frozen	3

Wool is classified as an industrial raw material and no import duties are applied. Trade between member countries is theoretically free of restrictions, but trade within the Community is affected by certain national restrictions—particularly in the case of France. France applies controls on supplies both from third countries and the new member states other than the Irish Republic. Other members of the Community are also subject to the arrangements if the meat or animals originated from third countries, the United Kingdom or Denmark. The French arrangements include:—

1. A virtual embargo on frozen mutton and lamb. An annual quota of around 3,000 tonnes is permitted which attracts special rates of duty.

2. Apart from pure-bred animals for breeding, sheep must not be imported at a liveweight of more than 44 kg (97 lb) or a deadweight of more than 22 kg (48 lb).

3. There is a system of import licensing. No licences are issued currently (July 1979) unless the domestic price of sheepmeat is above 19·85 francs per kg dw.

4. A special schedule of duties is imposed which varies according to

37

the domestic price. The duties range from a minimum of 5·50 francs per kg to a maximum of 8·50 francs per kg.

The application of these import controls, which discriminate against the United Kingdom has been challenged in the European Court of Justice and the decision on the case is expected soon.

Proposals for an EEC Sheepmeat Regime

In April 1978 proposals were presented for a Common Agricultural Policy for sheepmeat. A "light regime" was proposed with a basic price and a private storage rather than permanent intervention system. No MCAs would be applied and the current customs duties in third country trade would be replaced by variable import levies. Some discussion has taken place since then but no formal agreement has yet been reached. It would appear that any agreed regime would involve premium payments to compensate certain producers such as those in France for loss of income due to the introduction of a common policy for sheepmeat. With regard to third country arrangements, a system of voluntary restraint is likely to be agreed with New Zealand.

Pigs

Stock Resources

There are ten distinct established breeds of pigs in the United Kingdom and several other imported breeds are present in small numbers. Currently there are two principal breeds although some of the numerically smaller ones make a useful contribution to pigmeat production. Outlined below are the breeds, with an indication of their numerical importance, appearance, performance and contribution to pigmeat production.

BREEDS

One of the oldest established breeds in the United Kingdom and the most popular purebred pig, known in most other countries as the

Large White

Yorkshire, this is a white pig with erect ears. It has been bred for efficient production of lean meat over a wide range of liveweights. It is a prolific breed and the most efficient converter of feed into liveweight under national testing conditions.

British Landrace

The first Landrace were imported from Sweden as recently as 1949 and are now the second most common purebred pigs in the country. Developed in Sweden from the native European Landrace stocks including Danish Landrace blood, this breed was selected specifically for the Wiltshire bacon market, to be killed at about 91 kg (200 lb) liveweight. This is a lean pig with low amounts of backfat, but the Landrace is slightly less efficient as a converter of food than the Large White. The Landrace is a white pig, with "lop" ears, long body and rounded hams.

British Saddleback

This breed, formed in 1967 from an amalgamation of the Essex and Wessex Saddleback breeds, are black "lop" eared pigs with a white saddle over the shoulder. The Essex were seen most commonly in their county of origin and East Anglia in general, whereas the majority of Wessex were to be found in the south and south-west. The combination of the two breeds was a logical development because of their similarity and diminishing populations. Said to be a hardy breed suitable to outdoor conditions with good mothering ability, they tend to be early maturing, producing a relatively high proportion of fat and are unpopular with some sections of the trade because of a tendency to produce "seedy" cut. Compared with the Landrace and Large White, the feed conversion ratio and growth rate are poor and their use is now almost exclusively confined to providing crossbred dams with a white breed for use under outdoor, extensive systems.

Welsh

A breed almost identical in appearance to the Landrace with good prolificacy. Despite its name, the Welsh breed has found greatest popularity in the eastern counties where both purebred and cross-bred Welsh pigs have proved suitable for the pork, cutter and bacon markets. Under testing conditions, the carcase has a higher lean content than the Landrace but backfat measurements are higher.

Another black pig with a white saddle, but with erect ears. This breed has recently been imported from the USA and a small breeding population is now established. It is short, well muscled, with a high lean content in the carcase. In common with a number of American breeds, prolificacy is not very high.

Hampshire

None of these breeds is now of numerical importance, but staunch supporters of the breeds carry on with dedication the maintenance of these old British breeds. As economic forces and consumer tastes change, so the fortunes of breeds of livestock wax and wane. At present these breeds do not share national popularity, but it is probably wise to maintain small distinct populations of the rarer breeds in case they are needed to provide some particular characteristic which may prove to be lacking in breeds which are currently more popular.

Large Black, Tamworth, Gloucester Old Spot, Middle White and Berkshire

Small numbers of these imported breeds are under development and assessment by breeding companies and research organisations. Their value probably lies in their use as a component in a dam line or terminal sire line, rather than as purebreds.

Pietrain, Duroc, Lacombe and Poland China

Experimental evidence indicates that crossbred strains of pig perform better as mothers than purebreds. This superiority is expressed in more pigs born alive, more pigs reared, heavier piglet weights at 3 and 8 weeks of age plus a slightly faster growth rate to slaughter.

Crossbreds and Hybrids

Increasingly large numbers of crossbred or hybrid gilts are being kept for commercial production. The quality of carcase and efficiency of growth of the progeny of these crossbreds depends on the genetic quality of their purebred parents. The first cross gilt, resulting from crossing a Large White boar with a Landrace sow or vice versa, is probably the most popular crossbred, and generally gives excellent

41

prolificacy, good feed efficiency and a highly lean carcase in its progeny when mated back to one of the two parent breeds.

NATIONAL BREEDING HERD STRUCTURE

When a study is made of the national structure of breeding populations of pigs in different countries, it is clear that a form of pyramid generally exists. At the apex of the pyramid are a small number of prominent breeders who supply stock to a larger number of other breeders slightly further down the hierarchy. These breeders multiply the breeding stock which finds its way to the commercial herds forming the majority of herds at the base of the pyramid. Most countries in Europe have sought to exploit the natural pyramid structure by taking steps to ensure that herds forming the apex of the pyramid make the greatest possible progress in pig improvement. This is carried out through constructive breeding programmes.

Britain has also adopted this approach for a number of years, firstly under the auspices of PIDA and currently the Meat and Livestock Commission (MLC).

Pedigree

Pedigree means that the ancestry of stock is recorded accurately, normally by some central agency such as a breed society; it does not normally infer a specific potential of progeny. To maintain a "pedigree" herd of pigs means belonging to a breed society which is responsible for collecting the details of birth, allocating herd book numbers to adult breeding stock, ensuring that correct ear numbering procedures are maintained and/or specifying the correct breed type in respect of colour and conformation. Breed societies also have a responsibility for promoting the breed both nationally and internationally. Most, if not all, of the early breed improvement work was carried out by keen members of breed societies and to this day the majority of breeders involved in genetic improvement work are members of their respective breed societies.

Breeding Companies

In the mid 1960's a new development was seen in the pig industry when owners of poultry breeding companies began to take an interest in pig improvement. These poultry breeding companies had on their staffs trained geneticists who believed that poultry breeding techniques could be applied successfully to pigs. The interest generated by these "pioneers" fired the imagination of others, including meat processing companies and groups of traditional pig breeders. Thus emerged some thirty or more pig breeding companies to compete with the pedigree breeder for a stake in the breeding stock business. Companies started by poultry breeders tended to be tightly-knit organisations with nucleus lines of two or three breeds which were interbred to provide a crossbred sow for sale to commercial breeders. The companies also generally offered boars from their pure breeds which were used on the crossbred females. Multiplication of crossbred females normally remained under the direct control of the company. Other companies were less formally organised and often took the form of consortia of pedigree breeders seeking to co-operate to promote faster progress and better marketing. Breeding companies in general now command a very substantial part of the total female breeding stock market and an increasingly large part of total boar sales.

Role of Breeds

Britain is fortunate in having a number of breeds which are prolific, unlike some of the breeds recently imported from North America. Most crossbred commercial females have the Large White and Landrace breeds as their main components. The choice of boar to use on these crossbreds will vary. At present, for markets requiring a slaughter generation pig with a clearly defined level of fatness and conformation, the Landrace is normally used as a "terminal" sire. Where growth rate, feed conversion efficiency and a high total lean content are the requirements, the Large White boar is more commonly used. However, as European Market requirements begin to exert some influence on our home market, breeds with different characteristics will probably assume importance. In general, Europe tends to require a pig which is shorter, thicker and meatier with large hams. In this context, breeds such as the Hampshire and Pietrain may prove useful as components in a crossbred, terminal sire line used to produce the slaughter generation.

Improvement Programmes

The profitability of pig production depends on three main factors and improvement programmes aim to identify and select animals which fulfil the undermentioned requirements.

(a) A high level of sow productivity, i.e. the number of piglets reared per sow per year.
(b) The economical conversion of feed into pigmeat.
(c) The production of the lean carcases which are now demanded by the consumer.

CROSSBREEDING

The first factor of economic importance, sow productivity, is largely dependent on correct sow management in terms of feed, housing, health and other points of this type. Performance can, however, be improved by mating together two entirely separate breeds. The number of piglets born and reared, and their weights, are normally greater than that expected from either of the two parent breeds alone. This bonus is generally known as "hybrid vigour". If this cross-breeding process is taken one stage further and a first cross sow is used, a further boost to performance can be expected. The overall benefit to sow productivity expected from the use of first cross sows as compared to purebred sows is as follows:

Number of piglets born alive	+ 5%
Number alive at 3 weeks	+ 7%
Total weight of litter at 8 weeks	+ 11%

SELECTION FOR GENETIC IMPROVEMENT

The other two factors of economic importance, economy of growth and carcase quality, are to a larger extent under genetic control. If superior animals can be identified, these can be selected to become the parents of the next generation. However, the genetic make-up of an animal is fixed for life; it cannot get better however long the animal lives. For this reason, progressive improvement of genetic potential can be achieved only by continually introducing better parent stock. There are, therefore, advantages to be gained by ensuring that breeding stock are not retained for too long. As boars

44

achieve sexual maturity at about 6 months of age and gilts at a similar time, and having a gestation period of 4 months, it would theoretically be possible to "turnover" each generation at less than yearly intervals. In practice, however, this speed of turnover is not achieved nor is it necessarily desirable and intervals of about 18 months are more common.

The improvement scheme is based on the pyramid structure described earlier. Four different categories of herd are involved, each with a very special and important function in genetical improvement. The first and most technically involved are the "Nucleus" herd and "Breeding Companies". These two categories of herd spearhead the improvement work and form the apex of the pyramid. About sixty Nucleus herds and thirty Breeding Companies make up the category and it is to these herds that MLC currently direct the greatest part of their testing resources. The next category is the Reserve Nucleus herd, which, as the name implies, includes herds striving to enter the Nucleus category. The final category is the Nucleus Multiplier group of herds. These herds take stock from the apex of the pyramid—Nucleus or Company herds—and multiply them into numbers large enough to fulfil the demand from those herds forming the pyramid's base, namely Commercial herds which concentrate on producing meat animals.

THE MLC PIG IMPROVEMENT SCHEME

Earlier reference was made to MLC Central Testing and that these resources were primarily reserved for Nucleus and Breeding Company herds.

CENTRAL TESTING

These central testing facilities consist of four testing "stations", strategically sited to provide breeders in Great Britain with test facilities within a reasonable distance of their farms. The English stations are situated at Corsham (Wilts), Stotfold (Herts) and Selby (Yorks) and the Scottish station is situated near Stirling. Each station has facilities for testing groups of four litter mates consisting of two entire boars, one castrate and one gilt.

The principle behind pig testing is a simple one. Pigs vary in their ability to convert feed into lean meat efficiently. Some of this difference in performance is due to the effects of housing, type of ration, climate and pig health (environment), but some of the difference is due to genetic factors inherited from parents (heritability). The test identifies pigs with an inherited ability for economic conversion of feed into lean meat so that the best pigs may be selected to become parents of the next generation.

Central test stations are designed, built and managed in a way which helps to standardise climatic effects and differences in nutrition and housing, thus revealing genetic differences more clearly.

Pigs from Nucleus, Reserve Nucleus and Breeding Company herds are sent to central testing stations at a young age and reared to slaughterweight alongside contemporary groups. Each group of pigs from a breeder is in competition with groups from other breeders under similar conditions of environment.

The Test Group

Normally, a test group comprises two boars, one gilt and one castrated male, all from one litter. Each pair of boars is housed in a kennel type house. The boars lie together, but are fed separately so that feed consumption can be recorded accurately for each boar. The castrate and gilt (sibs) are housed in a fully enclosed building and fed together.

The pigs enter the testing station when their individual weights are between 18 and 25 kg (40 and 55 lb) and the test starts when each boar weighs 29 kg (60 lb) and the sibs total 54 kg (120 lb). The test is completed when each boar weighs 91 kg (200 lb) and the sibs' total weight is 165 kg (364 lb) at which time the sibs are slaughtered. Detailed carcase measurements are taken and a sample joint is fully dissected into lean, fat and bone. Feed consumption, liveweight gain and ultrasonically measured fat depths are recorded on each boar. This information is combined with feed consumption and growth records from the sibs, plus their carcase measurements, and expressed in a selection index as a points score for each boar. The average score is always maintained at 100 points, although the actual performance

46

of boars in the testing stations is improving year by year. Boars scoring below 90 points are slaughtered and boars over 90 points are offered back to the breeder.

These index points reflect the likely extra profitability of the progeny sired by one of the tested boars. It is not possible to put a precise value of an additional index point scored by a boar owing to the rapidly changing values of feed and pigmeat. However, in view of the number of progeny a boar may sire in his lifetime—it is likely to be a thousand and could be many more—an extra value per offspring of one pence per index point scored by their sire would total £10. Hence, the difference between an average boar of 100 points and one of, say, 120 points could amount to £200 over a boar's working lifetime. About 5000 boars a year enter test and about 2000 are returned for use on farms.

On-farm Testing

The importance of central testing lies in the accuracy with which breeding value can be estimated and the between herd or farm comparisons which can be made. The impartiality of the central test gives an authenticity to results which enables breeders to use these as a sales factor and buyers to discriminate on the basis of these results.

The disadvantages of central testing are perhaps less obvious at first glance. The cost of taking groups of pigs to testing stations, the loss of profit on those pigs slaughtered as part of the test or subsequently slaughtered on grounds of poor performance and the transport cost of good performing animals after test, add up to considerable sums of money. Central tests, with their high housing, management, data collection and computing costs, make it expensive for those operating central test stations. Additionally there is a health risk incurred when pigs from large numbers of different farms are brought together on a central station and subsequently a proportion returned to farms.

If a farm test could be devised which could measure, with accuracy on the live pig, the economically important characteristics, many of the disadvantages of central testing would be overcome. It is too

47

early yet to say that such an on-farm test has evolved, but considerable progress has been made. Most of the larger breeding companies operate their own test stations and most of the breeders in MLC's pig improvement scheme test replacement female breeding stock on their own farms.

The basis of on-farm testing sponsored by MLC is the measurement of backfat by echo-sounding of large groups of gilts or boars on farms. An operator calls every 3 or 6 weeks to measure the backfat depth at four points on pigs weighing 68–113 kg (150–250 lb) liveweight. Two of these fat measurements are combined in an index score with average daily liveweight gain figures calculated from the birth of the pig. The index score indicates the ranking order in terms of genetic merit for the pigs in the batch measured. Improvements are constantly being made to the index and it has already been shown to be of great value to breeders. The test, at present, is less accurate than central testing and considerably less factors are measured. Because of large variations between farms, and also between batches on the same farm, comparisons between pigs on different farms or in different batches can be very misleading and should not be made. However, about 70,000 potential breeding gilts and 10,000 young boars are assessed in this way each year.

Commercial Production Evaluation

Commercial Product Evaluation is a form of test devised by the MLC to provide objective information for commercial producers on gilts and boars being offered for sale by pig breeding companies. This type of test aims to evaluate the genetic level of company pigs in relation to either crossbreds from non-tested sources or purebreds from the improvement scheme or in relation to both according to availability of space. The test also aims to show the rankings of companies in the various characteristics of economic importance.

The test also evaluates both reproductive performance of gilts and the performance and carcase quality of their progeny under commercial conditions.

Each company is represented by 24 gilts and 7 boars and two litters from each gilt form the test. Thus, a total of 48 litters per company is obtained in any one year and samples of pigs are fed to three different weights on two different feeding systems. Two systems of feeding are practised, ad libitum and restricted, and slaughter weights are appropriate for light pork, 61 kg (135 lb), bacon, 91 kg (200 lb) and manufacturing pigs 118 kg (260 lb).

The test is undertaken at Celyn Farm, Northop, Flintshire, where accommodation for over 250 sows and a proportion of their progeny is available.

Care is taken to ensure that a truly representative sample of breeding stock on offer to commercial producers is obtained from each company.

Annual results will be published.

Market Requirements

Of all pigs slaughtered in the United Kingdom about 85% are sold privately with about 15% being sold by auction.

In the United Kingdom we eat about 61 kg (134 lb) of meat per head each year. Of this total about 12 kg (27 lb) is fresh pork including sausages and pies while 11 kg (25 lb) is bacon and ham. Most of the pork we eat is home produced while approximately 50% of our bacon and ham supplies are imported, mostly from Denmark.

The fresh meat market requires two distinct classes of pigs with a proportion of fresh meat being supplied from a third class.

The light pork pig, killed at 61 kg (135 lb) liveweight or less, is required in the south-east of England for small joints for purchase with bone-in. A particular requirement of this trade is the shape of the leg and the requirement is for leanness in order that joints can be sold whole without fat trimming. From a production viewpoint it is essential that a pig for this market grows fast and economically and that a premium is paid to the producer. Even if these three criteria are fulfilled the high overhead costs per score of finished pig make light pork production a very specialised job.

An increasingly large amount of fresh pork is produced from cutter pigs killed at about 64–82 kg (140–180 lb) liveweight and in some cases of even higher weights. Because of the amount of cutting which may be undertaken, the conformation of these pigs is less important and the requirements are for maximum lean content and a large well-shaped eye muscle. Many of the large multiple retail shops and supermarkets prefer these heavier pigs for their trade.

The Wiltshire bacon trade requires a pig killed in the range 82–95 kg (180–210 lb) liveweight and the conformation and fat distribution are important. The trade requires a long, lean pig with well-shaped hams, light shoulder and a minimum of fat along the centre line. However, an increasingly large quantity of bacon is made from the curing of individual joints as opposed to whole sides. When this type of operation is undertaken, both lighter and, more often, heavier pigs can be used.

The manufacturing trade requires a pig of over 100 kg (220 lb) live-weight and pigs of 118 kg (260 lb) are required by some companies. The precise conformation and level of fatness is less important for these pigs, though a well-shaped eye muscle and overall high lean content are desired.

Most of the foregoing specifications are satisfactorily met by white breeds either purebred or crossed. Breeds such as Hampshire and Pietrain are more commonly used as a component of pigs slaughtered at heavy weights.

Commercial Production

Some pig production was traditionally found on a high proportion of farms but as farmers, in general, have tended to limit the number of their enterprises, pigs are now found in fewer but larger units.

MAFF census figures show that 15% of agricultural holdings in England and Wales in 1977 had some pigs. Total pig production has risen over the years while it has been a regular feature of the annual statistics over the last decade that the number of holdings with pigs has fallen.

TABLE 1

Pig Numbers—England and Wales (June) (Source: MAFF)

	1965	1969	1973	1977	1978	1979
Total Pigs (000)	7979	6150	7297	6561	6503	6640*
Number of holdings with pigs (000)	94·6	63·9	46·6	29·6	—	—
Average herd size	65·3	96·2	156·7	221·6	—	—

* Provisional

Table 1 shows that although there is some fluctuation in total pig population, concentration into larger units has been consistent. Table 2 shows that there has been a similar trend in the size of breeding units; there has been only a very slight tendency for specialisation as 75% of the holdings with pigs have some breeding animals.

TABLE 2

Breeding Sows—England and Wales (Source: MAFF)

	1965	1969	1973	1977	1978	1979*
Number of sows and gilts (000)	756·3	732·5	829·1	705·0	709·0	722·0
Number of holdings with sows and gilts (000)	73·0	48·4	35·0	22·2		
Breeding units as % of total holdings with pigs	77	76	75	75		
Average herd size	10·4	15·1	23·7	31·8		

* Provisional

There are distinct regional variations in the size of units: those in the arable areas are larger than those found in the more western areas where farms are smaller. This is partly due to the availability of cereals and straw in the arable areas while on smaller grass farms the pigs have made more use of dairy by-products.

TABLE 3

Regional Pig Numbers—June 1977 (Source: MAFF)

	Total pigs (000)	% Total	Average herd size	Total Sows (000)	% Total	Average herd size
Eastern	1,748·2	26·6	367·3	185·4	26·3	51·4
South Eastern	845·6	12·9	331·6	95·3	13·5	51·8
East Midland	644·3	9·8	248·1	67·2	9·5	35·3
West Midland	586·5	8·9	170·4	64·0	9·1	25·2
South Western	874·0	13·3	141·5	93·7	13·3	20·7
Northern	439·8	6·7	191·8	48·7	6·9	28·2
Yorks & Lancs	1,277·1	19·5	287·8	132·1	18·7	38·3
Wales	146·0	2·3	43·6	18·8	2·7	7·2

Pig production falls naturally into two places; the breeding unit with mature animals producing weaners followed by the feeding stage in which the weaned pigs are taken to an appropriate slaughter weight. Both stages may take place on one unit or the weaners may be transferred or sold to a specialised feeding unit when they are in the weight range 20–50 kg (45–110 lb).

BREEDING UNITS

Systems of production in breeding units range from extensive outdoor systems involving little labour or capital through to closely controlled factory type techniques. When large areas of relatively poor land are available, sows may be kept in large batches; at farrowing they have cheap uninsulated shelters and little attention is given; weaning is on a batch principle, several litters being weaned at the same time. Coloured sows with some Saddleback in their breeding are favoured for this system. The majority of systems, however, involve much closer control of the sow. Where conditions are suitable, many producers prefer to run sows outside on grass during pregnancy but in poor weather more feed is needed; also, poor working conditions can be a problem in winter. The majority of sows are kept indoors in batches in yards often with individual feeders to ensure that each sow receives an adequate ration and wastage is reduced. More recently there has been a trend to keep pregnant sows as individuals, either loose or tethered, in stalls within environmentally controlled buildings.

Towards the end of pregnancy sows are moved into specialised farrowing quarters which limit the movement of the sow to avoid overlying the piglets and provide a safe warm area for the litter. The most commonly adopted method is to use a crate which ensures that the sow, when lying down, lowers herself gradually while piglets have a safe "creep" area at each side. The creep area is heated and there is space available for feeding of the piglets. Sows and litters normally stay in the crates for 2–3 weeks and are then moved to rearing pens, often grouping up to six sows and litters in a large multi-suckling pen. The mixing of sows and litters at this stage results in little fighting and reduces the stress effect resulting from weaning and mixing of pigs at a later stage. Weaning normally takes place at 5–7 weeks.

Weaning at 2–4 weeks is practised in a number of units to try to lower cost per weaner by boosting the sow's production of pigs per year. The piglets at this age need special conditions; very young ones are weaned into cases in a controlled environment house while those weaned at 3–4 weeks can be put in veranda type housing with a low-roofed solid-floored kennel and a slatted or mesh dunging area. Success with these systems requires that the piglets are separated from the dung and urine. From about 8 weeks of age the young pigs' basic requirements are good food and warm dry conditions. General practice is to run them in batches of up to sixty pigs in a weaner pool on ad libitum feeding. From this they can be drawn in level lots for sale or transfer to a feeding section. Sales may take place through auction markets, private contact or one of the groups organised by producers, by a feedingstuffs supplier or by a large scale buyer of finished pigs. Genetic improvement programmes do not benefit the weaner producer directly but do improve the feeder's margin. Often arrangements are made to pass some of this back to the breeding unit on condition that the breeder takes advantage of improved stock which is available.

FEEDING UNIT

It has been found that it is not normally advisable to put pigs into the finishing house until they are at least 32 kg (70 lb) in weight and most feeders continue the weaner pool system until this weight is reached. Systems of housing and management in feeding houses are numerous; some possibilities to be considered are:

1. controlled environment or natural ventilation
2. straw or slurry system
3. solid floor or slats
4. trough or floor feeding
5. wet or dry feeding
6. ad libitum or restricted feeding
7. manual or mechanical feeding

Many decisions are based on personal preference but availability of capital, cost of straw, muck disposal facilities, locality, availability of labour and type of output, all have an important bearing on the system to be adopted. It is not possible to recommend any system of housing to give the best results as management is a vital factor. Given the necessary confidence in a house, it will normally produce results which stand comparison with other units.

PROFITABILITY

The profitability of the enterprise is governed by the margin between costs and returns.

Costs consist of the fixed parts such as rent, depreciation and interest charges on loans and the variable parts of which feed forms over 90%.

As food is the major part of the cost of producing a pig, particular attention must be given to efficient usage at all stages. The breeding sow has a basic requirement of food of over 1 ton per year and variations around this figure are relatively small according to her litter production. As the number of weaners increases so the quantity of food fed per weaner decreases.

The effect of sow productivity on cost of feed per weaner

Number of weaners per sow/year	14	16	18
Sow feed including share of boar's food—tonne (cwt)	1·27 (25·0)	1·30 (25·5)	1·52 (26·0)
Creep—kg (cwt)	178 (3·5)	203 (4·0)	229 (4·5)
Average food per weaner—kg (lb)	103 (228)	94 (207)	86 (190)

With large litters weaned at 6–7 weeks the sow is normally given some additional food to aid milk production and, of course, the amount of creep feed is the same for each piglet.

The number of weaners per litter can be affected by management of the sow post-weaning, at service and at farrowing. Judicious use of crossbreeding to make use of the phenomenon termed hybrid vigour can increase productivity by 1·5 pigs per sow per year. Control of the sow's intake during pregnancy is best achieved by provision of individual feeders when sows are run in groups or by keeping sows in individual stalls in a controlled environment building.

The inability of the average sow to suckle adequately more than ten pigs coupled with advances in the production of foods suitable for young pigs, has led to much interest in reducing the suckling period to allow the sow to recommence the breeding cycle and produce more litters per year.

Based on a gestation period of 115 days and with an average of 5 days between weaning and service, lowering the age of weaning from 7 weeks to 3 weeks can theoretically produce an extra 0·25 litter per sow per year. It is found that the sows take longer on average to come on heat with 3-week weaning, and that the number of pigs born per litter is lower so some of the advantage is lost. In practice, weaning at around 5 weeks after farrowing is very popular.

In the post-weaning to slaughter stage food forms an even higher proportion of the cost of production. As the pig gets heavier, it requires more food to put on each successive pound of liveweight so, with heavier cutter pigs, worse overall food conversion is expected than for a light porker pig. Food conversion ratio (f.c.r.) is the average quantity of meal required to put on 1 kg of liveweight but the cost per tonne of the food also is very important.

The feeding of swill to pigs, once a widespread practice in urban areas, is considered to be a health hazard and recent legislation to control the processing and feeding of swill may further deplete the number of producers using this valuable food. Pigs which have been fed on swill may not be sold through auction markets but must be sold directly for slaughter.

FIXED COSTS

The portion of fixed costs which is debited to each pig produced is reduced by spreading these costs over as large a number of pigs as possible. This favours the larger herd which can also make the most of available labour.

The health of a herd has an important bearing on the economics of a unit but incidence and effects of disease are difficult to quantify particularly for chronic forms. Use of antibiotics as growth promoters is now strictly controlled and many of them can be used only under veterinary supervision.

In general terms, type of production is determined by availability of outlets having regard to the producer's assets, both of these may limit his choice. For example, the building available may not be suitable for breeding animals. Returns will depend on volume of sales, on quality and on marketing expertise. The larger the herd, the better equipped it is to provide level deliveries of pigs and to lower unit costs of transport and handling. The great majority of pigs for slaughter are sold on a deadweight basis with some form of quality payment. The producer has to decide whether payments offered for some aspect of quality justify any additional expense likely to be incurred to attract the premium. Severely restricted feeding to meet tight grading schedules may be uneconomic. It is the proportion of pigs in each grade which determines the average price and so governs the return. When marketing pigs it is important to know what specification is required and to understand the terms of any contract particularly on the financial responsibilities for other charges such as transport and insurance.

56

Cattle

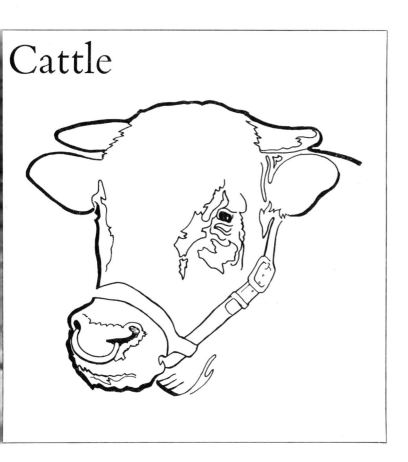

Dairy and Beef Cow Numbers

The national breeding herd comprises both dairy and beef cows. The dairy cows are all of dairy or dual purpose breeds—mostly Friesians, and the majority of their calves are artificially reared either as replacement breeding stock or for meat production. In contrast, the beef herd includes purebred beef breeds, crosses between beef breeds and beef/dairy crosses. The calves produced are suckled by their dams and reared for beef production for slaughter or as replacement breed stock.

There are approximately 13·7 million cattle in the United Kingdom of which about a third are breeding cows (4·8 million). One-third of the cows are beef cows, most of the 3·2 million dairy cows are Friesians.

From 1960 to 1975 there has been an increase of 31% in the number of breeding cows. Only 6% of this increase has been dairy cows and the remainder of the increase has been in the number of beef cows. Put another way, while dairy cow numbers have increased only slightly, the beef cow herd has more than doubled in the same period. Table 1 compares cattle numbers in 1960, 1975 and 1978.

TABLE 1

Breeding Herd and Total Cattle Numbers—June Census for UK
(000 head)

	1960	1974	1975	1978 (provisional)
Dairy cows	3165	3393	3242	3278
Beef cows	848	1887	1899	1584
Heifers in calf	823	1041	903	866
Total cattle and calves	11771	15203	14717	13655

The substantial increase in the national beef breeding herd has been largely the result of government policy and the decline in the profitability of milk production. In 1959 the government was determined that home beef production should be increased and, as an incentive, increased the rate of hill cow subsidy from £2 to £12, and calf subsidy from 75p to £9·50 for steer calves. At that time milk production was above the needs of the liquid milk market and therefore any incentives needed to be directed towards increasing beef cow numbers and the retention of calves from the dairy herd for beef production rather than any increase in dairy cow numbers.

The hill cow subsidy was again increased in 1965 to £13 per head and in 1966 the government, being anxious to arrest the decline in beef cow numbers outside the hill areas, introduced a further beef cow subsidy of £6·50. This new subsidy was for any cows kept wholly or mainly for the purpose of producing calves for beef which were not eligible for the hill cow subsidy.

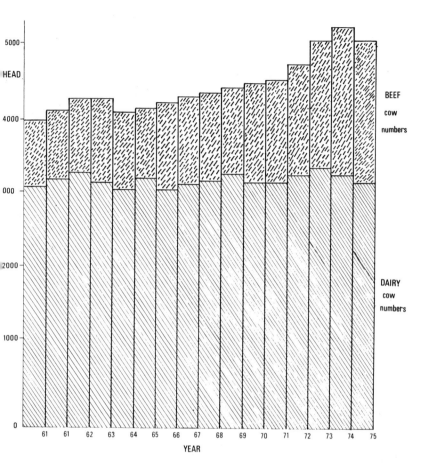

Fig. 1. *Breeding Cow population.*

Beef cow numbers have increased at a fairly substantial rate since 1966 while the government increased both the hill and beef cow subsidies every year until 1971. Between 1970 and 1975 the subsidies have remained constant at £18·75 per head for hill cows and £11·00 per head for beef cows. In 1977 the lowground beef cow subsidy was removed. At about the same time the calf subsidy was also removed. It was the combination of these two factors which led to the marked decline in the size of the national suckler herd. In 1976 a new scheme—Hill Livestock Compensatory Allowances was introduced to replace the Hill Cow Subsidy scheme. These allowances

59

are paid on an annual basis per cow and the current rate is £29 per cow.

Government policy as regards the dairy herd has been to maintain the herd at a level so as to meet the demand for liquid milk supplies. This has meant that any incentive payments have been linked to liquid milk consumption. The production of milk since 1960 has risen by about 10%. This has been brought about almost exclusively by higher yields per cow as the dairy cow population has remained almost static. Until 1965 milk production was in excess of requirements and this led to a small reduction in dairy cow numbers. Subsequent to 1965 there has been a small increase in the size of the dairy herd in step with the increase in demand for milk and milk products, but this trend appears to have been reversed since 1973 possibly resulting from the sudden increases in costs of production.

Bearing in mind the time lag from a policy decision to the availability of heifers for breeding it would appear that the cattle industry has responded to the demand for increased beef production in line with government objectives. It is therefore likely that future trends will also tend to follow government policy.

In England and Wales the number of holdings with beef cows has fallen from a peak of 63,000 in 1966 to 53,406 in 1977. Similarly in Scotland the number of holdings with beef cows has fallen from 20,000 in 1965 to 15,300 in 1977.

The average size of beef herds is small and in 1977 was 16·6 cows in England and Wales and 32·8 in Scotland (this includes pedigree herds which tends to reduce the overall average). The average herd size, however, is increasing. In England and Wales in 1977 12·9% of the beef cows were in herds of 100 while 12·2% were in herds of under ten cows. The comparable figures for Scotland were 29·8 and 4·5%, respectively. Full details are given in Table 2. There is little doubt that the concentration of production into fewer but larger herds will continue.

TABLE 2

Beef Cows and Holdings with Beef Cows by Herd Size 1977

Size of herd (beef cows)	England and Wales			Scotland			Great Britain	
	No. of cows (000)	% cows	% holdings	No. of cows (000)	% cows	% holdings	% cows	% holdings
1– 4	43·2	4·9	36·0	8·0	1·6	22·9	3·7	33·3
5– 19	199·4	22·5	36·6	45·8	9·1	28·3	17·6	34·6
20– 39	237·3	26·6	16·2	83·6	16·6	19·4	23·0	17·0
40– 59	152·9	17·2	6·1	84·1	16·7	11·4	17·0	7·1
60– 99	141·2	51·7	3·6	131·9	26·2	11·4	19·6	5·4
100–199	88·0	9·9	1·3	116·0	23·0	5·8	14·7	2·3
200 plus	26·6	3·0	0·2	34·5	6·8	0·8	4·4	0·3
Total cows	888500			503900			1392400	
Total holdings	53400			15300			68700	
Average herd size	16·6			32·9			20·3	

Home-produced beef accounts for well over 90% of total UK beef supplies. Figure 1 shows the sources of home-produced beef.

Total production averages about a million tonnes per year. Cow beef accounts for about a quarter of this production, a high proportion from the dairy herd because of the shorter average production life of dairy cows compared to beef cows. A very high proportion of

TABLE 3

Estimated Sources of Home-Produced Beef 1973

Breeding herd			Meat production (% total tonnage)
Dairy cows—Calves		—Beef×	21
		Dairy	20
	Cull cows		17
Suckler cows—Lowland 24%		—Calves	6
	Upland and Hill 76%	—Calves	19
	Cull cows		6
	Imported Irish cattle		11
		Total	100

the calves reared for beef are crossbred. The dairy herd contributes just under 60% of total home production.

Thus the major proportion of home-produced beef continues to be a by-product of the dairy herd, particularly in England, but only a small proportion of calves from the dairy herd are finished on dairy farms.

Sources of Calves for Beef Production

BEEF FROM THE DAIRY HERD

Calves from the dairy herd accounted for 41% of home-produced beef. Table 4 gives details of the breed distribution of the national dairy herd. Overall the Friesian breed accounts for about 85% of dairy cows and the Ayrshire, the second most numerous dairy breed, accounts for less than 17% of dairy cow numbers. The proportion of Dairy Shorthorns in the dairy herd in England and Wales has decreased from 23% in 1955 to about 1% in 1974.

TABLE 4

Dairy Herd Breed Distribution
(Cows and heifers in milk and cows in calf)
(Source: 1977 Dairy Facts & Figs. MMB)

Breed type	England and Wales 1974	Scotland 1975	N. Ireland 1976
Ayrshire	3·6	47·0	2·4
Friesian	81·0	27·9	91·3
Friesian/Ayrshire cross	★	24·1	★
Guernsey	2·8	} 0·7	★
Jersey	2·2		★
Dairy Shorthorn	0·9	★	5·4
Others	9·5	0·3	0·9
Total cows (000 head)	3086	299	258

★ Included in "others"

The Ayrshire as a breed is more common in Scotland compared with the Friesian but the percentage of Friesian/Ayrshire cross cows had increased from 2·5% in 1965 to 24·1% in 1975 at the expense of the purebred Ayrshire.

TABLE 5

Dairy Herd Breed Distribution—England and Wales 1955-74
(Source: 1973 Dairy Facts & Figs. MMB)

Breed type	1955	1960	1965	1970	1974
Ayrshire	16·3	17·7	15·7	9·7	3·6
Dairy Shorthorn	22·8	13·3	6·3	2·5	0·9
Friesian	30·7	44·5	64·2	76·3	81·0
Guernsey	4·9	6·1	5·7	5·2	2·8
Jersey	2·3	3·1	4·3	3·8	2·2
Others and crossbreds	23·0	15·3	3·8	2·5	9·3

In 1973/74, 80% of all dairy bull licences were for Friesians.

Information on the relative importance of the breeds of bulls used in Great Britain is obtained from the bull licensing statistics published by the Agricultural Departments. In 1974, 24% of all bulls were Friesian or Holstein and 30% were Herefords. An MLC estimate of the breed types of calves produced from the total breeding herd is given in Table 6.

TABLE 6

Estimate of Availability of Breeds or Types of Calf for Beef Production
(excluding female replacements for both dairy and beef herds)

Type of breed of calf	%
Friesian	31
Beef	27
Beef × Friesian	25
Friesian × Dairy	6
Beef × Ayrshire	3
Ayrshire	2
Others	6
	100

Figures for England and Wales from 1955 to 1974 (Table 5) show clearly how there has been a dramatic change in the structure of the dairy herd in favour of the Friesian, largely at the expense of the Dairy Shorthorn and to a lesser extent the Ayrshire. This has coincided with the increased demand for beef production as well as efficient milk production for which the Friesian is admirably suited.

The use of artificial insemination in dairy herds is well established and AI statistics, showing the relative use of bulls of different breeds, indicate the popularity of Friesian and Hereford sires (Table 7). Because AI accounts for over 60% of all dairy matings, these figures

TABLE 7

Inseminations in Great Britain by Breed of Bull 1976–77

Breed of Bull		England and Wales %	Scotland %
DAIRY			
Ayrshire		1·2	13·3
Friesian & Canadian Holstein		61·2	48·2
Guernsey		1·6	—
Jersey		1·5	0·7
Dairy Shorthorn		0·3	—
	Sub total	65·8	62·2
DUAL PURPOSE			
South Devon		0·3	—
Welsh Black		0·4	—
	Sub total	0·7	—
BEEF			
Aberdeen Angus		4·1	6·7
Beef Shorthorn		—	1·5
Blonde d'Aquitaine		0·1	—
Charolais		5·2	8·2
Chianina		0·1	—
Devon		0·6	—
Hereford		20·9	11·9
Limousin		0·5	—
Murray Grey		0·5	2·2
Simmental		1·1	4·4
Sussex		0·3	—
Others		0·1	2·9
	Sub total	33·5	37·8

give a reasonable assessment of national trends. During the past decade between 30 and 40% of dairy cows have been mated with beef bulls, and the majority of the crossbred calves from the dairy herd are Hereford Friesian steers and heifers.

Aberdeen Angus bulls are often used on dairy heifers to provide easier calving and a similar number of Charolais matings are made with mature cows. The advantage of these breeds, the Hereford and to a lesser extent the Simmental, is that they colour mark their calves and hence provide an easy method of breed identification when the calves are sold to commercial beef rearers and feeders. Several of the minority British breeds (South Devon, Devon, Lincoln Red and Sussex) provide useful crosses from dairy cows but, because of their lack of colour marking, have so far failed to make any marked impact in crossing with dairy cows.

BEEF FROM THE BEEF HERD

Calves from the beef herd tend to be produced as the result of two or three way crossing. In the UK the beef cow is normally crossbred. Traditionally this cow was produced by crossing beef breeds and frequently a third beef breed was used as a top-crossing sire. The most well-known crossbred cow used is the Blue Grey which is produced by mating a Galloway cow with a Whitebred Shorthorn bull.

A recent interesting development which has arisen because the supply of replacement breeding cows has been a major problem during the period of the marked increase in the size of the national herd, has been the increased use of beef cross Friesian cows for this purpose, particularly in the lowlands. This type of dam tends to have a better milk yield and produces higher calf growth rates than the traditional beef cows (see pp. 76 & 77). In addition, as a by-product of the dairy herd the breeding complications of maintaining two separate pure beef breeds to produce the crossbred commercial beef cow are avoided.

The bulls most widely used in suckler herds are those of the earlier maturing beef breeds such as the Aberdeen Angus and Hereford. Many of the calves produced are of attractive conformation but their small size and lack of growth potential makes them difficult to finish profitably. With more information becoming available on the performance of the different breeds, there is an increasing interest in the use of the larger breeds as sires such as the Charolais, Simmental, South Devon and the red breeds (Devon, Lincoln Red and Sussex). If calves are to be sold in the weaned calf sales, colour marking can still command a premium but if the calves are to be finished on the same farm or sold privately the choice of breeds is not limited to those which are marked in this way.

Irish Store Cattle

The importation of Irish store cattle has represented a useful addition to UK beef production. Since 1960 the number imported per annum has ranged from 500,000 to 700,000. In the past these cattle were predominantly beef cross Dairy Shorthorns but increasing numbers of beef Friesians and pure Friesians are now involved.

Pedigree Breeding Structure

There is no formal stratification of beef breeders as there is with pig breeders, but a hierarchical situation has developed whereby a rela-

TABLE 8

Pedigree Recording Statistics

	No. of MLC recorded herds June 1979	Average herd size (no. of cows)	No. of weighings (all sexes, all weighings) July 1978–June 1979	No. of bulls recorded★ July 1978–June 1979	No. of bulls recorded as percentage of bulls licensed 1975/6†
Aberdeen Angus	92	19	5688	763	47
Beef Shorthorn	10	21	323	119	58
Charolais	392	10	7660	807	100
Devon	34	25	3432	240	63
Hereford	247	24	16226	2342	34
Limousin	70	8	1180	121	100
Lincoln Red	22	47	2458	238	100
Simmental	119	9	2533	302	90
South Devon	77	23	4908	323	89
Sussex	37	37	4408	325	100
Welsh Black	45	25	3708	176	37
Overall	1145		52524	5756	66

★ Either 400- or 500-day weight
† Last year of licensing

tively small number of elite breeders tend to dominate breed development. These herds feature prominently in the export trade and in the supply of stud bulls to less influential pedigree herds. The latter group, much larger numerically, supplies the bulk of the crossing bulls required by commercial beef and dairy herds. In the major breeds the small groups of elite breeders normally do not rear and finish cattle commercially for slaughter. The number of cattle registered of the recently imported breeds increases year by year.

The average size of pedigree herds is small, and this makes selection and improvement difficult. The average size of MLC recorded pedigree herds by breed type are given in Table 8.

Beef Improvement Programmes

The beef improvement programmes being carried out in Britain are:

> Commercial beef recording.
> Recording pedigree stock on farms.
> Central bull performance testing.
> Progeny testing.
> Commercial breed evaluation.

COMMERCIAL BEEF RECORDING

The schemes for commercial producers are intended to provide physical and financial information on the various systems of beef production for management purposes. Records are collected of daily gains, slaughterweights, stocking rates, feed inputs, costs and returns. Study groups are organised for the different systems to enable beef producers to discuss their results with a view to making further improvements in efficiency.

RECORDING PEDIGREE STOCK ON FARMS

The MLC operate recording schemes for all beef and dual purpose breeds. The purposes of recording are to identify calves of rapid growth, to compare sires and dams by the growth of the calves, and to compare herd results with overall breed averages.

67

Cattle are weighed at intervals of 100 days from birth, and the weights processed to provide adjusted weights from 100 days to 600 days of age. Pre-weaning weights (100 and 200 days) reflect the mothering ability of the dam as well as the calf's own potential for growth. These weights are thus useful as a guide to culling cows within the herd and the selection of heifer replacements.

Post-weaning weights (300, 400, 500 and 600 day weights) reflect more closely the calf's growth potential and are used as a basis for selecting young bulls. The 400-day weight is taken as the main indicator of the calf's capacity for growth. The heritability of weight for age (see p. 75) is such that a bull 45 kg (100 lb) above herd average at 400 days will on average be expected to produce calves 11 kg (25 lb) heavier than average at the same age. Of course individual bulls may do better or worse than this.

The results are intended mainly for use by the breeder within his own herd, but when the processed records are returned to him his results are also compared with the breed average. Sire and dam progeny records are updated regularly and returned to breeders as an aid to selection.

About 1300 breeders now participate in this scheme. Details of number of herds by breed and number of weighings are given in Table 8, p. 66. The number of bulls recorded in 1973 are given as a percentage of bulls licensed; overall two-thirds of all beef bulls licensed have performance records available for them. A major problem with the information is to decide how much of the weight difference between bulls from different herds is genetic and how much due to variations in herd management. Buyers should seek bulls which are not only above breed average but also above the average performance of the herd in which they are reared. An indication of the variation in 400-day weights between and within breeds is given in Fig. 2. Performance differences between bulls within a breed can be larger than the differences between breeds. Thus the bull purchaser should decide on his breed and then ensure he selects an above-average bull of the breed. This variation which exists in weight for age is the basis for selecting for further improvement. Most Breed Societies have up-to-date lists of bulls with above-average weight

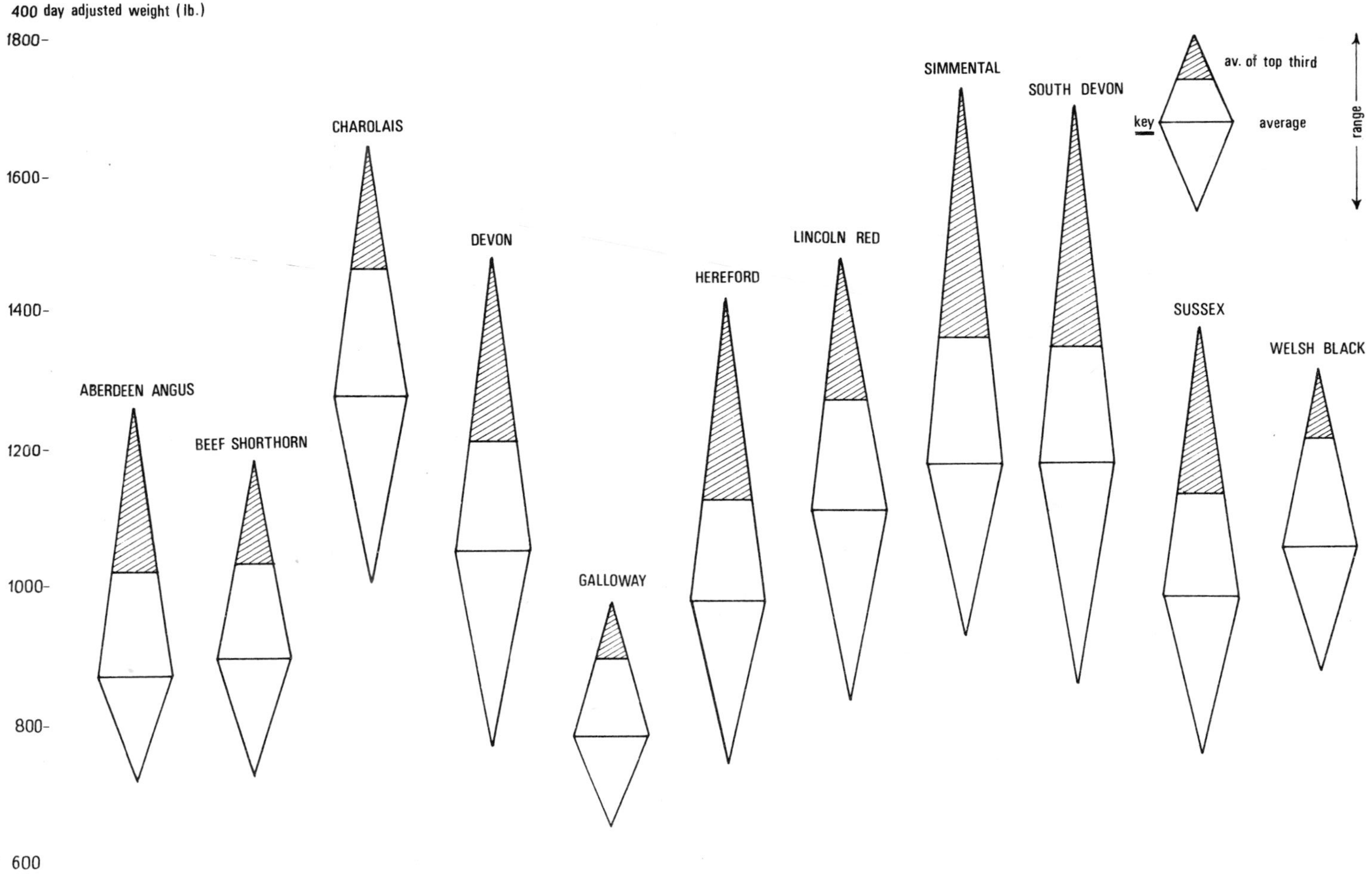

400 day adjusted weight (lb.)

1800–

1600–

1400–

1200–

1000–

800–

600

Fig. 2. Diagram showing variation of Bull 400 day weights by breed and within breed.

69

records and can usually guide buyers to the breeders with suitable bulls for sale.

CENTRAL BULL PERFORMANCE TESTING

Differences among animals referred to above are due to two major causes, genetic and environmental. The observed performance of an animal is a function of its heredity and the environment in which it is raised. When cattle are kept under uniform conditions, genetic differences can be identified and measured.

The structure of many beef breeding herds which are numerically small and have widespread calving patterns, often makes within herd comparisons difficult. Additionally, it is difficult to make comparisons between herds or between groups treated differently within the herd because of the large environmental differences due to location, management and nutrition. It is not possible to correct accurately for these differences.

Central performance testing stations are locations where animals are assembled from several herds to evaluate differences in performance under uniform conditions. The concept of performance testing relies on the fact that the important performance factors are heritable, i.e. are passed on to the next generation. Heritability estimates for characters of economic importance are shown in Table 9. These levels are such that selection on a performance testing would be effective, i.e. the performance of the animal itself is a reasonable

TABLE 9

Heritability Estimates for Various Beef Cattle Traits

	Median value
Post-weaning growth rate	50%
Yearling weight	60%
Feed conversion efficiency	40%
Subcutaneous fat thickness	40%
Eye muscle area	70%

(Preston and Willis (1970). *Intensive Beef Production*, Pergammon Press, Oxford, England)

guide to how well or otherwise its progeny will grow, and convert feed into liveweight.

The estimate of breeding value is based on the final weight for age, adjusted to 400 days. Weight for age is a summation of birthweight, gain to weaning (the start of the test) and test gain. Relatively little emphasis is placed on daily gain during the last period per se.

The performance tests are orientated towards evaluating terminal sires for crossbreeding. Bulls for performance tests are drawn from herds participating in the Meat and Livestock Commission pedigree recording scheme. There are five testing stations and bulls from the main beef breeds are tested. Details are given below:

Test centre	Maximum capacity (bulls per year)	Breeds tested
Aberdeen	60	Aberdeen Angus, red beef breeds
Holme Lacy, Hereford	160	Hereford, Welsh Black
Stockton on Forest, York	140	Charolais, Hereford
Stoneleigh, Warwickshire	120	Devon, Lincoln Red, Sussex
West Buckland, Wellington, Somerset	80	Charolais, South Devon

There are regular intakes throughout the year at each centre.

Qualifying standards of pre-test weight for age are set for each breed. Bulls start tests at 150–190 days of age and the duration of the test is 210 days so that the test terminates at approximately 400 days.

A complete diet in the form of cobs containing mineralised rolled barley and dried grass is fed ad libitum to facilitate feed recording so that estimates can be made of feed conversion efficiency. Ultrasonic measurements of backfat are also taken at the end of the test as are withers height measurements and conformation assessments.

Testing has a value only when it is linked to selection and the subsequent use of animals with superior breeding values. AI centres purchase many of their beef bulls from performance tests. During the last few years beef breeders in Britain and elsewhere have shown an increased awareness of the importance of good performance and they have also shown increased confidence in performance testing. This is measured by the quantity and quality of bulls being offered for testing in Britain and their subsequent successes in show and sale rings. Bulls that have done well in tests are being sought by breeders as herd sires. In order to foster this use of superior sires, the MLC has introduced Young Bull Proving Schemes, operated in conjunction with breed societies.

Breeds recently imported into Britain wishing to participate in performance testing must also agree to participate in some form of selection programme. The linking together of recording, testing and selection programmes is illustrated in Fig. 3.

The first organised breeding scheme for beef cattle in Britain was launched for the South Devon breed in 1970 and similar schemes have been introduced with most of the other established beef breed societies. The schemes are designed to promote the use of promising young bulls and to establish their breeding value through progeny testing.

High performance bulls are short-listed by the MLC and the final selection is made by the society concerned. The owner of the selected

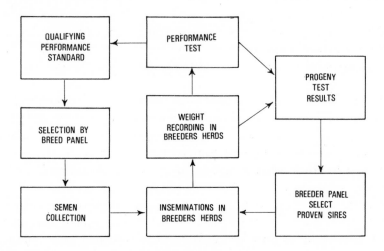

Fig. 3.   MLC Co-operative Breeding Scheme Flow Chart.

72

bull is offered a financial incentive by the MLC to collect a limited quantity of semen. This is then offered for free inseminations in breeders' herds. Thus breeders are able to try out promising bulls of different breeding in their herds, and at the same time the influence of these bulls is spread through the breed. Progeny records are collected on farms and young bulls bred in the scheme are submitted in competition with other bulls for performance testing. These schemes also enable bulls to be used in breeding herds which otherwise might be "lost" in commercial suckler herds or through export. A quantity of semen is stored on behalf of the owner at the MLC's expense until the progeny test result is available.

Performance testing relies on the performance of an individual animal being a reasonable guide to how well the offspring of that animal will perform. An alternative method of assessing the merit of an individual animal is by directly measuring the performance of its offspring—progeny testing.

PROGENY TESTING FOR BEEF

Progeny testing is much more expensive and it takes a much longer time to complete than performance testing. It is more accurate than performance testing but the advantage is small in the case of weight for age and is far outweighed by the fact that a much larger number of bulls can be performance tested than progeny tested, because of the difference in cost, and facilities required. The larger number of bulls tested means that greater selection can be applied and performance testing is therefore the major method of testing beef cattle. Progeny testing is used to compliment performance testing for bulls which go into AI where they will produce many times the number of progeny of a bull in natural service. Here the extra reliability is of considerably more importance.

The Milk Marketing Board of England and Wales have been operating a beef progeny testing programme for their AI stud at Warren Farm since 1965. This programme has involved the purchase of crossbred calves out of Friesian cows by teams of Aberdeen Angus, Charolais and Hereford bulls. The results have been used to form a completely progeny tested stud of Charolais bulls and the MMB expect that in the near future their Hereford and Aberdeen Angus

73

studs will also be completely progeny tested. The best of these bulls are offered in a special service at a slightly increased insemination fee.

COMMERCIAL BREED EVALUATION

It is evident that the many breeds and crosses available have different characteristics for beef production. Thus whilst the improvement of individual breeds is important, the reasoned substitution of one breed for another can be even more attractive as a short-term measure.

Considerable information is now available from commercial farms comparing the growth rates and slaughter weights of different breeds and crosses in the main systems of beef production. However, further information is required on carcase quality and feed conversion efficiency. The need for detailed carcase investigations and accurate recording of feed intake of individual animals require the control that is available only at specifically built experimental units. Two separate units were established in 1973 to cater for the two main sources of production—dairy bred beef and finishing of suckled calves. At each unit two systems of production are being undertaken so that the importance of breeds for different systems can be assessed. In the dairy-beef unit calves are reared for slaughter at 15 months and 24 months of age. These systems in commercial practice involve grazing and the feeding of mixtures of concentrates and conserved fodder. However, the need for accurate feed inputs requires the feeding of complete diets and excludes typical farm feeds. Therefore the growth performance typical of 15-month and 24-month beef production are being attained by the feeding of dried grass pellets.

Breeds and crosses from suckler herds are being compared in two typical systems of production—the overwintering and summer finishing of spring-born calves and the intensive finishing of autumn-born calves. Although the overwintered calves are fed typical overwintering rations, the finishing stages of both systems are based on dried grass feeding. Individual beasts are slaughtered at a constant fat cover as estimated by ultrasonic measurements of subcutaneous fat thickness. Carcases are subjected to MLC beef carcase classification and one side from each animal is subjected to retail cutting tests. The objective of the programme is to evaluate the different breeds

74

and crosses in terms of the efficiency of saleable meat production. The results of the programme should be used by the industry together with other information not investigated at the experimental units such as fertility, longevity, ease of calving to objectively appraise the use of breeds and crosses and to promote changes where necessary.

Performance of Breeds and Crosses

Breeds and crosses differ in growth rate and rate of fattening both of which have an important influence on slaughterweight and efficiency of feed utilization and the suitability of the cattle for different systems of production. Growth rate is one of the most important factors affecting profitability in beef production.

The information available on different breeds and crosses has shown constant variation between breeds for growth rate and slaughter weights. Additionally the results show:

That crossbred progeny performance ranks in a similar order to sires of the pure breeds.
That the breeds of crossing sires rank in a similar order in each main system of production.
That the breeds of crossing sires rank in a similar order regardless of the breed type of cow to which they are mated.

Approximately two-thirds of all cattle slaughtered for beef are sired by bulls of the beef breeds. The average weight for age of the various beef breeds is illustrated in Table 10.

Calves produced from the dairy herd are utilised for beef production in a range of beef systems which will be dealt with in detail later. Tables 11 and 12 show the performance of crossbred calves out of Friesian cows compared with purebred Friesian steers in three feeding systems. In each of the three systems the breeds ranked in a similar manner with the Charolais×Friesian showing the fastest growth rates followed by the Simmental and South Devon, which were

BREEDS AND CROSSES FROM THE DAIRY HERD

TABLE 10

| Breed | Breed Average Weight at 400 and 500 Days of Age for Bulls | | | |
	Average 400-day weight		Average 500-day weight	
	kg	lb	kg	lb
Charolais	579	1276	675	1489
Simmental	531	1171	627	1383
South Devon	530	1168	600	1322
Lincoln Red	498	1097	564	1243
Devon	477	1052	557	1228
Welsh Black	475	1048	564	1243
Limousin	455	1002	502	1106
Sussex	442	975	518	1143
Hereford	441	973	533	1175
Beef Shorthorn	410	904	486	1071
Aberdeen Angus	397	875	467	1029
Galloway	352	776	420	926

very similar. The breeds ranked in the same order as the pure breed 400-day weights. All of the beef cross Friesians grew faster than the pure Friesian with the exception of Aberdeen Angus×Friesians.

Table 12 shows that the quicker growing breeds tend to have the heaviest slaughter weights and that the ranking of the breeds for

TABLE 11

Comparison of Breeds and Crosses in Dairy Beef Systems
Daily gain (% above or below Friesian steers)

| Breed type | | 18-month beef | | |
	Cereal beef	Grazing	Yard finishing	Overall
Charolais × Friesian	+8·4	+9·9	+10·6	+10·4
Simmental × Friesian	+7·7	+7·2	+17·1	+10·3
South Devon × Friesian	+7·2	+8·7	+9·3	+8·9
Lincoln Red × Friesian	+6·4	+7·0	+7·6	+7·2
Devon × Friesian	+6·7	+6·9	+6·6	+6·7
Sussex × Friesian	+3·6	+3·4	+3·5	+3·4
Hereford × Friesian	+3·3	+5·7	+2·8	+3·1
Limousin × Friesian	+2·9	+3·6	—	+3·6
Friesian	0	0	0	0
Angus × Friesian	−12·1	−10·9	−12·2	−11·4

TABLE 12

Comparison of Breeds and Crosses in Dairy Beef Systems

	Slaughter weights	
	(% above or below Friesian steers)	
Breed type	Cereal beef	18-month beef
Charolais × Friesian	+9·1	+7·2
South Devon × Friesian	+7·8	+4·4
Simmental × Friesian	+1·8	+5·4
Devon × Friesian	+5·5	+2·4
Lincoln Red × Friesian	+4·8	+2·8
Friesian	0	0
Sussex × Friesian	−2·1	−0·9
Hereford × Friesian	−5·1	−10·4
Angus × Friesian	−14·2	−15·7

growth rate and slaughter weight were very similar. However, it is of particular interest that although the Hereford × Friesian grows quicker than the Friesian it is slaughtered at appreciably lower weights, i.e. it is earlier maturing.

Table 13, below, shows the performance of crossbred calves out of different types of dairy cows. The breeds of sire rank in the same order regardless of the type of dairy cow with which they are mated. Additionally the results clearly demonstrate that the Friesian is by far the quickest growing of the dairy breeds but that for beef production growth rate can be improved further by crossbreeding.

TABLE 13

Performance of Crossbred Calves Out of Different Cows

	Daily gain (% above or below Friesian steers)		
Sire breeds	Friesian cows	Ayrshire cows	Channel Island cows
Aberdeen Angus	−11·4	−25·3	—
Charolais	+10·4	−7·6	−22·2
Hereford	+3·1	−14·8	−30·8
Red breeds	+5·7	−10·0	—
Simmental	+10·3	—	—
South Devon	+8·9	−9·0	−24·6

BREEDS AND CROSSES FROM THE BEEF HERD

The choice of sire breed can influence calf performance considerably. Breeds which themselves are large, such as the Charolais, Simmental and South Devon, sire crossbred calves which have high weaning weights. The influence of sire breed on calf 200-day weights in lowland, upland and hill herds is shown in Table 14.

TABLE 14

Effect of Sire Breed on Calf 200-day Weights

	Lowland		Upland		Hill		Weight above (+) or below (−) Hereford × calves in same herd	
			Weight at 200 days					
	kg	lb	kg	lb	kg	lb	kg	lb
Charolais	240	530	227	501	205	452	+15·4	+34
Simmental	232	512	222	490	198	436	+12·7	+28
South Devon	231	510	221	487	200	440	+11·8	+26
Devon	225	497	215	474	191	421	+9·1	+20
Lincoln Red	222	490	214	472	189	416	+8·2	+18
Sussex	215	474	207	456	186	411	+5·4	+12
Hereford	208	459	194	428	184	406	0	0
Aberdeen Angus	194	428	182	402	176	388	−14·1	−31
Overall	221	487	210	464	191	421		

The order of ranking of the different sire breeds is not affected by herd situations but the differences between different breeds is considerably reduced under the difficult conditions of hill farms. The influence of sire breed on the growth of suckled calves to weaning is carried over into their post-weaning performance and slaughter weights.

One inevitable result of using large sire breeds to improve growth rates is an associated increase in birthweight with a consequent increase in the number of difficult calvings and calf mortality. Table 15 shows the effect of breed of sire on calf birthweights and Table 16 shows the results of a recent survey of calving problems in dairy herds by four breeds of sire.

78

TABLE 15

Sire Breed Effects on Calf Birthweights

	Lowland		Type of herd Upland		Hill		Weight above (+) or below (−)	
Sire breed			Calf birthweight*				Hereford × calves	
	kg	lb	kg	lb	kg	lb	kg	lb
Charolais	43·5	96	40·8	90	37·2	82	+5·9	+13
South Devon	42·6	94	40·4	89	36·3	80	+4·99	+11
Lincoln Red	39·9	88	38·6	85	36·7	81	+2·27	+5
Devon	39·0	86	38·1	84	35·8	79	+4·08	+9
Sussex	37·6	83	36·7	81	34·9	77	+0·91	+2
Welsh Black	36·7	81	35·4	78	32·2	71	−0·91	−2
Hereford	36·3	80	34·5	76	31·8	70	0	0
Aberdeen Angus	33·0	73	30·8	68	29·0	64	−2·27	−5
Beef Shorthorn	33·6	74	32·2	71	29·5	65	−2·72	−6

* Average of bull and heifer calves

TABLE 16

Effect of Breed on Difficult Calvings and Calf Mortality
(Source: Simmental and Limousin Steering Committee Report No. 1)

	Friesian heifers		Friesian cows	
	% difficult calvings	% calf mortality	% difficult calvings	% calf mortality
Simmental	8·9	11·6	3·5	4·4
Limousin	7·7	11·8	2·4	3·2
Charolais	—	—	5·0	5·5
Hereford	2·5	5·0	0·9	2·7

t is unwise to use a large sire breed on heifers and in any situation where calvings cannot be properly supervised. On the other hand he benefits on the subsequent potential of the calf (and value) means hat many farmers are coping successfully with the use of larger breeds. Within the AI stud it is possible for producers to select a bull which has been demonstrated to produce a low incidence of difficult calvings.

As a practical guide for suckled calf producers the number of reared calves per 100 cows required from the different sire breeds to equate with 100 Hereford cross calves weighing 204 kg (450 lb) each at 200 days old is as follows:

Aberdeen Angus	107
Charolais	93
Devon	96
Lincoln Red	96
Simmental	94
South Devon	94

Suckler Cow Performance

The main job of the suckler cow is to bear a live calf and to provide sufficient milk for its growth potential to be realised. In addition, good fertility, ease of calving and longevity are all important. Because of hybrid vigour, crossbred cows rear more calves to weaning than purebreds and have a longer productive life. Naturally, therefore, they are preferred in commercial practice.

The variation in calf weaning weights due to breed type of cow is much less than that due to breed of bull. Nonetheless, breed type of cow is very important. The beef cow carrying dairy blood typified

TABLE 17

Effect of Cow Type on Calf 200-day Weights

	Type of farm					
	Lowland		Upland		Hill	
Cow type	Calf 200-day weights					
	kg	lb	kg	lb	kg	lb
Aberdeen Angus × beef breeds	200	442	186	410	176	388
Blue Grey	203	448	196·4	433	190	418
Irish Aberdeen Angus crosses	202	446	192	423	188	414
Irish Hereford × Shorthorn	212	468	209	460	196·8	431
Aberden Angus × Friesian	204	450	200	441	195	429
Hereford × Friesian	217	479	210	464	198	437
Difference above(+) or below (−) of Hereford × Friesian and Blue Grey	+21	+46	+16	+35	+9	+20

Smithfield Market London, in early 19th century

Courtesy Superintendent London Central Markets

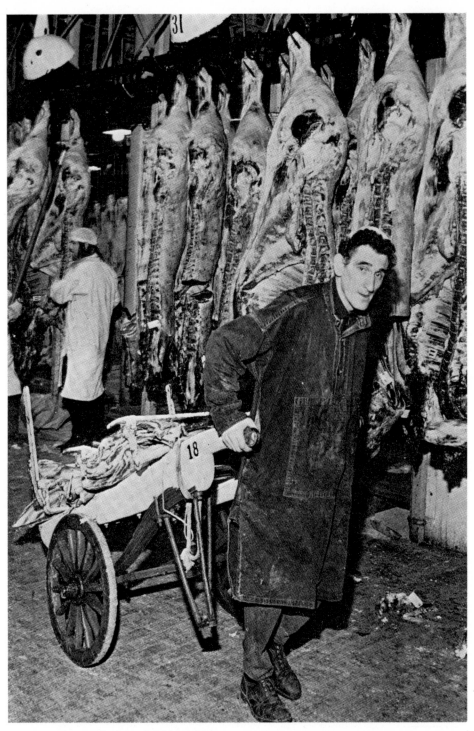

Meat Porter at work in Smithfield Market

Turkey stall at Smithfield Market

Comparison of breeds for efficiency meat production

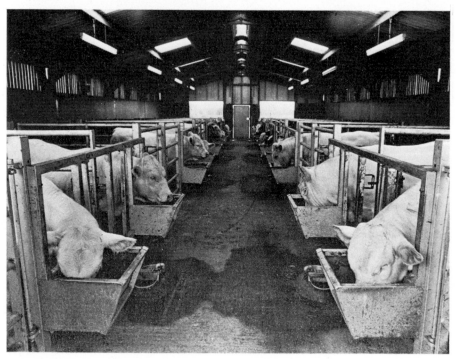

Charolais bulls on a central performance test

Courtesy Meat and Livestock Commission

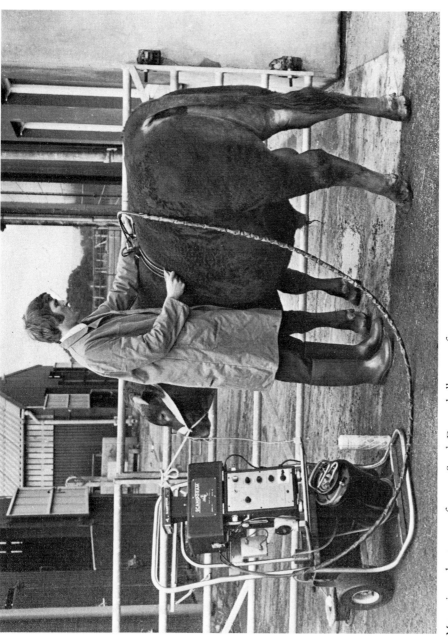

Measuring subcutaneous fat on a South Devon bull at a performance test centre

Courtesy Meat and Livestock Commission

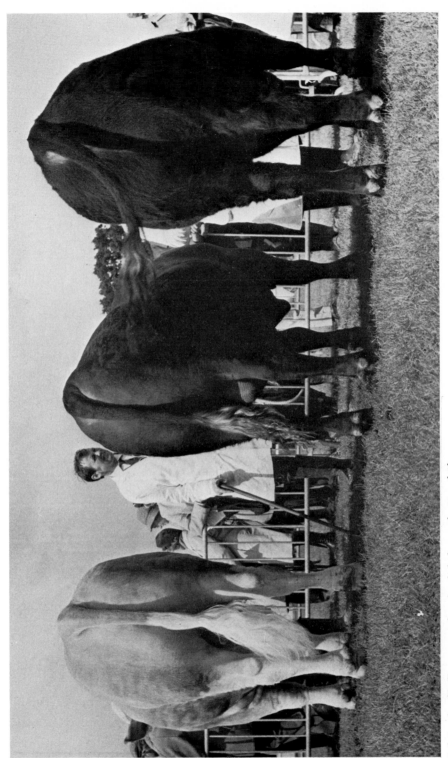

Judging beef bulls at a weight recorded class at the Royal Show

Hereford x Friesian steers at grass

Courtesy Warren Farm, The Milk Marketing Board

Sussex x Friesian reared at BOCM's Hobbs Cross Farm, Epping. Beef weight achieved in
44 weeks

Charolais x Friesian reared at same farm. Beef weight achieved in 34 weeks

Courtesy BOCM

by the Hereford×Friesian, which is widely used on all but the most difficult farms, produces calves with higher weaning weights than specialised beef cows such as the Blue Grey. On hill farms these differences are much reduced presumably due to the influence of the harsher environment (Table 17).

The more productive cow is an obvious choice if a large sire breed is used for crossing but the extra performance can only be gained at the expense of additional feed. Where feed is scarce the more resilient specialised beef cow is to be preferred. Indeed, on the toughest hill farms, the sparseness of grazing and the harsh climate restrict the choice of cow to the traditional hill breeds and their crosses.

There is still much to be learned about the lifetime productivity of the different types of suckler cow. They vary in mature size, and the number of cows which can be carried on a given acreage must be taken into account as well as calf performance.

In fact the difference in bodyweights between breeds and crosses of cow are not very dramatic. The danger is that the more productive cows sacrifice body condition to milk production, which makes it very much more difficult for them to be got successfully back in calf where the feed supply is limited as it is on many hill farms. Cows in the lowland are on average 38 kg (84 lb) heavier than the types of cows in hill herds.

Having chosen a breed type of cow, this must influence the choice of sire breed. Where a bull is mated to cows already carrying the blood of his breed, calf performance is often disappointing. Far better to use a bull of a third breed to produce triple cross calves suited to the needs of the finisher.

Beef Production Systems

Traditionally in the UK the production of beef was a leisurely process. Cattle were generally expected to make their growth at grass during the spring and summer and were kept throughout the winter

D

on low levels of performance. The age at slaughter was generally $2\frac{1}{2}$ to 3 years. However, the development of beef production systems in Britain over the last 15 years has resulted in a reduction in the age and weight at which cattle are slaughtered. This has been associated with the development of production systems with higher rates of lifetime liveweight gain (particularly due to the elimination of severe store periods but also due to better techniques of forage conservation and greater utilisation of concentrates) and a partial shift from grass to yard finishing. Beef production is still, however, often a secondary enterprise generally lacking adequate planning.

A recent MLC estimate of the relative importance of beef production systems in current use indicates that 6% of steers and heifers are slaughtered at less than 12 months of age. Of the 94% of steers and heifers which are slaughtered at more than 12 months of age about 49% are finished under 2 years of age, and 45% at older ages; 49% are finished in yards between December and May, and 45% are slaughtered off grass between June and November.

TABLE 18

Cow Bodyweights and Their Relationship with Calf Weaning Weights

Cow type	Lowland		Type of farm Upland		Hill		Comparison above (+) or below (−) weights of Hereford× beef cows		Weight of weaned calf produced per 51 kg (100 lb) of cow bodyweight	
	kg	lb	kg	lb	kg	lb	kg	lb	kg	lb
Aberdeen Angus×beef	485	1069	450	992	450	993	−24·5	−54	19·1	42
Blue Grey	453	999	468	1032	429	946	−21·8	−48	19·5	43
Irish Aberdeen Angus cross	476	1049	452	996	446	984	−23·6	−52	20·0	44
Irish Hereford× Shorthorn	475	1048	462	1019	423	933	−30·8	−68	19·5	43
Aberdeen Angus× Friesian	464	1024	452	997	445	981	−22·2	−49	19·5	43
Hereford× Beef	502	1107	481	1060	473	1043	0	0	18·1	40
Hereford× Friesian	501	1104	464	1023	45ʈ	994	+ 4·5	+10	19·5	43
Red Breed crosses	582	1282	526	1160	480	1058	+21·3	+47	16·8	37
Charolais crosses	690	1522	571	1259	499	1099	+40·4	+89	17·2	38

Systems of beef production differ according to whether they utilise the dairy or beef bred calf, and in the type and level of feeding particularly in the finishing stage. The production of suckled calves is a distinct system to which the finishing of the suckled calf may be carried out on the same farm as a separate enterprise. Complete systems using the dairy bred calf are those where the full production cycle—both rearing and finishing—takes place on the same farm. Partial systems are those where stores are purchased for either finishing or for re-sale as older, heavier stores. Complete systems differ in the feeds utilised and this influences the age at slaughter. Cereal and arable product beef systems involve housed cattle which make no direct use of land and are mostly slaughtered between 11 and 15 months of age. In grass/cereal beef systems cattle make use to a varying extent of grass for both grazing and winter feed and are mainly slaughtered between 15 and 24 months of age.

Since the early 1960's specific systems of beef production have been developed utilising the by-product calf from the dairy herd. The most intensive of these is cereal beef which was introduced in the early 1960's and for a short time increased in popularity. More recently higher cereal and before then higher calf prices have led to a marked reduction in the number of these units. The grass/cereal systems have increased most in importance and they have spread throughout the country.

CHOOSING THE SYSTEM

The choice of a system of beef production can only be made on an individual farm basis and, in particular, will depend on the physical resources available together with financial considerations:

PHYSICAL RESOURCES
   The acreage and quality of grassland, which determine the number and type of stock.
   The feeds available on the farm, including arable products and by-products.
   Fencing and field availability of water.
   Buildings.
   Labour.

FINANCIAL CONSIDERATIONS
   Rate of turnover.
   Cash flow.
   Initial capital investment.
   Peak capital requirements.

In several beef systems the production cycle exceeds one year. These clearly involve an appreciable delay before the investment in rearing and finishing is recouped from cattle sales—an important consideration where the full cycle of production takes place on the same farm. On the other hand, systems based on purchased store cattle have a rapid turnover and an even cash flow. Against this advantage, however, must be set the higher investment capital required to cover the accumulated costs and profit margins of previous owners, and the lower gross margins.

Whatever the system, it is necessary for the producer to know the likely profit margin and the capital required to achieve it. Return on capital can then be assessed and compared with likely returns from other uses of the resources available. The working capital necessary for beef production varies considerably from one system to another. For example, the working capital requirement for a suckler herd is low but the initial capital investment in cows is high. If suckled calves are finished on the same farm the cash flow is favourable because the calf subsidy in the autumn and beef cow subsidy in late winter partly finance wintering costs. Furthermore, the sale of finished cattle in spring provide cash to meet the costs of forage production.

Where batches of cattle overlap, capital must be available to meet the peak requirements when more than one batch is on the farm at the same time. In grass/cereal systems, the calf subsidy has fulfilled a useful purpose in that the subsidy received for one batch will partly finance the purchase of the next. Nevertheless, grass/cereal beef systems inevitably require more working capital than suckled calf production or cereal beef.

Cereal beef production has a relatively rapid capital turnover, and the cash flow depends partly on whether the barley fed is home-grown or bought-in. In this system overlapping batches can provide a fairly level output throughout the year.

In general, systems based on forage production involve buying and selling at fixed seasons and this can aggravate the problems of cash flow. The purchase of reared calves (3 months of age) instead of unweaned calves (7–10 days of age) can alleviate problems of peak capital requirements.

When introducing, changing or expanding a beef enterprise a major consideration to the producer must be to minimise the amount of working capital involved in operating the farm as a whole. The cash flow involved in the system of beef production should, therefore, be compatible with the cash flow of all the other farm enterprises.

Cereal beef is characterised by the lowest slaughter age—it is based on a high cost concentrate ration aimed at promoting rapid liveweight gains. Cattle are slaughtered at relatively young ages and light weights and a high rate of turnover is achieved. Because the cattle are housed throughout, the enterprise makes no direct use of land but it is extremely sensitive to changes in both barley and calf prices.

BEEF FROM CEREALS AND ARABLE PRODUCTS

MANAGEMENT AND PERFORMANCE TARGETS: Dairy bred calves are early-weaned, at 5–6 weeks, on to a concentrate containing 17%

TABLE 19

Physical Targets for Cereal Beef

Period	Gain kg (lb) per day		Feed conversion ratio kg (lb) concentrates per kg (lb) liveweight gain	
	kg	lb	kg	lb
Birth to weaning	0·59	1·3	—	—
Weaning to 3 months	0·95	2·1	1·45	3·2
3 to 6 months	1·22	2·7	1·95	4·3
6 months to slaughter	1·32	2·9	2·99	6·6
Overall	1·13	2·5	2·49	5·5

85

crude protein. At 10–12 weeks of age, weighing about 100 kg (220 lb), the calves are introduced to an ad lib diet based on rolled barley and containing 14% of crude protein. When the cattle are 6–7 months old and weigh 250 kg (550 lb), the diet can be further cheapened by reducing the crude protein level to 12%. This diet is fed ad lib until slaughter at 386–408 kg (850–900 lb) when the cattle are 10–12 months old. Carcase weights are in the range of 204–236 kg (450–520 lb) with typical carcase classes for Friesians of 2·3. A market premium is available.

Beyond the normal slaughter weight of 386–408 kg (850–900 lb) the feed conversion ratio deteriorates rapidly. This is an important factor because the use of a high cost diet makes efficient feed use essential.

TYPE OF STOCK: Late maturing breeds and crosses of high growth rate are best suited to the system because of their lower feed costs per kg (lb) of gain. Most producers use Friesian steers. Early maturing beef crosses and heifers become overfat at light weights on this high quality diet and are unsuited to the system. An increasing number of producers are taking advantage of the improved growth rate and feed conversion efficiency along with heavier slaughter weights obtained by using bulls.

TABLE 20

Bulls in Cereal Beef Production

	Bulls	Steers
Daily gain kg (lb)	1·3 (2·9)	1·18 (2·6)
Concentrates fed tonne (cwt)	1·9 (37·7)	1·85 (36·6)
Slaughter age (days)	333	328
Carcase weight kg (lb)	230 (508)	210 (464)
GROSS MARGIN	Extra 58% for bulls	
High priced cuts (%)	41·6	41·4

WHERE IT GOES WRONG: As with all housed young cattle, respiratory disease is a problem in the winter months. The incidence appears to be at a minimum in well-ventilated, draught-free accommodation. Bloat is a hazard, especially in cattle housed on sawdust or slats. The incidence can be minimised, without depressing performance, by feeding up to 1 kg (2 lb) hay per day.

Diets containing unnecessarily high levels of protein increase feed costs but do not improve growth rates. Home-mixed diets must contain adequate minerals and vitamins, particularly Vitamin A. Feed efficiency declines sharply after 363 kg (800 lb) so the timing of sale is important.

DEVELOPMENTS OF CEREAL BEEF PRODUCTION: The cereal beef system has shown a marked reduction in profitability in recent years as first calf, then cereal and protein prices increased. Attempts to make the system more profitable have centred on improving animal performance and reducing feed costs. In this context the benefits of bulls compared to steers cannot be ignored: improved growth (+13%), better feed conversion efficiency (+10%) and higher carcase weights (+10%). Costs of protein supplementation can be reduced by reducing the protein level in the ration as the animal increases in age, and replacing part of the protein by urea. Urea can be used to provide all the supplementary nitrogen from 6 months of age. However, the farm mixing of mineralised urea is not generally advised because of the dangers of toxicity. Attempts at reducing the costs of energy by the replacement of barley with ground straw have been less successful. Ground straw can be included at up to 20% of the ration for cattle over 227 kg (500 lb) liveweight without any marked effect on performance but the animal eats more of the ration. Above 20% inclusion the animal can no longer compensate by eating more of the ration and performance declines. The straw diet must therefore be much cheaper than the all-concentrate diet if a reduction in feed costs per kg (lb) of liveweight is to be achieved and in practice such rations are not so competitively priced.

Recently producers have been reducing feed costs by the increased use of arable products and by-products—maize silage, potatoes, swedes and turnips are all low cost high energy feeds capable of supporting good cattle performance. The interest in alternatives

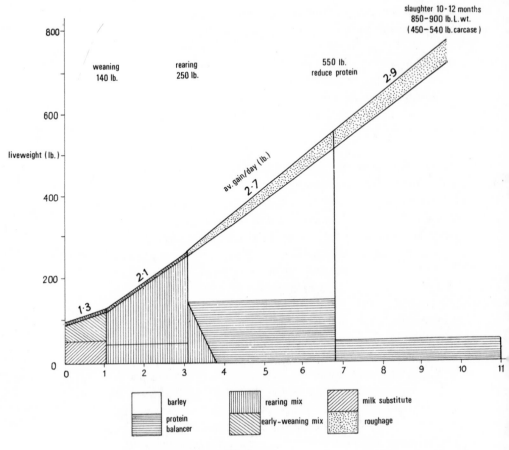

Fig. 4.   *Targets for cereal beef.*

lies in their high yields of feed per acre (Table 21). These feeds correctly supplemented will promote liveweight gains similar to those levels of gain in cereal beef systems and are particularly attractive to current producers of "barley beef" as well as being low cost feeds for finishing stores and suckled calves.

Turnips, swedes or potatoes can all replace cereals but growth is checked during the changeover period. For young growing cattle up to 50% of barley can be replaced by roots. In older cattle total

TABLE 21

Typical Yields of Different Feeds

	Fresh yield tonnes/hectare (tons/acre)		Yield dry matter tonnes/hectare (tons/acre)		Yield starch equivalent tonnes/hectare (tons/acre)	
	tonne	ton	tonne	ton	tonne	ton
Barley	4·5	1·8	3·87	1·55	3·10	1·24
Swedes	60·0	24·0	6·0	2·40	4·63	1·85
Turnips	51·25	20·5	4·63	1·85	2·38	0·95
Potatoes	30·0	12·0	6·38	2·55	4·63	1·85
Grass silage (a)	32·5	13·0	7·25	2·90	2·63	1·05
Hay	5·12	2·05	4·38	1·75	1·63	0·65
Grass silage (b)	37·5	15·0	7·50	3·00	5·25	2·00
Maize silage	45·0	18·0	8·75	3·50	5·75	2·30

replacement of barley is possible, but there may be restrictions imposed by feeding arrangements.

The use of maize silage is a very important development in beef production. Maize silage produces a high yield of low cost feed per hectare (acre), ideally suited to feed-lot beef production.

TABLE 22

Beef Production from Maize Silage

Physical targets	
Age at slaughter (months)	14
Liveweight at slaughter for Friesian steers kg (lb)	431–434 (950–1000)
Gain to slaughter kg/day (lb/day)	0·95 (2·1)
Stocking rate (finished beasts per hectare (acre))	5·25 (2·1)

It is a common system for bull beef production in the maize areas of Europe. The system applied to Friesian steers would use ad lib silage plus protein supplement from 12 weeks. Slaughter would be at 14 months at a liveweight of 431–454 kg (950–1000 lb). At a maize yield of 3857 kg per hectare ($3\frac{1}{2}$ tons DM/acre) in an average season, the stocking rate is 5·25 beasts per hectare (2·1 beasts/acre) maize.

The main problem with maize is its low protein content. Urea can be used as a source of protein in a concentrate supplement and studies are now in progress to incorporate non-protein nitrogen such as urea with maize during ensiling.

GRASS/CEREAL SYSTEMS

The term "Grass/Cereal Systems" embraces several methods of utilising dairy bred calves depending upon season of birth of calf, level of concentrate feeding and whether one or two grazing seasons are involved. At one end of the range is 15-month grass/cereal beef which utilises winter- and spring-born calves finished on a high concentrate diet after one grazing season at 13–17 months of age. At the other end of the range is grass beef, also generally using winter- and spring-born calves, but with a growing ration during the second winter and cattle being slaughtered at heavier weights off grass during their second grazing season at 18–24 months. For the most common method, 16–20-month grass/cereal beef, autumn- and winter-born calves are used. The aim is to finish the cattle out of yards at about 16–20 months of age. These systems are not rigid. For profitable production, they should be adapted to suit the time of calf purchase and individual farm resources.

Grass/Cereal 12-month Beef

In this system, calves born between November and March are grazed during the summer and finished out of yards on a diet almost wholly composed of concentrates.

MANAGEMENT AND PERFORMANCE TARGETS: Because a high cost diet is fed in the finishing winter, gains during grazing have a major impact on profitability. At the same time grazing management is made more difficult because the cattle are young and small and because there is little or no conservation acreage to control the early season flush of grass and to provide aftermath grazing from mid-season onwards. Because of these difficulties some producers feed supplementary concentrates throughout the grazing season but adopt high stocking rates.

Results are most successful where the system is integrated with another grass enterprise which helps to utilise the flush of grass in

90

TABLE 23

Physical Targets for Grass/Cereal 15-Month Beef

Period	Gain per day		Period	Stocking rate beasts/hectare (acre)
	kg	lb		
Purchase to turnout	0·77	1·7	Grazing: turnout to mid-season	10·0 (4·0)
Grazing	0·75–0·82	1·6–1·8	mid-season to yarding	5·0 (2·0)
Finishing winter	1·09–1·18	2·4–2·6	Overall (no conservation)	7·5 (3·0)

the spring and which may have a conservation requirement. A "follower-leader" technique can be successfully applied. This means that the young calves have first access to grass paddocks but do not have to eat the paddock "down" because older stock follow them.

TYPE OF STOCK: The later maturing Friesian is better suited to this system than the Hereford×Friesian, because the latter finished at too light a weight on the high concentrate finishing diet. Entire bulls fit well into this high cost, high energy finishing system.

WHERE IT GOES WRONG: Respiratory disease is a problem which is minimised in well-ventilated draught-free buildings but the rearing of winter/spring calves does create particular problems. Many producers feed excessive protein in the finishing diet—12% crude protein is adequate. Additionally producers must avoid delaying slaughter after feed efficiency has declined to unprofitable levels.

During recent years this system has increased in popularity. It involves calves born between July and January.

Grass/Cereal 15-month Beef

MANAGEMENT AND PERFORMANCE TARGETS: Calves are early-weaned and fed a rearing diet of hay and/or silage plus concentrates rationed at about 2·27 kg (5 lb) per day. The autumn-born calf should weigh at least 181 kg (400 lb) when turned out to spring grass. Later born calves usually weigh less and show poorer performance at grass, but a higher level of feeding in the rearing winter is not justified because it further depresses performance at grass. Daily gains of 0·68–0·77 kg (1·5–1·7 lb) are the required levels so that the calves are well grown but can still expect some compensatory growth when turned out to grass.

For the first 2 to 3 weeks after turnout calves should be fed supplementary concentrate at around 1·34 kg (3 lb) per day until they are accustomed to grazing.

Grass should be managed to provide a continuous supply of high quality herbage and this necessitates paddock grazing. In order to maintain performance, calves should be moved to a fresh paddock when they become restless rather than when the paddock is completely grazed. This usually requires topping to remove ungrazed herbage.

From mid-season, as herbage quality declines and infestation with parasitic worms increases, it is increasingly difficult to maintain performance. Integration of the grazing and conservation acreage is, therefore, essential. Some time during July the cattle are moved on to areas previously cut for silage or hay. At this time, on farms where parasitic worms are a problem, cattle are also drenched or injected with an anthelmintic. Regrowths on the paddocks can be conserved.

From late August, herbage quality and availability inevitably decline and supplementary feeding should be introduced to maintain liveweight gains. Any decline in performance at this time seriously increases winter feeding costs. If parasites are a problem paddocks should not be regrazed in late season.

TABLE 24

Physical Targets for Grass/Cereal 18-Month Beef

Period	Gain per day		Period	Stocking rate beasts/hectare (acre)
	kg	lb		
Purchase to turnout	77	1·7	Grazing: turnout to mid-season	8·75 (3·5)
Grazing	0·82–0·91	1·8–2·0	mid-season to yarding	5·0 (2·0)
Finishing winter	0·82–0·91	1·8–2·0	Overall (including conservation)	
Overall	0·82	1·8	454 kg (1000 lb) slaughter weight	4·25 (1·7)
			499 kg (1100 lb) slaughter weight	3·5 (1·4)

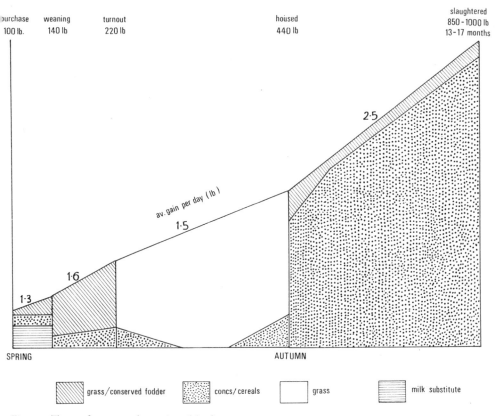

purchase
100 lb.

weaning
140 lb

turnout
220 lb

housed
440 lb

slaughtered
850 - 1000 lb
13 - 17 months

2·5

av. gain per day (lb)

1·5

1·6

1·3

SPRING

AUTUMN

grass/conserved fodder   concs/cereals   grass   milk substitute

Fig. 5.   *Targets for 15 month grass/cereal beef.*

The cattle should weigh between 295–340 kg (650–750 lb) at yarding for their second and final winter and be built up to a maximum intake of 2·72–5·47 kg (6–12 lb) of rolled barley per day, depending on the quality of the conserved forage available, weight at yarding, breed type and the planned weight at slaughter. In addition hay or silage should be fed to appetite.

In the most profitable units, silage made from wilted herbage to give a dry matter content of around 25% in the finished product, is invariably chosen as the conserved feed. Silage of this quality is eaten in large quantities, the total requirement of 2·54–4·06 tonnes (2½–4 tons) per head being dependent on the length of the feeding period and the level of barley feeding. During the finishing winter

93

barley is limited to about 2·72 kg (6 lb) for Hereford×Friesians but up to 4·54 kg (10 lb) for Friesians.

Cattle are slaughtered at 16–20 months weighing 408–522 kg (900–1150 lb) and produce carcases of 204–272 kg (450–600 lb) with carcase class usually 2·3 for Friesian and 3H·4 for Hereford×Friesian.

The target stocking rates are based on grassland fertilised at 200 units of nitrogen per acre.

TYPE OF STOCK: The Friesian and Hereford×Friesian are most commonly used for this system. Generally the Hereford×Friesian grows faster at all stages with an overall advantage of some 5%. This is usually offset, however, by the higher cost of the calf and, for this reason, many producers prefer the Friesian. The fact that the Hereford×Friesian finishes at rather lower weights than the Friesian may be an advantage or a disadvantage, depending on individual farm circumstances and season of birth.

The daily gain of heifer calves is up to 20% poorer than similarly bred steers, and they finish at lighter weights. But the lower cost of the heifer calf is a consideration. It has been demonstrated that entire bulls can also fit into this system and show similar advantages in terms of growth rates, carcase weights and gross margins as in cereal beef production.

Carcase Data from Bulls in 18-Month Grass/Cereal Beef

	No. cattle	Carcase weight kg (lb)	Carcase composition %				High priced cuts (%)
			Fat	Lean	Bone	Waste	
Bulls	26	286(630)	14·2	68·7	15·7	1·4	42·6
Steers	19	258(569)	19·7	62·9	15·7	1·7	44·0

The behaviour of bulls during grazing is different from steers with bulls tending to herd less closely.

WHERE IT GOES WRONG: Because calves are reared during the winter, respiratory disease is a particular problem on many farms. Again, well-ventilated, draught-free accommodation is essential.

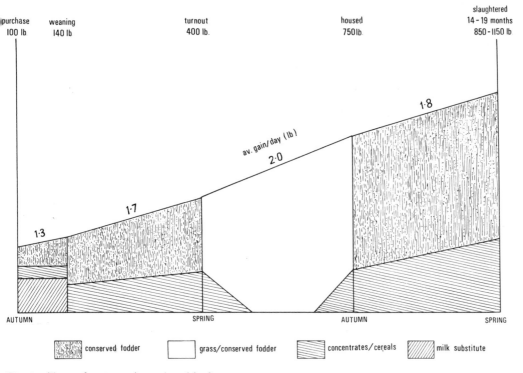

Fig. 6. *Targets for 18 month grass/cereal beef.*

The importance of good gains at grass has been stressed, and it is at this time that the greatest loss of potential performance occurs. Sound grazing management is essential, but the problem of internal parasites is worse when young susceptible calves are grazed on pasture grazed by cattle the previous year. Perhaps the greatest problem arises where calves are turned out year after year on to the same paddock near the buildings. This should be avoided. Winter finishing itself is profitable only where high quality conserved forages are available, and the amount of concentrate can be kept to a minimum. Where silage is being fed, barley supplemented with minerals and vitamins will be adequate. Some producers un-necessarily add to their costs by feeding over-expensive concentrates containing excessive protein.

Excessive concentrate feeding in the finishing winter often occurs because the silage is cut late, is not made quickly enough and is conserved without careful attention to detail (sheeting etc.). The penalties for poor conservation are severe in terms of extra concentrate feed costs.

Within the framework of the feeding system it is possible to vary the amount of cereal supplementation to achieve the required combination of gain, slaughter weight and slaughter date. If less cereals are fed, gains are inevitably reduced and this increases the weight at which the cattle reach a given degree of finish and the date on which slaughter takes place.

GRASS BEEF

Calves born between November and April require high standards of management if they are to be finished out of yards at 13 to 17 months of age, and even then this normally means finishing on an expensive diet dominated by concentrates. An increasing number of producers have, therefore, adopted a system in which these calves are finished off grass in their second summer at 18 to 24 months of age. Where autumn-born calves are fitted into the system the age at slaughter is 2 years.

MANAGEMENT AND PERFORMANCE TARGETS: Calves are early-weaned and turned out to grass as soon as possible. To maintain gains of 0·68 kg (1·5 lb) per day supplementary feeding is continued throughout the grazing seasons or until the calves reject it, in which case it is started again in late summer. They are yarded in the autumn and fed good quality conserved forage plus 1·36–1·8 kg (3–4 lb) concentrate per day. When turned out in the spring the cattle are well grown but in lean condition, and make effective use of grass. They are normally sold in the July to September period at 18 to 24 months of age when they weigh 431–454 kg (950–1000 lb). Carcase weights are in the range of 236–286 kg (520–630 lb) with normal class classification for Friesians of 3·3 and Hereford× Friesians 3L·4.

Liveweight gain in the winter should be in the region of 0·454–0·59 kg (1·0–1·3 lb) per day, which allows weight to be put on eco-

96

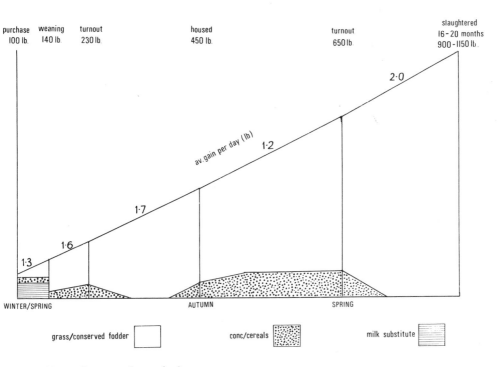

purchase 100 lb. weaning 140 lb. turnout 230 lb. housed 450 lb. turnout 650 lb. slaughtered 16-20 months 900-1150 lb.

2·0

av. gain per day (lb) 1·2

1·7

1·6

1·3

WINTER/SPRING   AUTUMN   SPRING

grass/conserved fodder   conc/cereals   milk substitute

Fig. 7.   Targets for 20 month grass beef.

nomically but still allows for compensatory growth when the cattle are turned out to grass.

TYPE OF STOCK: The earlier maturing Hereford×Friesian, or Aberdeen Angus×Friesian, are probably better suited to the system than the Friesian, as they can be more easily finished off grass. Heifers can be used because of the ease with which they can be finished, but they grow more slowly than steers.

WHERE IT GOES WRONG: Respiratory diseases which are a particular problem with winter calves are least problem in well-ventilated but draught-free buildings. Gains at grass in the first season are often poor especially for late-born calves. Careful grassland management and strategic use of concentrate supplements are the answer. A more recent development is a grazing system in which the first season

97

cattle graze ahead of second season cattle in the paddocks, and fresh young herbage is available for the young cattle.

A major fault in practice is excessive feeding in the growing winter which merely depresses gains at grass in the second summer.

INTEGRATION OF GRASS/CEREAL AND GRASS SYSTEMS: In large units there is sometimes a case for integrating two or even three grass/ cereal systems to even-out the cash flow of the total enterprise. The systems differ in their requirements for average and peak working capital and the peaks occur at different times. The 18 month grass/ cereal system provides the highest gross margin and is probably the easiest of the three systems to operate. Grass/cereal 15 month beef has the lowest peak capital requirements of the three systems because slaughter starts at about 13 months of age which means that cash is available for calf rearing. Similarly, cash income from the sale of quickly finished cattle can be used for financing the winter feeding of 18 month grass/cereal cattle. If a further group of cattle are finished at grass at 20 months, this means that there is a fairly regular level of cash income for 9 months of the year (December to September) which is available for reducing the peak requirements of the different systems. Obviously careful budgeting is required on an individual farm to assess the right balance of beef systems to optimise the gross margins per hectare (acre), the cash flow and the turnover of capital.

Beef Systems for Dairy Bred Calves
Effects of Changes in Feed and Calf Prices

The profitability of beef is governed to a large extent by calf price, feed prices and sale returns. Changes in the costs of the two main inputs—calves and feed—have a different effect according to the system of production. Consequently the relative profitability of systems change with changes in calf and feed prices. When calf and feed prices increase it is those systems which make the most use of grass that increase in relative profitability to other systems.

Changes in the prices of concentrate feeds substantially affect the profitability of all beef systems but the precise effect varies accord-

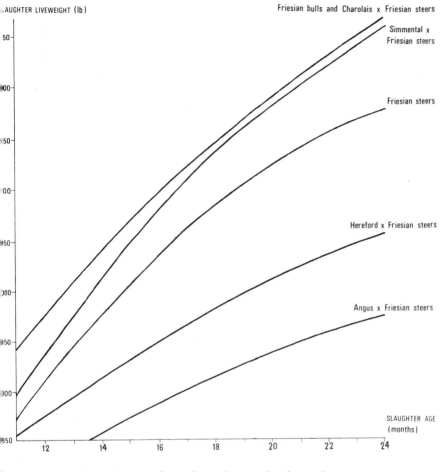

SLAUGHTER LIVEWEIGHT (lb)

Friesian bulls and Charolais x Friesian steers

Simmental x
Friesian steers

Friesian steers

Hereford x Friesian steers

Angus x Friesian steers

SLAUGHTER AGE
(months)

Fig. 8. *Average liveweight at slaughter and typical carcase classification for different breeds and crosses.*

ing to the individual system. Even in the grass/cereal and grass systems approximately 1016 kg (1 ton) of concentrates are being fed. Extension of the production cycle by feeding less concentrate per head per day in the yarded periods does not bring about a dramatic reduction in concentrate usage. For example the average producer with a 24 month grass beef system is feeding more concentrates than is the top third 18 month grass/cereal beef systems (see Table 25). The most efficient producers not only have cattle growing at above average rates but also feed less concentrates and so cost

TABLE 25

Concentrate Usage of the Main Beef Systems and
Effect of Increasing Concentrate Prices on Gross Margins per Head

	Concentrate usage kg (cwt)		Average Change in price of concentrates (£ per kg (ton))		Concentrate usage kg (cwt)		Top ⅓ Change in price of concentrates (£ per kg (ton))	
			10	20			10	20
			Effect on gross margins per head (£)				Effect on gross margins per head (£)	
	kg	cwt			kg	cwt		
Cereal beef (12 months)	1880	37	−18	−37	1727	34	−17	−34
Grass/cereal (15 months)	1270	25	−13	−25	1168	23	−12	−23
Grass/cereal (18 months)	1168	23	−12	−23	864	17	−9	−17
Grass beef (20 months)	914	18	−9	−18	762	15	−8	−15
Grass beef (24 months)	1016	20	−10	−20	813	16	−8	−16
Yard finishing of stores	610	12	−6	−12	559	11	−6	−11
Yard finishing of suckled calves	559	11	−6	−11	508	10	−5	−10
Grass finishing of stores	102	2	−1	−2	51	1	−1	−1
Lowland suckler herd—autumn calving	610	12	−6	−12	457	9	−5	−9
Lowland suckler herd—spring calving	457	9	−5	−9	305	6	−3	−6

TABLE 26

Average Concentrate Use Performance and Stocking Rate for Different Beef Systems

	Concentrate usage kg (tons)				Stocking rate beasts/hectare (acre)				Slaughter liveweight kg (lb)			
	Average		Top ⅓*		Average		Top ⅓*		Average		Top ⅓*	
	kg	tons	kg	tons	hectare	acre	hectare	acre	kg	lb	kg	lb
Dairy calves												
Cereal beef	1829	1·8	1728	1·7	—	—	—	—	392	864	395	871
15-month grass/cereal	1219	1·2	1118	1·1	7·25	2·9	8·00	3·2	432	953	430	947
18-month grass/cereal	1118	1·1	813	0·8	3·25	1·3	3·75	1·5	491	1082	491	1082
20-month grass	915	0·9	813	0·8	3·00	1·2	3·50	1·4	472	1040	490	1080
24-month grass	1016	1·0	813	0·8	2·50	1·0	3·00	1·2	484	1067	500	1100
Suckled calves												
15-month beef calf—autumn calving	915	0·9	711	0·7	1·75	0·7	2·25	0·9	417	920	421	929
2-year beef calf—spring calving	711	0·7	610	0·6	1·25	0·5	1·50	0·6	444	978	463	1020

* The top third of producers on the basis of gross margin

increases have less effect on their gross margins. At the present time s the reliance on concentrates decreases and more emphasis is put on grass and grass products the higher the gross margins per head increase and the lower is the overall stocking rate per grass acre s less reliance is placed on bought-in feed (Table 26). This is a result of both the effect of increases in concentrate prices and the effect on slaughter weight. As the amount of concentrates in the system decreases, not only is the liveweight gain reduced out the weight at which the animal is in a marketable condition ("finish") is substantially increased.

As the amount of concentrates in the system increase the higher the overall growth rate but the lower the slaughter weight. This can be illustrated by reference to three systems utilising the Friesian steer calf. Cereal beef production promotes gains of about 1·13 kg 2·5 lb) per day and cattle produce carcases of about 215 kg 475 lb). Similar steers reared in grass/cereal 18 month production would gain at 0·81 kg (1·8 lb) per day and would produce carcases of about 249 kg (550 lb). Extending the production cycle further to 2 years involves a reduction of growth rate to about 0·68 kg 1·5 lb) per day and the production of carcases of about 281 kg 620 lb). As the slaughter weight increases the effect of calf price on the cost of production decreases.

For the majority of producers the main response to higher concentrate prices will be a switch to systems with lower but more efficiently used concentrate inputs together with greater utilisation of grass and grass products. This means a considerable further improvement in grassland management is required.

The trend towards greater use of grass will result in accompanying changes in cash flow, capital investment and stocking rate. On one hand a longer production cycle means that capital is locked up for longer and the consequent higher requirements have to be met. However, the number of calves and the amount of capital required to stock a given acreage is increased. Two years represents the ceiling to the economic extension of the production cycle. Over 2 years' peak capital requirements have to cope with three batches of cattle on the farm at the same time, and the overall stocking falls markedly.

Grass/cereal systems cover the whole range of slaughter ages and weights which are economically viable now and will remain so into the foreseeable future. For the majority of cattle the range in age of slaughter will fall into a relatively narrow band of 15 to 24 months. With high concentrate prices the swing will be away from the 15-month systems to the 29-month systems by the incorporation of an extra grazing season and the substitution of a "finishing" winter by a "growing" winter.

Beef Systems for Dairy Bred Calves
Guides to Slaughter weights

There is bound to be a range of slaughter ages and weights within the broad description of a single system depending on the type of cattle and precise way in which they are managed on the individual farm. In fact there is a continuous range of slaughter age/weight combinations within which adjacent systems of production overlap. There is no clear division between the end of one system and the start of another. Results from units recorded by the MLC which produce beef from dairy bred calves have been used to construct the graph (Fig. 8) which shows a curve relating slaughter weight and slaughter age up to 24 months for the main breed types. The table below the graph shows the typical carcase classifications achieved by cattle for the three major identified systems of production.

There are considerable differences in slaughter weight at each age between the main breeds and crosses. For example, there is a difference of more than 100 kg (2 cwt) in slaughter weight at 18 months of age between the early maturing Angus×Friesian and the late maturing Charolais×Friesian. These differences between breeds and crosses in slaughter weight are the subject of close study in the MLC Beef Evaluation Units at Ingliston and Sutton Bonington (see p. 74). Also of special interest are the age/weight combinations of Friesian bulls which are considerably higher than Friesian steers and equivalent to Charolais×Friesian steers.

Information on carcase classification shows a tendency for both the fatness and conformation scores to increase as cattle are slaughtered

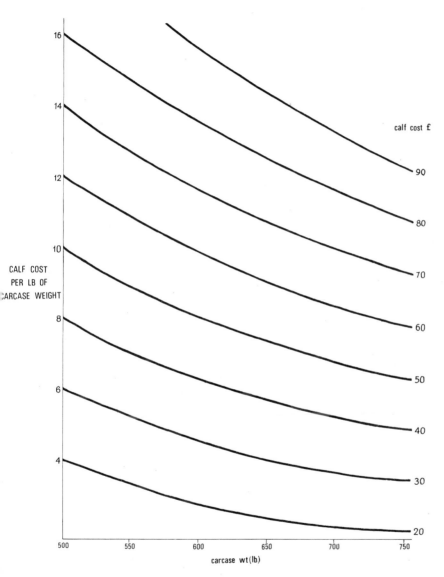

Fig. 9.   Relationship between calf cost per kg (lb) of carcase weight and calf price.

103

at older ages and heavier weights. Obviously at each age point slaughter at a lower level of fat cover would reduce the slaughter weight.

For cattle in dairy beef systems there is no case for slaughtering at less than 386 kg (850 lb) and higher slaughter weights are preferred. The graph clearly shows that, because the Angus×Friesian cannot attain a slaughter weight of 386 kg (850 lb) within the cereal beef system, it is unsuited to that type of production. Although a few Hereford×Friesian calves are reared for cereal beef this cross is also at a disadvantage in the system because it finishes at an excessively light slaughter weight.

Early maturing crosses such as the Angus×Friesian and Hereford× Friesian are used to their best advantage when reared in systems of beef production which have a high dependence on forage and, as a result, operate at moderate lifetime growth rates. Under these conditions lean beef development predominates and the cattle can be taken to higher slaughter weights without the danger of becoming over-fat. Also if early maturing cattle are to be reared in, say 18-month beef production they should be fed a lower level of concentrates during the finishing winter than later maturing Friesians. These considerations are important when planning a beef production system for the individual farm.

The graph has possible usefulness in the marketing of cattle. If the breed type and age at slaughter are known then it is possible to predict the average slaughter weight of a batch of cattle at the typical carcase classification. Of course there will be variation in slaughter weights between individual cattle of the same age. But the total variation is unlikely to be greater than 32 kg (70 lb) when cattle are slaughtered for cereal beef and 45 kg (100 lb) for 24-month beef. This is good enough to suggest that the graph could be used to aid forward buying or selling of finished cattle.

INCREASING SLAUGHTER WEIGHT

When there is a strong commercial interest in beef production it manifests itself through increased calf and store prices. Calf supply is the major factor limiting the size of the home beef production ndustry. When calf prices are high, therefore, other things being equal, producers benefit from higher slaughter weights because the overhead cost of the calf is spread over a greater volume of production (Fig. 9).

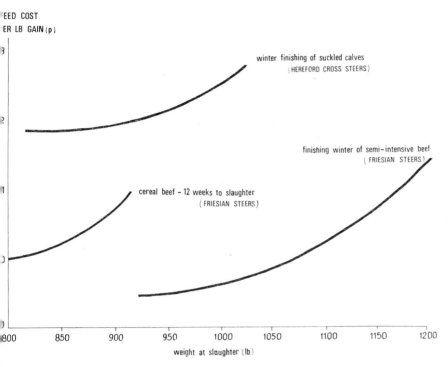

Fig. 10.   *Effect of slaughter weight on feed cost per lb of gain during the finishing period.*

Currently most producers finishing cattle out of yards endure a period when, to achieve the required amount of fat cover on the carcase, liveweight gains are made at a cost per kg (lb) of gain greater than the average value of unit liveweight. Within a system of production the average feed cost per kg (lb) of gain during the finishing period increases at an accelerating rate as slaughter weight increases; Fig. 10 shows the effect of increasing slaughter weight on overall feed costs per kg (lb) of gain during the finishing period.

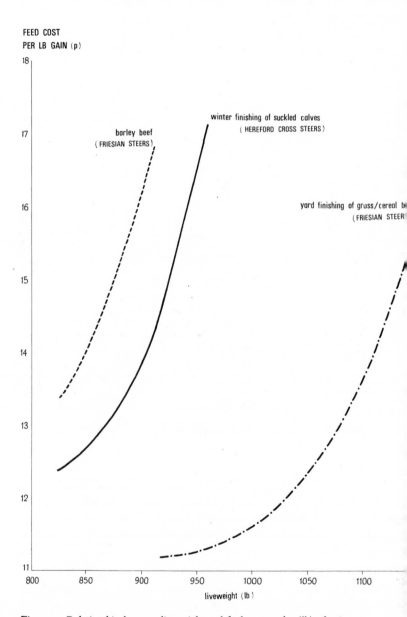

Fig. 11. *Relationship between liveweight and feed cost per kg (lb) of gain.*

The increase in feed costs per kg (lb) of gain as cattle get bigger and fatter is demonstrated more dramatically by an examination of short-term changes (Fig. 11). These diagrams show that there are clearly strict upper limits to the weight at which a margin over feed costs can be obtained. Thus minor feed modifications within a system of production offer little scope for increased slaughter weights. Potential slaughter weight is affected by the type of animal (see p. 102) and by making use of the variation within breeds, breed differences and sexual status of the animal, it is possible to substantially increase slaughter weights. For example, replacing Friesian steers by entire bulls utilises the effects of sexual status. If Friesian bulls are replaced by Charolais×Friesian bulls this adds the increased potential slaughter weight of that cross. The higher growth rate and later maturity of Friesian bulls compared with steers in grass/cereal 18 month beef production allows the production of a beast 54 kg (120 lb) heavier at slaughter, the still higher growth rate of the Charolais×Friesian and its very late maturity allows a further increase in slaughter weight of 113 kg (250 lb). However, this will result in some extension of the production cycle and a reduction in the overall stocking rate as additional hay or silage will be required. The net effect is that increasing slaughter weights does not significantly affect the profitability of beef production to the individual producer, although it is an obvious way of increasing the national supply of beef if this is considered desirable.

Finishing Suckled Calves and Stores

Cattle sold in store markets are derived both from beef and dairy herds. Their overall performance depends on the number and severity of winter store periods they have passed through. Suckled calves are finished at 16 months weighing around 408 kg (900 lb) and in this case no store period is gone through. With cattle slaughtered at 2 to $2\frac{1}{2}$ years, one or two such periods are involved; with those slaughtered at 3 to $3\frac{1}{2}$ years, two or three periods. Weights are typically of the order of 454 kg (1000 lb) with a range from 386 kg (850 lb) to 522 kg (1150 lb). Carcases range from 195–295 kg (430–650 lb) and typically have a carcase class of 4·4. A fat score of 3 L or H is preferred in most markets but can be difficult to achieve with early maturing crosses without a sharp reduction in carcase weight.

Despite the wide range of feeding systems, there are only two really important profit factors. First, skill in buying and selling stock and secondly, low feed costs per kg (lb) of gain.

Both yarded and grazing systems are used, depending on the time of year and weight of cattle involved.

YARDED SYSTEMS
Winter Finishing
Systems

The types of feed used vary greatly between farms but for winter finishing in yards, two types of diet are capable of producing economic gains. In the first, high-quality conserved forage is fed with a minimum supplement of rolled barley—up to 3 kg (6 lb) per head per day. In the second, cheap arable by-products are fed with higher levels of balanced concentrates.

For efficient production and even marketing it is essential that feeding groups should be matched for weight and type. Steers and heifers should be separated and managed differently—heifers requiring lower levels of concentrates.

Over-wintering
Systems

Many suckled calves reared in hill and upland herds are, of necessity spring-born and as a result they are too small to finish over the first winter, as also are lighter weight dairy bred stores. These cattle are therefore overwintered on a low cost feeding system and then finished the following summer at grass.

A level of performance is required which keeps the cattle growing during the winter with liveweight gains put on economically yet still allows compensatory growth to be expressed at grass. This can be accomplished with a liveweight gain of 0·50–0·59 kg (1·1–1·3 lb) per day, which gives the best overall economic performance (Table 27).

TABLE 27

Relationship Between Winter and Summer Gains

Winter gain		Following summer gain		Total gain winter+summer	
kg/day	lb/day	kg/day	lb/day	kg	lb
0·23	0·5	1·04	2·3	184	405
0·23–0·32	0·5–0·7	0·95	2·1	184	405
0·32–0·41	0·7–0·9	1·1	2·2	204	450
0·41–0·50	0·9–1·1	0·9	2·0	204	450
0·50–0·59	1·1–1·3	0·86	1·9	211	465
0·59–0·68	1·3–1·5	0·68	1·5	197	435
0·68–0·77	1·5–1·7	0·54	1·2	191	420
0·77	1·7	0·45	1·0	184	405

GRAZING SYSTEMS

For grass finishing, good grassland management is essential. The produce is able to adjust stock numbers to grass production, and with heavy stores, it is often possible to complete two or more production cycles within one grazing season.

The lighter weight stores are more heavily stocked at grass and are kept growing during the grazing season to ensure efficient grassland utilisation. They are then either sold in the autumn as medium or heavy stores or finished in yards during the winter. Fertiliser N associated with increased stocking rate almost invariably increases profits.

Slaughter weights range from 431–567 kg (950–1250 lb) with carcase weights of 204–318 kg (450–700 lb). Typical carcase class for Hereford crosses 3 L or H 3H 4.

An assessment of the price that can be paid with different feed costs and sale prices to leave different levels of profit can be made using a calculator. Such calculators are produced by the MLC. An hour spent by the prospective purchaser in the farm office applying the calculators to the farm situation before going to market, or before designing the winter diet, will pay dividends.

Suckled Calf Production

Herds in suckled calf production are kept under a wide range of conditions on lowland, upland and hill farms and cows are calved at various seasons. Consequently, no precise description of the system is possible. But, whatever the conditions, the overall objectives are the same: cows with a long productive life and a calving interval of 1 year, low calf mortality, and good calf weaning weights.

Management and Performance Targets

The calving season affects costs and returns. In general, the choice of calving season is influenced by individual farm circumstances, especially the availability of winter feeds and buildings. Whatever policy is adopted, only where the calving season is short can herd feeding levels be adjusted to avoid gross over- or under-feeding of individual cows.

Winter feeding of winter/spring-calving herds should maintain cow bodyweight and condition until the last month of pregnancy. The feeding level is then gradually increased to provide for the rapid growth of the unborn calf and subsequent lactation. Generally, hay or silage is fed, supplemented with low levels of concentrate in late pregnancy. For a 559 kg (11 cwt) cow, 27–32 kg (60–70 lb) medium quality silage per day should be sufficient to maintain weight. Self feeding of good quality silage, resulting in intakes above this level, is economically undesirable. In the last month extra feed should be provided by 2 kg (4 lb) concentrates (in winter) to allow for growth of unborn calf and prepare for lactation.

Saving in winter feed costs is possible with spring-calving cows if these are in good body condition at the start of the winter. Additional protein is only required, even during the period of maximum cow productivity, when cows are wintered on low quality roughages such as poor hay or straw.

Some farms use arable by-products, a typical daily diet being 7–9 kg (15–20 lb) straw and 2–3 kg (4–6 lb) barley, supplemented with urea, vitamins and minerals. This is a relatively low cost diet, and makes no demands on the grassland acreage.

110

After calving, feeding is maintained at the immediate pre-calving level for 2 to 3 weeks, and then increased, as the calf takes more milk, until grazing becomes available.

Autumn-calved cows need higher levels of winter feed to maintain milk production. A typical daily diet would be 32 kg (70 lb) of medium quality silage and 3 kg (6 lb) of barley. Body condition must be maintained at a level sufficient to ensure that the cow conceives 3 months after calving. In housed herds, calves are often creep-fed on good quality silage and concentrates from a few weeks of age.

Grazing management should be designed to maintain milk production in winter/spring-calving herds and to provide good quality grazing for autumn-born calves. On marginal land, grazing management is often influenced by the need to conserve the better, flatter areas. On such farms the integration of sheep and cattle grazing is also important. On lowland farms and improved upland areas. paddock grazing is an effective means of maintaining herbage quality. But from mid-season, herbage quality and calf performance generally decline. Fresh grazing can be provided by the aftermaths on previously conserved areas. Worm parasites are not a major problem in single suckled calves.

Creep feeding can improve calf performance. With autumn-born calves, creep feeding a total of 152 kg (3 cwt) barley will improve weaning weights by up to 23 kg (50 lb). But as the calves grow larger it is difficult to allow them access to a creep while excluding smaller cows. Some producers wean early, graze the calves on high quality aftermaths and use the cows to eat down rougher areas.

Because milk contributes more to the growth of spring-born calves, creep feeding can be delayed until later in the season. But in the last few weeks before weaning, a total of 51 kg (1 cwt) barley can be expected to increase weaning weights by up to 9 kg (20 lb).

Creep feeding of calves prior to weaning also has the advantage of conditioning them for future diets and guarding against any check in growth rate that may occur as a result of weaning.

TYPE OF STOCK: Aberdeen Angus, Hereford, Charolais and to a lesser extent Simmental bulls distinctively colour-mark their calves. When calves are marketed at suckled calf sales this is an important selling point. If calves are finished at home or sold by private treaty, however, colour-marking is less important and the choice of breeds from which bulls can be selected is wider.

WHERE IT GOES WRONG: Excessive winter feed costs, which account for over half the variable costs, often lead to low profitability. Minimising costs at this time depends on a short calving season to facilitate rationing, and the efficient use of grassland. Adequate fertiliser levels, combined with good conservation and intelligent grazing management can do much to reduce costs and, at the same time, improve performance.

Poor overall reproductive performance due to high rates of barrenness and calf mortality is a major loss of profitability. Good conception rates depend on the proper timing of feed inputs so that cows are in good condition at bulling. Proper feeding also increases the viability of the calf but stockmanship at and after calving is also critical. Abortion caused by brucellosis is an important source of low calving performance, particularly among newly purchased heifers. Where it is a problem, veterinary advice should be sought. An increase in the beef cow and hill cow subsidies to brucellosis-accredited herds is designed to encourage herds to participate in the eradication scheme.

Grass staggers or hypomagnesaemic tetany can be an important cause of cow mortality. It can occur at any time of the year and is most common in cows when they are milking heavily. The incidence tends to rise as the intensity of grassland management is increased. The main preventative measures include applying potash fertilisers in the autumn, rather than in the spring, feeding 57 g (2 oz) calcined magnesite per day in the concentrate or sprinkled on silage, or dosing with magnesium bullets.

A recent development is the production of bull beef.

Hereford Bull

Typical Hereford

Lincoln Red Bull

Courtesy The Lincoln Red Cattle Society

Friesian Steers

Courtesy The British Friesian Cattle Society

Group of Sussex Cattle

Courtesy The Sussex Cattle Society

Simmental Bull

Murray Grey Bull

Courtesy The Murray Grey Beef Cattle Society and The Sir William Angliss College, Melbourne

Devon Bull

Charolais Bull

Courtesy The British Charolais Cattle Society

Galloway cattle

Belted Galloway Bull

Output from a suckler herd can be increased by placing a second purchase calf with the cow and double-suckling. But good stockmanship is essential. The introduction of disease through the bought-in calf is a risk, and acceptance of the calf by the cow must be ensured. But perhaps the greatest danger lies in over-estimating the milk production of the cow, and producing two small calves in place of a single good one. The technique applies only to summer or autumn calving periods.

TABLE 28

Bull Beef Productions in Suckler Herds

No. of herds	Bulls		Steers		
	5		5		
Weaning weight of male calves kg (lb)	291	(641)	272	(599)	
Overall stocking rate cows per hectare (acre)	1·75	(0·7)	1·75	(0·7)	
Gross margin per cow					Extra 5% for bulls
Gross margin per hectare (acre)					Extra 3% for bulls

TABLE 29

Targets for Double Suckling

Calving interval (days)	365	
Calf gain per day (kg (lb))	0·9	(2·0)
Calf weaning weight (kg (lb))	261	(575)
Stocking rate (cows/hectare (acre))	1·50	(0·6)

The usual practice in suckler herds is to purchase all replacements either as bulling heifers or as heifers in calf.

Traditionally, the sources of heifer replacements have been such crosses as the Shorthorn cross hill breed (e.g. Whitebred Shorthorn cross Galloway to give the Blue-Grey) or a beef breed cross Dairy Shorthorn often imported from Ireland. In recent years, however, during the expansion of the national herd, both these sources of

HEIFER REARING— SUCKLER HERD REPLACEMENTS

E

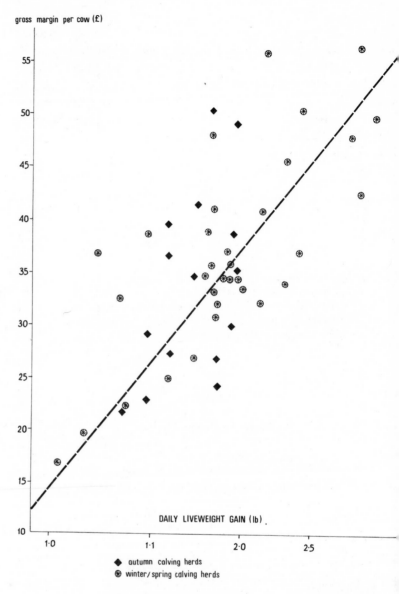

Fig. 12. *Relationship between liveweight gain per day of calf and gross margin per cow.*

supply have diminished or become less reliable, and an increasing tendency to retain home-bred heifers has aggravated the problem.

The use of alternative replacements from the dairy herd, such as the Hereford×Friesian, has rapidly increased in popularity and these crosses have proved to be very reliable with excellent mothering qualities.

As a result of these changes, the rearing of brucellosis-accredited heifers for suckler herds is becoming an important system of beef production. The brucellosis eradication scheme has resulted in a substantial premium being available for accredited stock.

Suckler herds require heifers to calve at between 2 and 2¼ years of age. To achieve this target heifers have to be well grown to reach the right weight at bulling—approximately two-thirds of the optimum calving weight. Targets are set out in the following table:—

TABLE 30

Suckler Herd Breeding Cow Replacements

Cattle targets (Hereford×Friesian heifers)

Age at bulling	15–17 months
Weight at bulling	340 kg (750 lb)
Age at calving	24–26 months
Weight at calving	454–499 kg (1000–1100 lb)

Target gain (kg (lb)/day)

	Autumn-born calf		Spring-born calf	
	kg	lb	kg	lb
Rearing winter	0·65	1·4	—	
First grazing season	0·68	1·5	0·59	1·3
Growing winter	0·59	1·3	0·59	1·3
Second grazing season	—		0·82	1·8
Overall	0·65	1·4	0·65	1·4

Importance of Cattle Performance on Profitability

Although commercial beef producers are increasingly accepting the desirability of planned systems there are wide ranges in performance and in the financial returns between farms. Within all beef systems there is a close relationship between physical performance and profitability. The factor which has the greatest effect on gross margin per head is the growth rate, and the higher average daily gain within a system of production the higher the gross margin:

System	Effect of an increase of 0·045 kg (0·1 lb) per day in daily gain (£)	
	kg	lb
Cereal beef	+5·55	+5·00
Grass/cereal 15-month beef	+6·12	+5·50
Grass/cereal 18-month beef	+8·03	+7·22
Grass 20-month beef	+7·98	+7·18

Increasingly, therefore, with higher feed prices, it will be important for beef producers to obtain the highest rate of gain within systems to minimise feed costs per kg (lb) of gain.

In suckler herds other cattle performance factors are of equal importance. The ability to produce a calf every year is essential to the economics of suckled calf production.

In all systems of production there is a close relationship between growth rate, feed conversion, cost of feed per kg (lb) of gain and profitability. In the longer term, therefore, producers must ensure that the right breeds and crosses have been selected. Whenever possible high performance strains within the breed should be used. Management should extract the maximum growth potential from the cattle with the available feed. The factors which particularly influence growth rate and with it the related characteristics of lateness of maturity and slaughter weight are breed type, potential within a breed and sexual status particularly, whether male calves are castrated or left entire. These effects are at least partly additive.

Sheep

Systems Of Production

Sheep play a role in a whole range of farming systems, nevertheless they are more predominant in certain types of farming—approximately 44% of breeding ewes are found on hill farms, 17% on upland farms, and 39% on lowland farms. The proportion in the first two categories has increased in recent years at the expense of the lowland flocks. This has happened partly as a result of government policy implemented through subsidies and price support but it also reflects the lack of alternative farming enterprises in the hill and upland areas. In the hill and mountain farming areas sheep production is commonly a specialised and fairly stable single farm enterprise. In upland areas it is associated with cattle rearing or dairying.

In lowland areas sheep are usually part of mixed farming systems and over recent years have suffered from competition not only from other types of livestock production but also from arable crop production. Approximately 60% of the breeding ewe population is found on livestock rearing farms. Slightly less than half of these being farms where sheep form the main enterprise.

Although dairy farming is an increasingly specialised and larger scale operation, sheep flocks are still run in conjunction with dairy herds on a great many farms. There are something like 1 million ewes on farms in the "mainly-dairying" category, this is approximately one-fifth of the most productive ewes in the country.

Flock size varies widely, some flocks having well over 10,000 breeding ewes. The overall flock size is 179 breeding ewes with hill flocks averaging about 210 ewes and lowland flocks about 130 ewes.

Sheep are often only secondary enterprises on lowland farms. Sheep production is still predominantly an extensive operation with low inputs and low unit output and returns. However, sheep can, and do, play a useful and financially valuable role in lowland farming systems. They often act as a complementary enterprise utilising resources (e.g. unploughable grassland or arable by-products) which would otherwise not be used. They also enhance the farm rotation adding appreciably to cereal yields. Although there is very wide variation in the profitability of flocks, the average margin per acre from traditional systems has been and still is to some extent, low compared to alternative farm enterprises. The advantage which sheep enjoy is their low requirement for working capital, relative to other livestock enterprises. Requirements for fixed capital are also usually lower, in particular housing is not essential for most sheep flocks.

Lowland spring lambing flocks can be broadly categorised into three distinct types:

Extensive grassland flocks run on traditional low cost lines, generally utilising second or third class land on permanent or semi-permanent pasture with low levels of fertiliser application.

Intensive grassland flocks higher cost systems associated with reseeded young grassland many practising a rotational or other controlled grazing systems.

Arable flocks utilising arable by-products such as stubble grazing winter cereal grazing, sugar beet tops etc. for significant periods, and in some cases additional areas of roots and catch crops grown specifically for the sheep flock. There is obviously considerable overlap between the three types of flock described.

On upland farms the type of production is very dependent on the quality of the land involved. In some cases it may be good enough to enable lambs to be sold direct for slaughter, but a high proportion of lambs are sold as stores for subsequent finishing on lowland farms. In addition many upland farms specialise in the production of cross-bred ewe lamb replacements for lowland flocks. On upland farms the number of alternative enterprises is lower than on lowland farms and sheep production is generally of much greater importance.

Lamb slaughterings follow a distinct seasonal pattern increasing steadily from May to November and then declining gradually. As the season advances after May there is a continual increase in the proportion of heavy lambs. These changes partly reflect regional differences in breeds and systems of production. Just over a quarter of the lambs are sold as hoggetts—lambs carried over from one year to the next. Within this slaughter pattern four main systems of lamb production can be broadly distinguished as follows.

Early out-of-season production in which the ewes are lambed in December and January to provide lambs for sale at the time of high early spring prices. Except in areas favoured with a mild climate and very early grass growth it is normally a high feed cost system. It accounts for about 2% of the annual lamb crop slaughterings.

Production of grass in which flocks are managed with the object of selling all lambs fat by the end of the grazing season. Marketing takes place from May onwards, lowland flocks predominate. This system accounts for about 50% of the annual lamb crop slaughterings.

Production off grass plus forage crops in which the ewe flock is managed on a low cost system. A proportion of the lambs are sold off grass but the bulk are finished and sold off forage crops from the end of the grass grazing season to the end of the year, either on the same farm or after passing through a store lamb market to a feeder's farm. A significant proportion of the purchased hill lambs are involved in this type of production system which accounts for 20% of the annual lamb crop slaughterings.

Hoggett production is an extension of lamb production involving the finishing of spring-born lambs over the winter period and following spring on a variety of forage crops. It accounts for approximately one-quarter of the annual lamb crop slaughter. Most of the hoggetts are sold before the beginning of the next grazing season.

The lack of any precise plan or target is commonplace in sheep production. This reflects itself in fat-lamb flocks having very variable success in the proportion of lambs finished off grass. A further example of this lack of precision is that many flocks are built up or maintained on the basis of opportunity buying of replacements and selling of cull ewes. The implications of this are not generally appreciated by commercial flockmasters. Many flocks contain "mixtures" of breeds and crosses of ewes which differ by as much as 50% in bodyweights which complicates feeding and stocking rates. Additionally the weights of lambs at slaughter are determined to a large extent by the weights of the parent breeds, consequently there is often a wide range of weights and fatness in the lamb carcases produced from a single flock.

PERFORMANCE FACTORS DETERMINING PROFITABILITY

There is a very close relationship in practice, in all types of production, between physical performance and the financial results achieved. Improvement of physical performance is important in both lowland and hill flocks, the difference lies in the level of performance at which improvement is aimed.

The output from the ewe flock consists of lambs and wool, of which the former is by far the most important. In lowland flocks wool

TABLE 1

Relationship between Physical and Financial Performance

| | Extra profit for an increase of 0·1 lambs reared per ewe | | Extra profit for an increase of 0·1 ewes per ha in stocking rate | |
| | (£ per ewe) | | (£ per ha) | |
	Lowland flocks	Upland flocks	Lowland flocks	Upland flocks
1974	+1·16	+1·05	+2·42	+2·05
1976	+2·15	+1·97	+5·14	+3·63
1978	+2·24	+2·09	+1·89	+2·12

accounts for about 10% of the financial output per ewe, however, in hill flocks where lambing percentages are lower the contribution of wool to financial output increases to 18% of ewe output. The output of lambs depends on the number of lambs sold, their average weight and the price realised per pound.

The number of lambs reared per ewe has a significant effect on profitability. The factors which determine the number of lambs available for sale are:

* Number of lambs born per ewe lambed.
* Number of barren ewes.
* Number of ewe deaths pre-lambing.
* Number of perinatal lamb deaths.
* Number of later lamb deaths.

The extent of the variation between recorded flocks in these factors is summarised in Table 2 which shows that although the number of lambs born per ewe lambing is the biggest single source of variation in the number of lambs for sale from the flock, each of the separate wastage factors has a significant effect.

Individual flocks may have a problem with a single factor but generally it is more often the case that a low performance in one factor is associated with an equally poor performance in several

Number of lambs reared per ewe

others, thus the loss of production is accumulative. Whilst differences between breeds and crosses account for some of the variation between flocks in these vital performance factors differences in the management and feeding of the ewes are equally important and it is to these that the commercial farmer must direct his main attention.

TABLE 2

Recorded Lowland Flock Performance Levels

	From	Range To	Average all flocks
Number of lambs born per ewe lambed	1·0	1·91	1·41
Number barren ewes as percent ewes to ram	0	20·0	8·4
Number ewe deaths as percent ewes to ram	0	17·2	4·2
Number of lambs dead or died at birth as percentage of all births	1·0	34·5	8·0
Number of lambs died subsequent to birth as percent of lambs survived birth	1·0	14·1	5·5

The causes of recorded barrenness are numerous and include sexual immaturity, fertilisation failure and embryonic and foetal mortality. No single remedy is therefore available, but nutritional deficiencies and the incidence of certain infectious diseases such as enzootic abortion and toxoplasmosia are particularly important. Losses from the diseases which cause premature loss by direct damage to the conceptus, might be appreciably reduced by simple management procedures such as avoiding non-essential seasonal movements of stock which are liable to spread infection in breeding flocks. Lamb mortality levels are generally high. Lamb mortality up to weaning, including lambs dead at birth, runs at an average of about 20 to 25% of all lambs born, though in well managed flocks the losses can be as low as 7 to 10%. Much of the loss is avoidable by improvement in husbandry and nutrition of the lambing flock. Perinatal mortality usually accounts for well over half of the total lamb losses.

Improved control of microbic and parasitic diseases has in recent years greatly reduced wastage in the pre- and post-weaning stages but improvements in the timing of routine veterinary programmes could reduce losses still further. Medicines and veterinary costs comprise on average about 10% of the variable costs per ewe, but there are flocks in which these costs are considerably higher without any complementary increase in the gross margins per ewe.

Due to the seasonal variation in prices the value of the slaughtered lamb and consequently the output per ewe is affected by the date of sale. Early lamb production systems with their higher feed and forage costs are particularly sensitive to changes in the price received for the lamb. Rapid growth rate to slaughter weight is an important factor in determining the returns from the production of early out-of-season lamb and lambs off grass. Growth rate is of less importance in late lamb and hoggett production off forage crops where the final carcase weight obtained is relatively more important.

The weight of lambs sold can have a large effect on the gross output of the ewe. If the weight of the carcase is increased it can improve the profitability because the relatively large fixed input of ewe maintenance cost is spread over a greater total output. However, this is modified to some extent by price. Higher prices during the early part of the year substantially reduce the weight of carcase required to produce any given return.

Table 3 shows the effect of carcase weight, lamb price and number of lambs sold per ewe on lamb sales per ewe. It is striking that an increase in the average carcase weight per lamb of 2 kg in a flock of 100 ewes is equal to an increase of about ten in the number of lambs reared per 100 ewes. Lamb sales per ewe of about £44 at a price of 160p per kg can be obtained from 1·80 lambs reared per ewe and a carcase weight of 15 kg or 1·40 lambs reared per ewe and a carcase weight of 20 kg.

The weight at which lambs are fit for slaughter is largely determined by the mature weights of the parent breeds, modified to some extent by the growth rate and type of rearing of the lambs. Change in lamb carcase weights therefore mainly presuppose changes in the parent breeds. Most slaughter lamb is produced from lambs which are relatively immature. The main criterion for their acceptability as a

125

TABLE 3

The Effect of Carcase Weight, Lamb Price,
and Number of Lambs Sold per Ewe on Lamb Sales per Ewe

Deadweight (kg)	No. of lambs reared per ewe: Price per kg	1·3	1·4	1·5	1·6	1·7	1·8	1·9
	p							
15	160	31·2	33·6	36·0	38·4	40·8	43·2	45·6	48·0
	180	35·1	37·8	40·5	43·2	45·9	48·6	51·3	54·0
	200	39·0	42·0	45·0	48·0	51·0	54·0	57·0	60·0
16	160	33·3	35·8	38·4	41·0	43·5	46·1	48·6	51·2
	180	37·4	40·3	43·2	46·1	49·0	51·8	54·7	57·6
	200	41·6	44·8	48·0	51·2	54·4	57·6	60·8	64·0
17	160	35·4	38·1	40·8	43·5	46·2	49·0	51·7	54·4
	180	39·8	42·8	45·9	49·0	52·0	55·1	58·1	61·2
	200	44·2	47·6	51·0	54·4	57·8	61·2	64·6	68·0
18	160	37·4	40·3	43·2	46·1	49·0	51·8	54·7	57·6
	180	42·1	45·4	48·6	51·8	55·1	58·3	61·6	64·8
	200	46·8	50·4	54·0	57·6	61·2	64·8	68·4	72·0
19	160	39·5	42·6	45·6	48·6	51·7	54·7	57·8	60·8
	180	44·5	47·9	51·3	54·7	58·1	61·6	65·0	68·4
	200	49·4	53·2	57·0	60·8	64·6	68·4	72·2	76·0
20	160	41·6	44·8	48·0	51·2	54·4	57·6	60·8	64·0
	180	49·1	52·9	56·7	60·5	64·3	68·0	71·8	75·6
	200	54·6	58·8	63·0	67·2	71·4	75·6	79·8	84·0
21	160	43·7	47·0	50·4	53·8	57·1	60·5	63·8	67·2
	180	49·1	52·9	56·7	60·5	64·3	68·0	71·8	75·6
	200	54·6	58·8	63·0	67·2	71·4	75·6	79·8	84·0
22	160	45·8	49·3	52·8	56·3	59·8	63·4	66·9	70·4
	180	51·5	55·4	59·4	63·4	67·3	71·3	75·2	79·2
	200	57·2	61·6	66·0	70·4	74·8	79·2	83·6	88·0
23	160	47·8	51·5	55·2	58·9	62·6	66·2	69·9	73·6
	180	53·8	58·0	62·1	66·2	70·4	74·5	78·7	82·8
	200	59·8	64·4	69·0	73·6	78·2	82·8	87·4	92·0

marketable produce is that they should have reached an appropriate
level of finish, this is normally when the lamb has reached about
half its potential adult bodyweight. The time taken for a lamb to
reach half its potential bodyweight is influenced by the quantity and
quality of the feed available, both through its dam and directly from
grazing and supplementary feeding. With a knowledge of the adult
bodyweight of the parent breeds it is possible to specify a range of

Thus:

	kg	lb
Breed weight of ram (Suffolk)	91	200
Breed weight of ewe (Clun Forest)	73	160
	————	————
	164÷4	360÷4
Slaughter weight of lamb	41	90
Carcase weight	20	45

crosses to produce lambs whose carcases fit into specified weight classes that are required by the market.

Table 4 shows how information on some breeds can be combined to define the appropriate weight class of the fat lambs each cross would be expected to produce.

Note: Each figure shown is the average for males and females combined and the carcase weights would be either below or above according to sex. Wether or entire male lambs have a potential adult bodyweight that is between a fifth and a quarter higher than their sisters. When all lambs are to be slaughtered, as in fat-lamb production, the difference in potential adult bodyweight will be reflected in

TABLE 4

Bodyweight of Parent Breeds and Lamb Carcase Weight (K.O. 50%)

Carcase class	Breed		Mature adult bodyweight						Crossbred fat-lamb			
	Ewe	Ram	Ewe breed		Ram breed		Crossbred fat-lamb		Bodyweight at slaughter		Carcase weight	
			kg	lb	kg	lb	kg	lb	kg	lb	kg	lb
11–15 kg	Welsh Mountain	Southdown	45	100	61	135	53	117	26	58	13	29
25–34 lb	Cheviot	Southdown	64	140	61	135	62	138	31	69	15	34
16–20 kg	Scottish Blackface	Dorset Down	70	155	77	170	74	162	37	81	18	40
35–43 lb	Welsh Halfbred	Dorset Down	73	160	77	170	75	165	38	83	18·5	41
20–24 kg	Romney Halfbred	Suffolk	77	170	91	200	84	185	42	93	21	46
44–52 lb	Mule	Suffolk	84	185	91	200	87	193	44	97	22	48
24 kg plus	Scottish Halfbred	Oxford Down	88	195	105	235	97	215	49	108	24	54
53 lb plus	Suffolk × Scottish Halfbred	Oxford Down	88	195	105	235	97	215	49	108	24	54

differences in carcase weights of entire male lambs, wethers and ewe lambs.

The production of lambs within the lower carcase weight class fits in well wherever the land or conditions limit the potential for production. Because larger breeds and crosses are more costly to purchase and feed and cannot be stocked as heavily as lambs of the smaller breeds, it is important for the commercial producer to ensure that they produce a high output of lamb carcase per ewe. For this purpose they should be mated to rams of large breeds so that carcases in the higher weight range, which are not overfat, can be produced.

The general principle of predicting carcase weights from the breed weights of the parents also applies to store lambs. It is not quite so reliable as in the case of fat-lambs which grow quickly and unchecked to slaughter at an early age, because store lambs have a more variable growth pattern.

The rate of fat deposition in lambs appears to vary with their growth pattern, especially as they get older, and this can have an effect on the weight at which they are fit for slaughter. For instance, it appears that store lambs which have been severely checked in early life may fatten rapidly when given generous feeding and be ready for slaughter at a lower weight than would be predicted from the parent breed weight formula. Conversely, if the restriction in early life is not too severe and store lambs simply grow slowly but steadily they will probably not reach the desired level of fatness at the predicted slaughter weight.

Store lamb systems are very flexible and can permit the benefits of crossbreeding, particularly in relation to target carcase weights, to be fully utilised. For example, they can allow the potential of the smaller breeds and crosses of ewe to be exploited by crossing with a large breed of ram. Because early growth is not so important in store lamb systems the disadvantages of the smaller ewe breeds—lower lamb birthweights and growth rates—are not critical. The weight of lamb carcase produced per ewe is the main factor determining the economic and biological efficiency of this type of lamb production; within reasonable limits the time taken to produce the carcase weight is of lesser importance.

The following example illustrates the effect of using rams of different breed sizes on ewes of small breed size for different systems of production.

TABLE 5

Effect of Ram Breed Size in Crossbreeding with Small Ewe Breed

System	Ewe			Ram			Lamb		
	Breed	Breed weight		Breed	Breed weight		Carcase weight		Age at slaughter
		kg	lb		kg	lb	kg	lb	days
Fat-lamb production	Welsh	45	100	Southdown	62	135	13	29	140
Store lamb production and finishing	Welsh	45	100	Suffolk	91	200	17	38	270

Because the production cycle for store lambs is long it is an essential part of such systems that both ewes and lambs are fed on low cost feeds.

Trade requirements for lambs vary in weight and fatness and it is possible to meet these differing requirements by exploiting the known variations in bodyweights of the British sheep breeds. Acceptable carcases in four weight ranges can be produced in practical and efficient systems by crossbreeding fat-lamb sire breed rams on various currently available types of ewe.

The seasonal variation in the price per kg (lb) of carcase has to be taken into account considering the slaughter weights of lambs.

It is important for the producer to choose the weight at which the lambs are sold for slaughter very carefully so as to ensure that the potential of his ewes are realised.

On the cost side there are opportunities for useful reductions in most sheep flocks and the first to be considered is the flock replacement

Controlling costs of Production

cost. Flock replacement cost is the cost of maintaining a flock of a given size taking account of ewe deaths and the ageing of the flock. It accounts for 34% of the direct costs of production. The factors which control the cost of maintaining the breeding stock are the number of replacements required annually and their mortality rate. In practice, replacement and selling policies are flexible, varying with circumstances and particularly with market prices. The average levels of ewe mortality in MLC recorded lowland and upland flocks are 3·6% and 4·5% respectively but with considerably higher rates recorded in individual flocks. The normal breeding life of ewes in commercial flocks is between 4 and 5 years.

The length of breeding life in the flock can be increased from ewe lambs and by reducing the rate of premature culling. The true extent of premature culling may be partially masked by economic opportunism in disposing of cast ewes in a favourable market.

Careful attention to the health and feeding of the flock could allow a high proportion of the ewes in many flocks to produce an extra lamb crop. Providing the over-age ewes do not suffer an abnormally high death rate or cause the value of the cull to drop too markedly, then retention reduces the cost of maintaining the breeding stock.

The practice of mating ewe lambs reduces flock costs since the average breeding life is normally increased and it also increases the number of lambs available for sale. It is estimated that currently 26% of ewe lambs which are retained for breeding are put to the ram in their year of birth, but there are considerable breed and regional variations in the practice. There is no reason why a higher proportion of lowland ewe lambs should not produce a lamb crop in their first year with proper attention to husbandry techniques. The results show that in practice the number of lambs reared per 100 ewes lambs put to the tup is disappointingly low—78 in lowland flocks, 58 in upland flocks, due mainly to the low conception rates since the numbers born to each ewe lamb which lambed was satisfactory. Selection of well-grown lambs, adequate feeding and separate treatment of the main flock are the keys to improving performance.

Good body condition of the ewes at tupping is the first essential in ensuring a good lamb crop. To ensure that ewes are in good body condition, it is important to take positive action by ensuring that they are weaned in good time and have access to an adequate supply of high quality feed, such as aftermath grazing, at least 6 weeks before tupping is due to commence. Reliance on the rapid flushing of lean ewes immediately before tupping is not satisfactory. A routine condition scoring of ewes about 6 weeks before tupping provides a useful guide to the level of feeding required to get them to the right weight for tupping.

For the first 3 months following tupping the feed requirements of the ewe are low because pregnancy does not make any significant demands during this period. Providing ewes are in good condition at tupping and they are not housed or heavily stocked, supplementary concentrate feeding is not generally required.

During the final 8 weeks of pregnancy the lamb foetus grows very rapidly, making about 75% of its growth, with consequent heavy demands on the ewe. This critical period of heavy demand on the ewe normally coincides with late winter when the available supply and quality of feed from grazing is at a very low level. In these circumstances it is necessary to provide some supplementary concentrate feed in steadily increasing amounts.

In practice it is impossible to feed precisely to meet the varying demands of the individual ewes in late pregnancy because the flock will normally lamb over a period of 5 to 6 weeks. In addition the proportion of ewes carrying triplets, twins and singles varies not only between flocks but also within the same flock from year to year. In the case of most lowland flocks not more than 25% of the ewes should be carrying singles if a satisfactory number of lambs are to be available for sale. This suggests that feeding in this very important phase should be geared to meet the requirements of the twin-bearing ewe.

The result of adequate feeding in late pregnancy is that ewes give birth to strong lambs and have a good supply of milk available to suckle them. These two factors are extremely important in ensuring that lamb losses are minimised. They also ensure that lambs are able to take advantage of the ewe's milk supply and grow rapidly and efficiently.

Within any flock there is considerable variation at any one time in the requirements of individual ewes, and it is obviously wasteful to give the same level of feeding to all ewes. Efficiency in the utilisation of feeds is achieved by dividing the flocks into groups of ewes with similar nutritional requirements. This is simplified where steps are taken to ensure a tight lambing period, in practice the average spread of lambing in recorded flocks is 62 days. Flocks can be divided according to expected lambing date by the use of coloured markers with the rams at mating.

Similarly the feed requirements of lactating ewes with twins are in the order of 25% higher than ewes suckling singles. Thus there is a strong case for separation and giving a high plane of nutrition to the ewes suckling twins. This can be achieved in a number of ways—concentrates are fed at a high level for some weeks after lambing or a high intake of digestible herbage is ensured, either through a lower stocking rate or by grazing on better quality pasture.

Date of lambing is important, putting the tupping date back to obtain a late lambing to coincide with the growth of grass enables a saving in concentrate feeds to be made. Where grass is available at the time of lambing, concentrate feeding for ewes with singles can be cut-out completely, and that for ewes with twins cut back considerably. Later lambing will tend to increase the proportion of lambs sold at the time of lower prices, and/or increase the proportion of store lambs. Where the aim is to finish lambs in the autumn and over the winter on forage crops this is not of critical importance and lambing later is generally more profitable.

HILL SHEEP PRODUCTION

In hill flocks the important components of profitability are the same as in the lowland and upland flocks. Thus it is just as important to improve the number of lambs reared per ewe, but in the hill situation a sensible target would be 100 lambs successfully reared per 100 ewes to ram, compared to 180 lambs reared per 100 ewes to ram as a target for a lowland flock. Wool forms a higher proportion of the output in hill flocks and therefore is economically more important.

Length of life of the ewes is particularly important because, at the low reproductive rates commonly encountered, the flocks could not otherwise be maintained. Similarly ewe deaths are of much greater significance than in lowland flocks. Indeed in years of adverse weather conditions ewe mortality can be a serious cause of financial loss.

Average performance levels and targets for hill flocks are given in Table 6.

TABLE 6

Average Performance Levels and Targets for Hill Flocks

Ewe performance (per 100 ewes put to ram)	Average	Range	Target
Empty	9	3– 18	6
Died	6	2– 15	4
Lambed	86	75– 94	90
Live lambs born	90	70–120	103
Lambs reared	85	50–115	95

The main alternative sheep enterprise to a lowland breeding flock is a store lamb finishing enterprise during the autumn and winter. Where forage catch crops or arable by-products are used, this type of enterprise provides a comparative margin per acre and return on working capital employed.

Because of advances in techniques for growing forage roots reliable crops can be grown and fed without a high labour input. These techniques include precision drilling, pre-emergence residual herbicides, direct drilling and electrified netting. The use of roots as a break crop, particularly in the cereal growing areas, coupled with the reduction in the labour requirement has brought a halt to the downward trend in the root acreage. The increasing interest lies in

FINISHING STORE LAMBS

their high yield per acre and consequent lower cost per unit of energy.

A major factor controlling profitability is the feeder's margin (difference between purchase and sale price). An assessment of the price that can be paid with different feed costs and lamb prices can be made using a calculator of the type described earlier.

Stock Resources

The number of breeding ewes in the UK fluctuates around 12 to 13 million. Included in this total there are some fifty purebreeds and more than three hundred crossbred types derived from them. The fifty purebreeds have been classified into seven groups. A detailed description of each of the breeds and crossbred types is unnecessary but an examination of the breeding structure of the industry shows that there are eight breeds which have a major influence on the industry and these merit a short description. The breeds are:—

> Scottish Blackface Border Leicester
> Welsh Mountain Bluefaced Leicester
> North Country Cheviot Teeswater
> Swaledale Suffolk

These eight purebreeds together account for about 50% of the total ewe population, and crossbred ewes derived from them account for a further 25% and are the dams of a high proportion of the lambs slaughtered each year.

Scottish Blackface

The Scottish Blackface is numerically the most important breed with 2·4 million breeding ewes. It is native to the hill and mountainous areas of Scotland, although considerable numbers are also found in Scotland on the lower ground, crossed with Border Leicester rams or rams of the Down breeds. The ewes found on the low-ground are mostly those which have been drafted out of hill flock after producing four or sometimes five lamb crops under the harsh conditions of the open hills. The performance of the breed enerally depends on the conditions under which the ewes are kept. Under good hill conditions the ewes weigh an average of 54 kg (120 lb) prior to tupping and under upland conditions will produce 150 live lambs per 100 ewes, while under hill conditions this figure is reduced considerably to less than 100 lambs per 100 ewes.

An important crossbred ewe, the Greyface, is produced when a Border Leicester ram is crossed with Scottish Blackface ewes. There are approximately 214,000 breeding ewes of this crossbred, which are normally crossed with Down rams on lowland farms to produce fat lambs.

The North Country Cheviot is found in various areas of Scotland, but is generally associated with the extreme north of the country. A relatively large breed, weighing, under good conditions, 73 kg (162 lb) on average prior to mating, it is capable of producing 170 live lambs per 100 ewes. Numerically there are over half a million breeding ewes and along with the South Country Cheviot, which is a smaller type found in the hill areas, when crossed with the Border Leicester an extremely popular crossbred is produced—the Scottish Halfbred. There are over half a million breeding ewes of the crossbred being used for fat lamb production.

North Country Cheviot

The Pennine range is the base for this, the numerically largest English hill breed, which in size and conformation is similar to the Scottish Blackface. At tupping the average bodyweight of mature Swaledale ewes is about 48 kg (105 lb) and like the Scottish Blackface it produces about 150 lambs per 100 ewes under good conditions, but only 100 lambs per ewe under the more extreme hill and moor-land conditions. There are some half a million breeding ewes of the Swaledale breed and when crossed with the Teeswater and the Bluefaced Leicester two important crossbred ewes are produced—the Masham and Mule respectively. These two popular crossbreds account for 8% of the National ewe flock (717,000 breeding ewes).

Swaledale

Three other Pennine hill breeds can be mentioned here—the

135

Dalesbred, Lonk and Roughfell; they are comparable to the Swaledale in performance differing only in conformation and the environment under which they live.

Welsh Mountain

This breed, native to the hills and mountains of Wales, is a small hardy breed which like the other hill breeds is much more productive in a lowland environment. At tupping the mature ewes weigh an average of 36 kg (80 lb) and under hard hill conditions produce under 100 lambs for 100 ewes. On the low ground the number of lamb per 100 ewes may increase to around 130.

Numerically the Welsh Mountain is the second largest breed with nearly 2 million breeding ewes (16% of the National flock). Many of the draft ewes are mated with the Border Leicester ram to produce the Welsh Halfbred ewe. This is another popular crossbred ewe 300,000 of them are utilised for commercial slaughter lamb production.

Suffolk

The Suffolk is the major crossing ram breed and has a significant influence on commercial fat lamb production. Suffolk rams are used for crossbreeding to sire lambs for slaughter out of most of the common ewe breeds. Half the lambs slaughtered annually are sired by Suffolk rams. This extensive use of the Suffolk ram as a terminal sire is due to the suitability of its progeny to meet market requirements under various systems of production.

The average bodyweight of the Suffolk ewe at tupping is 82 kg (189 lb) and the average number of live lambs born per 100 ewes is 154. The fertility of the breed and its body size are important as the Suffolk ram is also the sire of crossbred daughters which are eventually used in commercial fat lamb production. There are nearly one million breeding ewes, sired by Suffolk rams, used in fat lamb production.

Border Leicester

This breed has been mentioned previously in relation to its crosses with major hill breeds. Its importance in the sheep industry is not due to its production as a pure breed, but more to the characteristic

136

which it transmits to its crossbred daughters. Numerically, the breed is small with only 12,000 ewes found throughout the British Isles. The majority of Border Leicester flocks are small and many are run in association with dairy herds.

The important characteristics of the breed are its size; on average pre-mating ewe bodyweight of 82 kg (180 lb), and its fertility—175 live lambs born per 100 ewes. These are the characteristics which the Border Leicester is expected to transmit to its offspring.

The performance characteristics of these two breeds which are found in the north of England and used mainly for crossing on the Pennine hill breeds are similar in most respects to those of the Border Leicester. Although numerically they are very small breeds, they have a disproportionate influence on the national ewe flock through the large number of their crossbred daughters which are used for slaughter lamb production.

Bluefaced Leicester & Teeswater

TABLE 7

Important Crossbred Ewes (Source: MLC Flock Recording Schemes)

Name	Sire breed	Dam breed	Number of ewes	Average* bodyweight at tupping		No. live lambs born per ewe lambing
			(000s)†	kg	lb	
Scotch Halfbred	Border Leicester	Cheviot (North and South)	590	80	176	1·71
Masham	Teeswater	Swaledale	406	71	157	1·77
Welsh Halfbred	Border Leicester	Welsh Mountain	304	58	128	1·51
Mule	Bluefaced Leicester	Swaledale	311	73	162	1·78
Greyface	Border Leicester	Scottish Blackface	214	70	154	1·83
Suffolk crosses	Suffolk	Welsh Halfbred	31	68	151	1·55
	Suffolk	Scotch Halfbred	389	80	176	1·64
	Suffolk	Clun	130	72	158	1·52
	Suffolk	Other breeds	293	—		—

* Performance characteristics quoted for mature ewes
† Estimated figures from M.L.C. Sheep Survey 1971

Crossbred Daughters of the Eight Major Breeds

The number of possible combinations which can be produced from the eight breeds already mentioned are numerous, but only a few of them become significant. Information is presented in Table 7 on the more important of the crossbreds. The sire and dam breed for each crossbred are shown, and also the estimated number of breeding ewes of each. It can be seen from the average bodyweights and litter size figures presented for each crossbred that the differences between them are not large.

Structure of National Sheep Flock and the Role of Stratification

STRATIFICATION

The traditional breeding pattern is a stratified crossbreeding system related largely to altitude of grazing. (Fig. 1) The system starts with purebred flocks of the hill breeds in the harder hill areas, followed by the drafting of aged ewes to lower ground for mating with rams of the Longwool crossing breeds. The result of this system is the utilisation of extensive areas of mountain and hills which would otherwise be unproductive. In isolation, the output from hill flocks kept under difficult climatic conditions would be negligible but within the stratification system they make a significant contribution as a source of breeding stock for the productive lowland flocks. It is

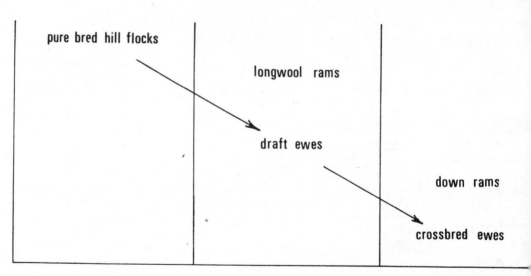

Fig. 1.   *Traditional stratification of the sheep industry.*

138

overnment policy to keep the hill and upland areas of the country in livestock production and to that end some financial assistance is provided in the form of direct subsidies on breeding ewes.

The second stage of the system involves the mating of the draft hill ewes to Longwool rams to produce crossbred ewes. This usually occurs in the upland areas where the production of these ewes is often a more viable proposition than the production of finished lambs for slaughter. The crossbred ewes resulting from this stage in the statification system are then moved to lowland flocks where they are mated to rams of the Down breeds to produce slaughter lambs.

This traditional stratification system in its simple form has been an accepted part of the British Sheep Industry for many years, but recent information from an MLC Sheep Breed Survey indicated that the stratification system is only the framework of a more complicated structure.

The breeding pattern of the industry is illustrated in Table 8. Apart from the expected relationships between hill ewes mated pure, hill ewes mated to Longwool rams (draft ewes) the most striking point from this table is the large number of draft hill ewes which are mated to Down rams. A diagrammatic representation of the breeding pattern involving the main purebred and crossbred group is present in Fig. 1.

TABLE 8

British Sheep Breeding and Production Pattern

	Ewes	Slaughter lambs	
	Proportion of	Proportion of total	
Breed type	total ewes	carcase meat	
		Purebred	Crossbred
	%	%	%
Hill breeds	57	18	16
Other breeds	14	6	11
First crosses	20	—	33
Other crosses	9	—	6
All	100	24	76

Sheep Improvement

The MLC sheep improvement activities fall into two main lines of work:—

Defining the performance characteristics of the main types of sheep available for commercial production and identifying their requirements for optimum economic performance in practical production systems.

The selection and multiplication of breeding stock with genetically superior performance for appropriate economic characteristics.

The first part of the programme is covered mainly by the operation of a whole-flock recording scheme in which information about the performance of the principal breeds and crosses is collected under a wide range of commercial farm conditions. In addition to the performance of the sheep, records of the management applied to them and the physical conditions on the farms concerned are collected, together with details of all financial transactions relating to the flocks. These comprehensive records provide the means for estimating and demonstrating the effects of the main production variables on the performance of the sheep under farm conditions. This whole-flock recording covers sheep in each of the main identifiable production systems in the lowland and upland sectors of the industry in a total of 850 recorded flocks. To supplement the information collected in the whole-flock scheme, formal controlled comparisons of breeds are being undertaken in a limited number of large commercial flocks. These comparisons are currently concerned with the evaluation of ten ram breeds as the sires of slaughter lambs from three breeds of ewe.

The selection and multiplication of improved breeding stock which forms the second main line of work in the sheep improvement programme is operated mainly through individual performance recording in purebred flocks. This type of recording combines performance and pedigree records in such a way that they can be used to base selection on an appropriate combination of information about the performance of individuals and their relatives. The scheme

TABLE 9

United Kingdom Supplies of Mutton and Lamb
Units of 000 tons (1018 metric tons)

	1969	1970	1971	1972	1973	1974	1975	1976	1977	1978
Opening stocks	14·2	25·7	28·7	23·3	24·1	24·0	22·2	16·7	18·3	15·2
UK production	203·5	223·4	226·0	216·4	231·6	248·5	259·5	242·9	223·2	228·4
Imports	362·0	326·0	347·7	326·0	261·4	209·4	243·9	225·8	218·9	225·9
Total supplies	579·7	575·1	602·4	565·7	517·1	481·9	525·6	485·4	460·4	469·5
Exports and re-exports	8·2	10·4	15·2	22·7	26·9	26·1	33·5	32·6	44·6	41·4
Closing stocks	25·7	28·7	23·3	24·1	24·0	21·8	16·7	18·3	15·2	32·1
Domestic consumption (by difference)	545·8	536·0	563·9	518·9	466·2	434·0	475·4	434·5	400·6	396·0
Home produced meat as % of domestic consumption	37%	42%	40%	42%	50%	57%	55%	56%	56%	58%

directed at the breeds where genetic improvement can yield the best return for the industry. These are primarily the ram crossing breeds; firstly the longwools which sire crossbred ewes that are the dams of a high proportion of slaughter lambs; and secondly, the terminal sire breeds which have a dominant position as the sires of the slaughter lambs.

Self contained upland and lowland ewe breeds are the second category which can yield a useful return from selection on performance records and these are also covered in the scheme which currently includes some 300 flocks with 25,000 breeding ewes.

The possibility of individual ewe recordings is limited in the important ewe breeds because of the physical difficulties of controlling matings and collecting records. In these as in some of the ram breeds where the flock size and length of flock life do not permit the effective application of selection on performance records for breed improvement, an alternative approach of establishing a small number of development flocks is being adopted. In these flocks individual ewe recording is complete and the maximum possible selection pressure is exerted on the appropriate economic characteristics of the sheep.

Because of some basic biological differences central performance testing does not have the same value in sheep improvement as in the beef cattle and pig programme; but some basic work is being done to ascertain if modified testing methods can provide a means for adapting performance testing techniques for sheep.

Artificial insemination is also a technique which has not hitherto been used in sheep improvement but with development of the sheep programme it could fulfil a useful role and some basic development work on sheep AI is therefore being carried out.

Quality Variations

Meat is a very variable commodity. Animals vary a great deal in form and composition, while the edible part of the carcase varies in appearance and in the eating satisfaction it gives. Indeed, this last and vital factor is, to a degree, unpredictable. Unless one knows about the background of the animal, the way it has been treated before and after slaughter, and the expertise of the butcher and cook, there will always be something of a question mark over how any particular piece of meat will eat. This is not to say that despite the influence of these variations on eating quality, some subtle, some gross, the consumer does not gain a high degree of satisfaction from this staple component of the diet of the developed world. There is a wide range of "quality" which is acceptable to consumers, many of whom these days show a surprising lack of discrimination. Equally there are those who can detect and gain great satisfaction from interesting texture and finer flavour, and who are prepared to pay to increase their chances of getting it.

PRIORITIES

The order of priorities of today's consumer is sometimes the despair of the traditional meat trader who judges meat at the end of the day by what it tastes like. But flavour must be the characteristic of fifth importance to consumers for the simple reason that they cannot appreciate flavour until they have purchased the meat (and that they will not do unless price, leanness and appearance are right), and until they have found it tender. Appearance of the meat displayed to the prospective purchaser is, therefore, vital. It is primarily a matter of colour and "bloom", or the impression of freshness which is so difficult to define or measure.

SELECTION OF MEAT

So selection of meat—whether animals, carcasses or cuts—by experienced traders is essentially a matter of narrowing the range of variability. This is done by taking account of a wide range of factors from breed type, area or even farm of origin, knowledge of the feeding system, knowledge of the place of slaughter, fatness, colour, age, etc. As a general principle, it can be said that in any particular state of the market, the greater the degree of discrimination exercised by the buyer, the higher the premium he will have to pay over the average market price.

But in selecting animals, carcasses or cuts for his trade, the butcher does not only have regard to likely eating qualities. He is concerned at all times with yield. The proportion of the animal or meat purchased that finally goes across the counter plays a key role in the economics of the meat business, and any trader has to weigh the advantages of better yield against any possible decrease in the average eating characteristics.

Weight loss

Along the marketing chain from farm to retail shop, weight is lost from the body in various forms. There is weight lost in handling

144

Suffolk

Hampshire

Courtesy Animal Photography Ltd.

Swaledale

Southdown

Dorset down

Welsh Mountain

Courtesy Animal Photography Ltd.

Scottish blackface

North country cheviot

Courtesy Animal Photography Ltd.

Middle white

British Landrace

Courtesy Institute of Meat

Gloucestershire old spots

Large black

Courtesy Institute of Meat

Berkshire

British saddleback

Courtesy Institute of Meat

Tamworth

Welsh

he live animal from farm to abattoir (gut fill and possibly carcase wastage), during the dressing operation in the abattoir (skin, guts, etc.), during the cooling of the carcase at the abattoir and the transport of the meat to the shop (evaporative loss), during storage in the shop, in the process of retail preparation (bones and fat trim) and on display. Weight may also be lost during cooking, although this will generally be in the home. Meat which is processed in many ways before sale to the consumer may, of course, gain in weight, but in the processing field generally, weight is under much tighter control than is usually possible in the fresh meat trade.

This chapter considers some of the many factors influencing composition, yields and eating qualities of cattle, sheep and pigs. Since this book is not a textbook on meat science, the amount of explanation is kept to a minimum. Readers are referred to "Meat Science" by Professor R. A. Lawrie if they are interested in the scientific basis of variation in the eating characteristics of meat.

Inherent Variations Between Animals

Animals can differ in the appearance, eating quality and yield of the meat they produce for many reasons, of which the most important are sex, state of maturity, breeding and feeding.

SEX

The national breeding herd produces male and female progeny in almost equal proportions on average (although the ratio of males to females can vary widely from year to year in a particular herd and from herd to herd). A proportion of the females have to be retained to replenish the breeding herd which produces an annual crop of culled breeding females of relatively advanced age, as well as a small number of mature males which have been used for breeding. The percentage sex make-up of the annual slaughter of each species in Britain is shown below. The figures for the proportion of young entire males are estimates based on available information.

Entire males convert food into lean meat more efficiently than females and also more efficiently than castrated males. But traditional

F

	Sheep	Cattle	Pigs
	%	%	%
Mature males and females	10	24	3
Young males—noncastrated	5	1	1
castrated	48	52	50
Young females	37	23	47

extensive systems of production required male animals to be held for so long that if left entire they would have become sexually mature and difficult to manage; hence, from time immemorial it has been the practice in Britain and other major producing countries to castrate male animals. This also has the effect of slowing down growth, reducing efficiency and making the carcase fatter. Up to quite recently this last was a desirable feature of meat production since the vast majority of the population was engaged in physical labour and needed the energy provided by fat meat. Further, the consuming population was not concerned with becoming over-weight and had not heard the medical criticism of the possible dangers of diets high in saturated animal fats.

But now, the meat industry is based on systems of production geared to the slaughter of young immature animals, coupled with an ever increasing demand for leanness and a growing awareness of the importance of using scarce resources like animal feed as efficiently as possible. For all three reasons, the entire male should now be reinstated as the primary production method over the castrated male. Indeed, in countries where the fattening of young animals for beef is a relatively new practice (as opposed to the traditional practice of rearing some calves for veal, slaughtering the rest and relying on cows for beef), a large proportion of male cattle are raised as entire on production systems from which they are sent for slaughter at 12, 18 or even 24 months of age.

BEEF

Although bull beef production is developing in Great Britain (approximately 3% of all clean cattle finished for beef in 1978), most meat traders still do not have experience of this type of meat. Among those who have used young entire bulls, their advantage of superior leanness and yield of saleable meat, excellent thickness of flesh in the oin, rump and round and the absence of the coarse flesh of the mature bull are clear, but they are to some extent counterbalanced by the following disadvantages:—

As the bull develops male characteristics, the muscles of the neck and shoulder develop faster than those of the rest of the carcase; this is not significantly advanced at 12 or 18 months of age, and the yield of high priced cuts from the bull carcase is probably only about 1% less than that of the steer of similar age.

For some trades, the level of finish (i.e. external fatness) attained by young bulls is not considered adequate to ensure acceptable quality.

Young bulls seem more prone than steers (to a degree as yet not known) to "cut dark"—that is the lean meat when exposed to the air is darker than the fresh cherry-red preferred by both trade and consumer. This is believed to be due to the effects of a greater susceptibility to pre-slaughter stress.

The first of these disadvantages is small in economic terms compared with their superior yield of saleable meat, and the low level of finish may become attractive to more and more traders as consumer demand for lean meat intensifies. The full exploitation of bull beef production in Britain, therefore, depends on the extent to which methods are developed for controlling its colour.

PORK

With the entire male pig (boar), the risk is of a different nature but the production advantages are clear. With young boars, 8 kg (18 lb) of lean meat can be produced from 45 kg (100 lb) of feed as compared with some 7 kg (16 lb) in the case of castrates—a $12\frac{1}{2}\%$ improvement in total efficiency. Yet there is a substance present to varying degrees in the fat of a small proportion of young boars which, on heating, produces an odour obnoxious to some but by no means all consumers. This is quite distinct from the odours in the flesh of mature boars which are well known to the meat industry and require meat from sexually mature boars to be used in very diluted amounts in

147

manufactured meat products. Considerable testing work has shown that pork and bacon from young boars up to about 73 kg (160 lb) carcase weight is equally acceptable to consumers as control gilt or castrate products. A proportion of meat traders have come to accept pork and bacon from young boars as providing excellent lean meat, and it seems likely that as confidence in young boar meat builds up within the industry, such meat is likely to contribute an increasing proportion of the nation's pork and bacon supply.

LAMB

In the case of lambs, ram lambs possess the production efficiency advantages of bulls and boars compared to their castrated counterparts, but the advantages appear less marked, particularly when lambs are slaughtered off the ewe. Ram lambs taken through to the hoggett stage may create management problems when foraging. However, from a meat quality standpoint, there is little evidence that meat from ram lambs, in comparison to castrates of the same age, suffers any of the quality problems associated with the other species.

Change from the historical practice of castration to the more efficient and rational method of raising males entire is likely to be a slow process in Britain. However, legislation which has had the effect of retarding change is progressively being dismantled. This will allow the freer operation of market forces.

The differences between the bred female and the young female arise essentially from the age difference with its associated changes.

AGE OR STATE OF MATURITY

As the animal grows, clear changes take place in the characteristics of the lean and fat, and these affect appearance and eating qualities. Briefly, these may be summarised as follows:

The amount of connective tissue within the muscles and on the outer surface of the muscles (gristle), increases and changes its chemical composition to become much harder.

Chemical changes also occur within the muscle fibres.

The combined effect of these changes is that the lean meat tends to become tougher.

The lean meat becomes darker.

The intensity of flavour tends to increase.

The water content of the lean meat decreases as a proportion of the weight of the meat. While this might have the effect of making the meat become drier and less juicy after cooking, any such tendency would be counterbalanced, to some extent, by an increase in fat within the muscle (marbling).

The chemical composition of the fat changes, so that it becomes harder and, after cooking, has a tendency to solidify more quickly as it cools.

In the case of grazing animals, the longer these are at grazing, the greater the extent to which the fat yellows.

The bones become harder and more brittle and change in colour.

These age changes are well demonstrated by the progression from veal (lean meat pale in colour, white fat, very tender, relatively flavourless), through intensively produced cereal beef ("barley beef"), through normal steer and heifer beef (at 18 months to 30 months of age) on to cows of medium age to old cows (with dark, rather dry flesh, tough, strongly flavoured, yellow fat). The darkening of the flesh from pork pigs to pigs of manufacturing weight, and from suck lambs through to hoggets and ewes is well known.

The rate of these changes is not greatly influenced by the way the animal is fed—it is essentially a function of age, not chronological age but physiological age, that is the state of maturity of the animal.

Beef animals are slaughtered at a relatively advanced state of maturity compared with pigs and sheep.

Bone and Cartilage

Traditionally, the state of bone and cartilage development is used as an indication of the maturity of a carcase and its meat. In cattle there is a fairly clear pattern of change in the colour of the bones and in the ossification of the cartilages. For example, in the young animal up to about 2 years of age, the pubic symphysis or "aitch bone" can be split easily by cutting with a knife through the cartilage. Later in life, the cartilage becomes ossified and the resulting bone will have to be sawn through. The spines of the dorsal vertebrae (or chine bones) also show quite marked changes with age. In the young animal, the spines are largely white and cartilaginous but they are gradually ossified and become redder, leaving tips of cartilage which eventually become ossified. Most cattle slaughtered at around two years of age or less have soft, red spines terminating in white cartilage which is also soft.

Another example of the use of bones to indicate maturity is in sheep, where the "break" or "straight" joint is exposed if the front feet are removed between the toe bones and the cannon bones. In lambs the saw like appearance of the exposed cannon bone is more pronounced than in mature sheep, and in older sheep this joint cannot be broken as easily as in lambs.

Teeth

In the live animal, and also in the carcase, an approximate indication of the state of maturity is provided by the pattern of tooth development. Cattle lose their central calf incisors and start to produce permanent incisors at an average of about 22 months of age. However, there is considerable variation about this age and a common range for the first permanent incisors to start erupting is 18–26 months. When both central permanent incisors have erupted, the animal becomes what is called a two-toothed animal. Later in life it becomes four-toothed, six-toothed and then develops a full mouth of permanent incisors. It follows that most commercial cattle will be slaughtered with either calf teeth or two permanent incisors. In the past, the state of development of the molar teeth has tended to be ignored when assessing maturity, but recently they have attracted more attention because they can be used to give a more precise guide to maturity. For example, the first pair of molars on the lower jaw start erupting at around 5 months of age and are fully

150

up by 1 year, at which time the second pair of molars are erupting. Further details are contained in *The Ageing of Cattle by their Dentition*, by A. H. Andrews, published by the Royal Smithfield Club, 1975. The location of the molars in relation to the incisors is illustrated in Fig. 1.

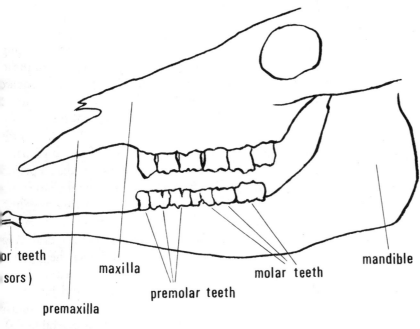

or teeth
sors)

premaxilla

maxilla

premolar teeth

molar teeth

mandible

Fig. 1. *Upper and lower jaw showing location of molars in relation to the incisors.*

So far as sheep are concerned, they lose their first lamb incisor teeth at about 15 months of age and achieve a full mouth of permanent incisors by about $3\frac{1}{2}$ to 4 years of age. As with cattle, the molars can be used to provide an indication of development at a younger age. As an example, the first pair of molars start erupting at about 3 months of age for the lower jaw and 5 months of age for the upper jaw.

Apart from changes in the characteristics of the lean and fat which take place as animals grow, and which have been discussed above, changes take place in the shape and carcase composition of animals.

Shape and Carcase Composition

151

One of the most obvious characteristics which one observes in growing children and animals is that their shape as well as overall size changes. Most noticeable is that the proportion which the head and feet form of the total body changes; early in life the head and feet form a greater proportion than they do later in life when the trunk takes on more rapid growth. Such changes are illustrated in Fig. 2.

Within the growing body there are simultaneous changes taking place in the proportions of the body tissues like muscle or lean meat, fat and bone. If animals have satisfactory health and an adequate supply and quality of food, growth of these major components of the carcase has three phases. In the very young animal the skeleton grows more rapidly than in any subsequent period. This allows for the development of the brain and for teeth development in preparation for having to forage for food. This phase is followed by a period in which muscle growth predominates to support the skeleton, and in the third phase the rate of fat growth reaches a maximum. Thus, during normal growth as any individual animal gets heavier it will put on more fat. Fat starts to be laid down early in life and in the pig much of it is subcutaneous (i.e. under the skin and over the external muscles). In the baby piglet it appears to act as an insulating layer. Fat is also laid down around the kidneys, stomach and intestines. Later in life, fat is also laid down between the muscles (intermuscular or seam fat) and within the muscles (intramuscular or marbling fat). While some of the fat laid down has an essential part to play in the animal's development, much of it is merely a store for the energy fed which exceeds its requirements for simply maintaining itself and for producing further bone and muscle growth.

Figure 3 based on results from a major beef research project, published by the Royal Smithfield Club, 1966 illustrates how, during normal growth, the composition of the familiar Hereford×Friesian steer changes. It shows that much of the food later in life is being used to deposit fat. The significance of this is considerable since it takes some six times as much food to produce 1 kg (1 lb) of fat as 1 kg (1 lb) of lean meat. When some of this fat may have to be trimmed in order to make the meat saleable (thus devaluing the fat

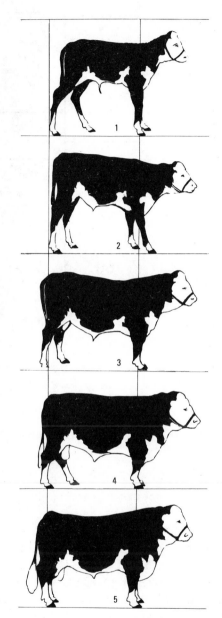

Fig. 2.   Changes of proportion of Hereford Cattle (J. Hammond 1935)
. Bull—2 days   2. Bull—5 weeks   3. Bull—13 months   4. Bull—22 months
. Bull—5 years.

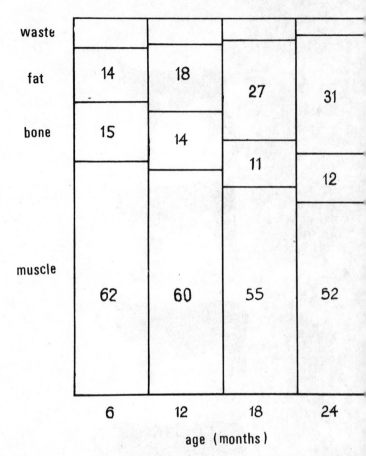

waste

fat	14	18	27	31
bone	15	14	11	12
muscle	62	60	55	52

| 6 | 12 | 18 | 24 |

age (months)

Fig. 3. *Percentage composition of Hereford × Friesian steer carcases of different ages.*

from saleable meat price to only a few pence), the full impact of th
cost of producing excess fat in a carcase can be appreciated.

It now seems likely that by about two months of age in pigs and eigh
months of age in cattle, the proportion of total muscle or of tota
bone found in different parts of the body remains fairly constan
thereafter, whatever the overall growth rate of the tissues may be
The distribution of the subcutaneous and intermuscular fat depots ir
different parts of the carcase seems more variable. Thus, while th
overall growth rate of each of the tissues varies from animal t
animal within any given species, so that, at any particular weight

they may vary quite markedly in the proportions of lean meat, fat and bone, the distribution of muscle and bone between different parts of the body tends to be quite similar. One of the consequences of this is that the distribution of carcase weight in different commercial joints is primarily influenced by variations in fat distribution. However, because of the relative constancy of the distribution of lean meat and bone, it has been found that when carcases are jointed in a standard way, the variation in joint proportions is less than is often supposed to be the case by some breeders and meat traders. Table 1, p. 192, shows the similarity with which the total muscle in the body is distributed in the high priced cuts of beef cattle varying widely in type. Similar evidence could be produced for the other species.

It has already been observed that as animals grow their shape or conformation changes and that the most dramatic changes are in early growth. However, if one considers animals around the stage in growth where they are slaughtered, there may be quite marked differences in their overall shape or conformation. In other words, they differ in the thickness of lean meat plus fat in relation to the size of the skeleton. To take extremes as examples, it is possible to have one carcase of poor conformation and another of very good conformation and yet they could be of similar carcase weight and have the same overall proportions of lean meat, fat and bone. What happens, in effect, is that in one case the lean meat is laid down over long bones and in the other over shorter bones. It appears that quite small differences in bone length can produce quite marked differences in conformation. The better conformation carcase tends to make the lean meat look thicker and, other things being equal, this can be advantageous in terms of visual appeal.

Fat can also play an important part in influencing conformation since it has the effect of filling in the indentations where the muscles meet or insert, as well as accumulating to give the appearance of thickness on the rump, loin, brisket, etc. In addition, seam fat can affect shape by subtly adjusting the relative muscle positions.

The subject of conformation will be discussed further, later in the chapter.

An animal's composition and shape is determined by its genetic make-up and also by the rearing environment, an important feature of which is nutrition or feeding. These aspects will be considered in turn.

BREEDING

Animals of the same species vary in their inherent capacity to grow and develop, and in their efficiency of converting food into lean, fat and bone. Generally speaking, animals of a particular breed tend to be more alike than animals from different breeds but, nevertheless, the variation in growth capacity within each breed is considerable and the average growth and the typical conformation is capable of improvement by selective breeding.

There is a general tendency within the livestock industry to talk of "early maturing" and "late maturing" breeds, but this sometimes leads to confusion since the "late maturing" types grow faster than the "early maturing". Maturity, in this sense, refers to the development of fat. If this happens relatively early and at light weights, the breed is said to be early maturing, while if fat is developed later and at heavier weights we have a later maturing breed.

Early Maturing Breeds

In pigs, the early maturing breeds once in favour for pork production, such as the Berkshire and Middle White, have virtually disappeared in response to consumer demand for leaner meat and producer demand for more economic lean meat production. Similarly, some of the more extreme early maturing types of sheep (such as the Southdown) no longer contribute to commercial production in purebred form for the same reason. The prime examples of early maturing cattle breeds—the Angus and Galloway—retain, at the present time, an important role in commercial crossbred production, and in certain parts of the country where they are traditionally favoured for their good grazing attributes and the quality of the meat they produce.

Breeds also differ in the muscle to bone ratios of the carcase, their dressing percentage and in muscle thickness. As has been indicated already, there are generally very small differences in the distribution

156

of muscle by weight through the carcase, and the differences between breeds that we can see in conformation reflect differences in fatness, in bone dimension and in lean meat thickness rather than "yields of high-priced cuts". Muscle to bone ratio is not related to rate of maturity; for example, among the early maturing beef breeds, the Aberdeen Angus has a high muscle to bone ratio while the Welsh Black has a muscle to bone ratio around the average. Yet among the later maturing breeds, the French Charolais and Limousin have a high muscle to bone ratio compared to the South Devon which has a low to average muscle to bone.

Later Maturing Breeds

Demand for faster growth (because of its relation to feed efficiency and profitability in the utilisation of grass, cereal or concentrate feed), and for leaner meat will accentuate the swing towards later maturing types. The implications of such a swing has caused some concern in the meat trade because it is feared that eating quality will suffer. Traditionally, marbling fat is believed to be important in influencing the flavour and tenderness of beef. However, extensive research both in the USA and elsewhere, particularly on beef, has thrown considerable doubt on the scientific basis of this. Marbling has been found to be much less closely related to tenderness than is traditionally supposed to be the case, and it now seems likely that, provided a carcase has a certain minimum level of external fatness, which can be low in relation to most beef produced today, its lean meat can reasonably be expected to have sufficient intramuscular fat to ensure juiciness and influence flavour. Those who like eating more fat than most modern consumers desire may wish to buy beef from the fatter, earlier maturing breeds, but they should be aware that they are doing so because they like the taste of fat and not because it is likely to improve the eating quality of the lean meat. In addition, they ought to be prepared to pay the premium required to cover the cost of producing such beef. While there is support for the view that beef from an Aberdeen Angus steer reared in the north of Scotland may produce more tender beef than, say, the more common Hereford×Friesian steer, due to differences in the structure of the lean meat, the detailed tests carried out so far have not shown that these are sufficient for a trained taste panel to detect differences between the breed types in flavour, juiciness or overall acceptability.

It seems, therefore, that differences between cattle breeds in quality are largely related to other differences such as state of maturity.

It would be incomplete to leave the discussion of the influence of breed on meat quality without referring to pigs, where breed does appear to influence the incidence of abnormalities in the lean meat and fat. For example, the Pietrain breed (imported from Belgium) is noted for its high incidence of pale, soft, exudative (so-called PSE) meat and, amongst the most common breeds in this country, the Landrace has a higher incidence than the Large White. This condition arises where formation of acid in muscle after death is unusually rapid. The effect is to damage the muscle structure so that the meat becomes paler and drips more than it would do if the rate of acid development were slower. The incidence of the condition in this country at present is not such as to cause great concern, but it is important that it is closely monitored and steps taken to correct any increase, otherwise it could lead to substantial losses in the marketing of pigmeat in its various forms.

There is also some evidence that the Hampshire breed imported from America is more prone to produce soft fat than our own breeds, but this may be related to specific sires.

FEEDING

In order to understand the way in which feeding can affect variability in carcase characteristics, it is important to appreciate how animals use food. Some of the food actually absorbed from the gut is needed just to maintain the animal in a living condition and to keep it warm by furnishing the cells with supplies of essential nutrients, including protein. This is called the maintenance part of the diet. Food fed in excess of maintenance requirements allows animals to grow, and its allocation to different functions depends on the animal's stage of growth and on how much food it has actually absorbed in excess of its maintenance needs.

When the supply of energy is unlimited, the demands of all the tissues will be met and any excess energy will be deposited as fat. If the amount of energy is lowered progressively, fat ceases to be laid down but the other tissues continue to grow.

158

Animals given free access to the same diet vary widely in their ability to convert food into lean meat as efficiently as possible. To take extremes as examples, some may grow fast but be fat, while others may be lean but take a long time to achieve a marketable weight. These differences are generally due to the differences in their genetic make-up and clearly neither of these extremes is desirable.

There are genetic differences which influence how much food an animal will eat—that is, its appetite—and this being so, growth rate will be affected. Since animals which grow slowly need to maintain themselves for a longer period of time, their efficiency of converting food into weight gain will tend to be poorer. There are also genetic differences in the capacity of an animal to develop more lean meat, i.e. in its lean meat potential. In some animals, for example in the so-called early maturing breeds, like the Aberdeen Angus, the timing of the growth phases is such that if the animals receive energy from their food which is over and above their requirements for lean growth, they use it to produce fat at lower weights than in animals where the growth phases are more extended. It follows that since it takes more energy to produce 1 kg (1 lb) of fat than 1 kg (1 lb) of lean, the later maturing breeds should tend to have a better efficiency of converting food into lean meat at commercial slaughter weights and their carcases also tend to be leaner.

Genetic Differences

Current evidence does not lend support to the idea formulated some years ago that the form of an animal and the development of the muscles in particular can be manipulated by feeding. An animal whose growth has been retarded, e.g., during a store period, is likely to be of similar shape to its normal counterpart of similar weight. However, when feed conditions are so favourable that fat development is enhanced or if feed is so severely restricted that growth of bone continues at the expense of other tissues, the shape of an animal may be altered, but usually to a minor degree. It appears, therefore, that under store conditions animals will lose fat and muscle but only under very severe restriction will bone growth be retarded. Evidence available indicates, for example, that if store

Store Conditions

159

conditions are imposed on reasonably fat cattle, there is about an equal weight of both muscle and fat used up in loss of liveweight and only a small actual weight loss from bone. However, if there is further weight loss from the carcase this would have to come increasingly from the muscle tissue, as the fat tissues sink to an insignificant proportion of the remaining carcase.

Quality of the fat

Feeding can exert quite a marked effect on the quality of the fat produced in the carcase. For example, the flavour of lamb is affected markedly by the type of pasture on which lambs are finished. Lambs reared on a wide variety of different crops can produce widely differing flavours in the fat, rape, for instance, producing a nauseating aroma and flavour, and oats a pungent flavour. Cattle and pigs do not appear to have the same ability to transfer flavour components from the diet into the fat. However, grass fed cattle produce a yellower fat than those fed on concentrates, and the contrast is particularly evident in spring and early summer when cattle first go out to grass when pigment concentration in the grass is at a higher level than later in the season. In pigs, feeding diets containing unsaturated fats, e.g. maize, can markedly affect the hardness of the fat. The feeding of certain types of swill in the diet of pigs can also produce soft fat which is undesirable for most uses.

Yellow fat in sheep occurs occasionally. The incidence is probably less than 1%. However, when it occurs it presents a problem since joints of markedly different colours do lead to consumer resistance. There is no evidence to suggest that yellow coloured fat is nutritionally inferior to whiter fat and, indeed, the opposite may hold true. The cause of yellow fat in sheep does not appear to be nutritional but due to a recessive gene.

Environmental Effects on Quality

It has already been noted that meat is a very variable commodity, not only in terms of how much lean meat and fat there is in the joints sold across the counter; but also in the eating characteristics of that meat. Some meat is very good as judged by its appearance

160

nd on how it eats, some meat is indifferent in both respects. While t is possible for the housewife to assess the appearance of a joint before she buys it, this does not generally provide a reliable guide to ow much satisfaction it will give when eaten. Of course, the way in which particular cuts are cooked can have a very important influence n eating quality and much needs to be done to educate the consumer n some of the basic rules in cooking meat and on how best to cook articular cuts. However, leaving aside any dissatisfaction which may arise due to poor cooking practices, there is a great deal of ariation in the potential eating quality of meat before it is cooked, nd it is clear that this is a source of some dissatisfaction which could vell contribute to adverse long-term trends in meat consumption. A issatisfied consumer may turn to more reliable and consistent lternatives, even though they would not get the great satisfaction vhich they could have from some of the best meat on the market.

Much of the variation which exists in eating quality is not due to nherently poor quality meat in the animals from which the meat ame, though, as observed earlier in the chapter, older animals, for xample, tend to have less tender meat than younger animals. A najor source of the variability arises from a poor understanding of he impact which environmental effects can have on quality. Apart rom the influence of these on final eating quality, such variation roduced can result in financial losses at all points along the dis- ribution chain from abattoir to consumer. These effects will be onsidered under the following headings:—

Pre-slaughter treatment, including transport and stress.

Refrigeration and ageing.

Freezing.

The way in which animals are handled before slaughter can have a narked effect on the quality of meat produced. Careful handling of nimals is important at all stages from farm to abattoir. For example, are should be exercised in collecting animals from pens or fields eady for loading on the farm and in loading on to the transporter

PRE-SLAUGHTER TREATMENT

at the farm. Penning arrangements on the transporter should be such as to avoid the mixing of strange animals and the vehicle should be driven carefully and be well ventilated. At the abattoir, care should be taken in unloading and the animals should be penned to avoid mixing previously established groups of animals. In moving animals from pens in the lairage to the point of stunning and sticking there should be as little excitement caused as possible. Sticks should not be used to drive animals and electric goads used with discretion.

Dark Cutting

If care in handling is not exercised at these various stages then, apart from physical damage which may be caused to the skin and under lying tissues resulting in financial loss, the animals may become stressed and this can produce poorer quality meat. For example, the phenomenon of "dark cutting" in beef appears to be due to some form of stress resulting either from a sudden change of temperature such as the first frosts in the autumn after the warmth of summer, or due to bad handling in the process of getting the cattle to the abattoir. It was noted earlier that the full exploitation of bull beef production in this country depends on controlling the tendency for bulls to be more prone to produce dark cutting meat than steers. Control can be exercised through the adoption of better handling procedures such as those referred to above.

After slaughter, the energy store in muscle (called glycogen) is converted to lactic acid. The amount of this acid produced depends on the amount of glycogen present at the time of slaughter. If an animal is very exhausted or stressed at death, then its glycogen will be depleted and little acid will be formed. This is the situation which gives rise to "dark cutting" in beef and which produces a condition in pigs called dark, firm and dry (DFD) meat. If cured, such pigmeat produces "glazy bacon". Apart from its less attractive appearance, low acid meat (called high pH—pH being a measure of acidity) provides a more favourable environment for bacteria to multiply and this can lead to a significant spoilage problem, particularly where the meat is pre-packed.

The rate of formation of acid in muscle, in addition to its amount, is also important. Reference was made earlier in this chapter to PSE meat which occurs mainly in pigs if acid is quickly formed in muscle after death. The incidence of such meat is influenced by inherent differences between animals and the way in which they respond to conditions of stress.

Much can be done to minimise the incidence of the PSE and DFD type conditions by adopting the handling procedures outlined at the beginning of this section.

REFRIGERATION AND AGEING

Modern distribution of meat cannot avoid delays in transport and holding while bringing meat from the abattoir to the home. Refrigeration must be applied to carcases at the abattoir to prevent risk of growth of bacteria and ensure ample storage life for all subsequent distribution operations. Rapid chilling is highly advantageous bacteriologically while, at the same time, reducing both weight loss from evaporation and drip loss from cut surfaces of joints when the meat is prepared for sale and consumption. However, rapid chilling can produce meat of extremely tough eating quality from an otherwise satisfactory carcase and this is why the way in which meat is refrigerated can have such a marked effect on eating quality.

Cold Shortening

The changes which take place in muscle after death, particularly in beef and lamb, are such that rapid chilling can readily produce a condition known as "cold shortening". If beef and lamb is subjected to rapid chilling shortly after slaughter, the muscles contract and they go through rigor mortis in this shortened state. The meat so produced, when cooked, is very much tougher than if the meat had been chilled less rapidly. The condition was first identified in New Zealand lamb where it is subjected to rapid freezing after slaughter. In beef the problem is less acute since the larger bulk of the carcase cannot be chilled as quickly as lamb. Nevertheless, cold shortening can occur in beef and tends to affect some of the muscles near the surface of the carcase, particularly where the muscles are unprotected by an insulating layer of external fat. Cold shortening is not a practical problem in pork because the rate of acid formation after death in relation to time and temperature is more rapid than in beef

163

and lamb and also because the fat cover on pigs reduces the rate a which muscle cools.

Prior to the widespread adoption of refrigeration for cooling pi carcases after slaughter, bone taint in hams was common. Bone tair is spoilage in the deep muscles of the leg arising from enhance microbial growth due to slow cooling. With modern refrigeratio practice which produces a deep muscle temperature of about 4°C, 2 hours after slaughter, bone taint does not occur. Since pork does no develop cold shortening, very rapid chilling can be practised whic has the advantage of significantly reducing carcase weight loss i cooling.

Because of their small size, lamb carcases are frequently coole without refrigeration. This safely avoids cold shortening, but it not currently acceptable under EEC requirements for meat which to be traded between EEC countries. Slavish adherence to thes requirements leads to carcases with some cold shortening and, henc increased toughness. If carcases are cooled at a temperature of 15°C for about 6 hours before further chilling, then most of the leg an loin avoid the risk of cold shortening. Such a procedure seems essen tial to maintain quality, but it could conflict with EEC requirement

Chilling Methods

Chilling methods used for beef vary greatly, ranging from ver efficient blast chillers to unrefrigerated cooling halls. With blas chillers there is real danger of cold shortening of some of th expensive parts of carcase and, with cooling halls, there is a risk o bone taint in the deepest parts of the carcase, especially if the anima has been stressed before slaughter and the meat has a low acid leve. A compromise seems to be to cool in an air temperature of 4°C s that the deep muscle temperature reaches below 7°C in less tha 48 hours.

After carcases have been chilled at the abattoir to the desire temperature, it is important that, as far as possible, the temperatur is not allowed to rise during the process of distribution to the reta shop otherwise there may be spoilage, colour and drip problem:

Apart from any effect which cold shortening may have on the tenderness of meat, meat which is cooked within a few days of slaughter is generally tougher than meat which is held for a longer period of time. This is due to the effect of ageing or conditioning. When the muscles of the carcase go into rigor mortis, or stiffening of muscle, after death, this is associated with a decrease in tenderness. After rigor is complete, changes take place in muscle which gradually improve the tenderness again. This process is commonly called ageing and associated with it is the development of fuller flavour. The effects seem particularly important in beef. The rate at which ageing takes place increases with increasing temperature, but in practice it is undesirable to age meat at a high temperature because of the deleterious side effects this produces. An example of the effect of temperature, has been to show that ageing beef for 2 days at 20°C gave the same degree of tenderising as 14 days at 0°C, and that the benefits of ageing were more marked with beef which was initially tougher.

The longer meat is aged, therefore, the more tender it becomes, but the advantages of holding meat beyond about 7 days at normal chill temperatures seem quite small with beef from young animals. With beef from older animals, such as cows, there are probably advantages to be gained from a longer period of ageing unless the meat is going to be largely comminuted. It is important that the significance of ageing is fully appreciated, particularly for beef, as one of the major attributes which the consumer looks for in beef is its tenderness and meat which is insufficiently aged is likely to give rise to dissatis-faction.

It is, perhaps, of interest to record that some of the best eating quality in meat is achieved by cooking the meat almost as soon as the animal is slaughtered and eviscerated, and before the onset of rigor mortis leading to increased toughness. Such a practice is common in, for example, rural areas of South America but, for obvious reasons, it cannot be practised to any significant extent in this country. However, in recent years, interest has been shown in various countries in the possibility of cutting carcases into cuts while the carcase is still hot. This so-called hot boning procedure requires very careful bacteriological and temperature control but, if it could be made to work successfully, it could lead to reductions in chill space and weight loss during cooling.

FREEZING

There are significant advantages in freezing meat. These include discouraging adverse microbial changes and prolonging the useful storage life of meat, leading to the easing of meat supply and distribution problems. Offsetting such advantages is the tendency for frozen meat to drip on thawing, and this produces financial losses at various stages up to the cooked end-product as well as making the meat look somewhat unattractive in colour.

When meat is frozen, ice crystals form within the muscles and cause damage to the muscle structure. As the time taken to freeze meat increases, structural damage increases as large ice crystals are formed. When thawing takes place, the damaged muscle structure is unable to hold some of the water it contains and so drip ensues. To minimise structural damage and, hence, drip loss, it is important to freeze meat as quickly as possible. The smaller the piece of meat being frozen, the less damage will be caused. Thus, freezing of individual portions of meat will be more efficient from this point of view than freezing large cuts. If meat is uncovered as it is frozen it will freeze more rapidly than if it is wrapped. Certain wraps can double the time to freeze over that of corresponding unwrapped meat. One of the practical implications of the foregoing is that freezing units should not be overloaded with meat during the actual freezing process. While the meat may appear superficially to freeze satis-factorily, it is inevitable that overloading will slow down the freezing process, leading to larger ice crystals, greater structural damage and thus, more drip on thawing.

Once the meat is frozen, it is important not to allow the temperature to rise too near to freezing point until the meat is to be finally thawed, otherwise larger ice crystals will be formed in the muscles producing more structural damage.

While freezing inhibits microbial growth, long periods of storage below freezing point may produce considerable rancidity in the exposed fat. The fats of beef and lamb are relatively resistant and

166

may still be good even after 18 months storage at $-10°C$, but the softer fat of pigs becomes rancid rather more quickly.

Ageing of meat, which was discussed in the previous section, does not proceed during freezing and, since frozen meat is normally cooked within a short period after thawing, it is desirable, therefore, for meat to be adequately aged before freezing.

Development work is being carried out to discover rapid methods of freezing meat which minimise the weight loss on thawing and retain the colour of fresh beef. If such methods could be developed, it would improve the image of frozen meat in the public's eye and not attract the kind of discounts currently encountered.

Grading and Classification

It will be evident from earlier sections that there is considerable variation in carcase quality. This partly reflects variation in the demand for meat of different qualities and has led over the years to wholesalers in this country, who buy on a deadweight basis, developing their own carcase grading systems to encourage production of the kind of carcase they want and discourage those which are not required. Such grading systems are particularly well developed for pigs, where there is a long history of deadweight purchase, but they are being used increasingly for cattle and sheep.

Carcase grades are effectively groupings of carcases by characteristics **GRADES** which are considered of most importance to those trading in them, and to which prices are attached to encourage the better type of carcase and discourage the poorer. Weight and fatness are among the characteristics incorporated in forming the grades which might number three or four. The problem with such grading systems drawn up by individual wholesalers to suit their own particular requirements, and those of their customers, is that while they each know the type of carcase which should fall in each grade it is difficult for this to be communicated clearly to others. It makes it particularly difficult for the producer who needs to know, for

167

example, the type of carcase he should be producing and to assess where he can sell his cattle to greatest advantage. It can also create problems for the retailer comparing prices quoted by various wholesalers for different selling grades if the retailer is personally unable to go and inspect the carcases.

The diversity of demand in this country makes it impossible in the short to medium term to envisage acceptance of national grading schemes for cattle, sheep and pigs in Britain. This is in contrast to the national grading schemes which have been operating in countries such as New Zealand and Denmark, which rely on a strong export trade for much of their home production. It also contrasts with national grading schemes which have been operating in a number of countries primarily for internal trading purposes. Perhaps the most well known of these schemes is the Beef Grading Scheme operating in the USA and a few of the main features of the scheme will be given.

USA GRADING

The United States Department of Agriculture's beef grading procedure is, in fact, in two parts: determination of the "quality" grade and then determination of the yield grade. Carcases submitted for quality grading must be well chilled and "ribbed" (i.e. partially quartered to reveal the eye muscle). The grader is required first to assess the "palatability—indicating characteristics of the lean" and then conformation. He assesses the degree of marbling from the eye muscle and the maturity from the state of ossification of the carti-lages, the colour of the bones, the colour and texture of the lean. These are then combined into one of the following preliminary grades according to a set method: Prime, Choice, Good, Standard, Commercial, Utility, Cutter and Canner. The basis of the com-bination of characteristics to form the preliminary grade is that as the animal becomes more "mature" (i.e. physiologically older), there is a need for more marbling to compensate for the deleterious effects of increasing age on the eating characteristics of the meat. At each level of maturity, the higher the degree of marbling, the "higher" the grade, but as the carcases become more mature, a higher level of marbling is required to achieve a particular grade. The preliminary grade is then modified by conformation according to clearly set-out rules. The principle of these is that poor conformation may cause a carcase to be down-graded relative to its meat quality, but not in general vice versa. Thus the expected eating characteristics of the meat is the dominant factor in determining the "quality" grade.

The USDA introduced yield grading as a supplement to the original quality grade, but yield grading must now be applied to every carcase which is quality graded. The purpose of yield grading is to separate carcases into five classes according to their expected yield of trimmed 35 mm ($\frac{3}{8}''$) external fat, de-boned, high priced retail cuts. The estimation of yield involves the fat thickness over the eye muscle with an adjustment for unusual distribution, eye muscle area, carcase weight and the kidney, heart and pelvic fat percentage. In practice this is carried out by trained graders subjectively, although the standards are written in terms of actual measurements.

One important way of overcoming the problems in this country associated with different traders wishing to operate their own grading systems is to make use of carcase classification schemes. The potential value of such schemes in improving the efficiency of meat marketing accounts for the Meat and Livestock Commission being required, when it was established, to develop and operate carcase classification schemes in GB.

Carcase classification schemes simply provide a method of describing the characteristics of carcases which are of prime importance to the retailer cutting up and preparing meat for sale. Each characteristic is described separately without attributing relative importance or cash value differences to them. However, given the framework of the classification information, the buyer or seller can draw up his own grades by grouping the separate classification descriptions together and he can then attach prices to these different groupings.

Classification, then, provides a common language for describing important carcase characteristics in a way which all can understand. It helps producers to appreciate more clearly what the market requires. The regular feedback of classification results to producers

CARCASE CLASSIFICATION SCHEMES

linked to wholesalers' buying schedules, enables them to know more clearly the types of carcases they are producing and those in demand by retailers. This provides a much sounder basis for livestock improvement than occasional demonstrations or observations in the market place.

It is important to emphasise that classification is not grading as it does not put carcases into a pre-determined order of merit. But wholesalers can use classification to define their buying and selling grades, to check on their buying methods and to assist in allocating carcases to retail outlets. This in turn facilitates price comparisons by producers and traders.

Classification can help retailers to describe their carcase requirements, compare suppliers' quotations, order with confidence, confirm specification on delivery and improve cutting and costing. These benefits of classification to the retailer are likely to become more important in future as the industry becomes more competitive and the retailer finds he has less time to go to the auction market to buy animals or to the wholesaler to buy carcases. He will find it is more important to spend his time in the shop where his skills are better employed, provided he can have confidence that the independently applied classification scheme gives him a satisfactory basis for his specifications.

Pig Carcase Classification

The MLC started the development of carcase classification scheme by first preparing one for pigs, since more was known about variations in pig carcase characteristics and methods of assessing these under abattoir conditions than for beef and sheep.

The method of classification introduced in 1971, and which, in June 1975 involved the classification for some 500 participating management units and 70% of the total kill in GB, describes carcases by the following five factors:

CARCASE WEIGHT
BACKFAT THICKNESS (see below)
VISUAL ASSESSMENT: applied to certain carcases: scraggy, deform

med, blemished, pigmented, coarse-skinned, soft fat, pale muscle, partially condemned. These are classed as Z.

LENGTH: at the option of the participating trader.

CONFORMATION: at the option of the participating trader. Carcases with poor conformation are classed as C.

Backfat thickness is determined by one of the following three methods, each of which involves using an optical probe or introscope to measure the thickness of fat over the eye muscle on the intact carcase or side without depreciating the value of the carcase. The optical probe works by a system of mirrors inserted in the probe which enables the operator to see the division between the eye muscle and the fat over it. When this division is found, the depth of fat can be measured with a gauge built into the probe.

METHOD 1 (used mainly for pork and cutter pigs)
Fat thickness is measured with an optical probe at two fixed points over the eye muscle—4·5 cm (1·8 in) (P_1) and 8 cm (3 in) (P_3) from the mid-line of the back at the head of the last rib. The measurements are added together to describe the degree of fatness.

METHOD 2 (used mainly for bacon pigs)
An optical probe measurement of fat thickness is taken at 6·5 cm (2·5 in) (P_2) from the mid-line of the back at the head of the last rib. This measurement, together with measurements in the mid-line at the maximum shoulder and minimum loin fat positions, indicates the degree of fatness.

METHOD 2A Method 2 without the mid-line fat measurements. Used by traders (utilising heavy hogs, for example) who only wish to describe fatness by a single fat thickness measurement.

The position of the optical probe measurements over the eye muscle is illustrated in Fig. 4. The optical probe enables fatness over the eye muscle to be measured on the intact carcase or side without depreciating the value of the carcase. It has been established that such measurements provide a better index of carcase leanness or per cent saleable meat in the carcase than do measurements taken in the midline.

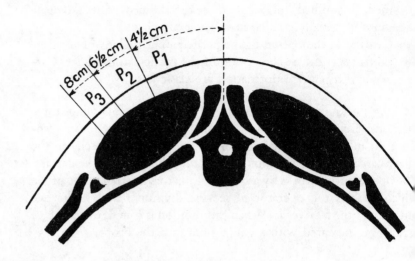

Fig. 4. Location of optical probing positions at the level of the head of the last rib.

An important feature of the scheme, so far as retailers are concerned, is that carcases have the probe fat measurements recorded on the carcase so that they can see for themselves the fatness of the carcase they are buying. Pigs classified by Method 1 have the sum of the two measurements recorded in a triangle on the trotter as shown below on the left. Those classified by Method 2 have the probe measurement marked in a triangle on the belly or foreleg as shown on the right.

Where a carcase is classed C or Z, the appropriate letter is recorded just outside the triangle.

The key factors described by the scheme are weight and fatness. Retailers are primarily concerned to know weight (because of relevance to joint or chop size, among other things) and the amount of fat and lean meat. Leanness is of dominant importance these days but it will be apparent that this is not measured directly but indirectly through fat measurements. This is possible because, among carcases of a given weight, the proportion of bone is not very variable; this means that a carcase with a high fat percentage must have a low lean meat percentage and vice versa. The measurement

172

f backfat thickness used in the scheme are related to total carcase atness and so to total carcase leanness. It may seem strange that the cheme does not incorporate a measure of lean depth since many have onsidered this provides a useful guide to the value of carcase. The eason why such a measure is not included is that no sufficiently recise procedure has yet been developed for measuring lean depth n the intact carcase. If it could be measured more accurately, then : could be included as a characteristic separate to fatness. This would hen give a measure of lean meat thickness, a characteristic of some mportance to retailers, but it has to be recognised that such a neasurement is of very little value so far as indicating the overall :anness of a carcase is concerned. Fat measurements are still best for his purpose. Until such time as it is possible to include a direct neasure of lean depth, the C and Z (scraggy) classes may be used to ssist in picking out those carcases with poor lean thickness amongst arcases of similar weight and fatness.

ig. 5.   *Probe fat measurements recorded on carcases.*

Recently, the EEC pig grading scheme has been introduced into this ountry for the primary purpose of providing a UK reference price o Brussels each week. The grades are defined in terms of ranges of er cent lean in the carcase, and carcases are put into these grades by ombining the P_2 fat measurements with weight.

After extensive consultation with industry representatives, a beef arcase classification was introduced in October 1972. In June 1975

Beef Carcase Classification

173

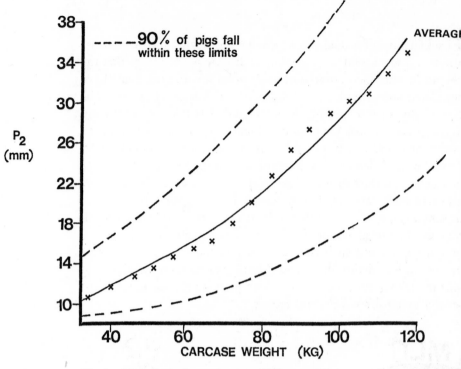

Fig. 6.   *Variation in fatness for pigs of different weights.*

the scheme was being carried out for some 500 participatin
management units throughout the country and covered about 45%
of total clean beef kill. While this level of application may seer
encouraging, the extent of its use as a basis of trading leaves muc
scope for improvement. There is reluctance amongst some whole
salers and retailers to exploit its potential. Wholesalers' salesme
and retailers' buyers are concerned that classification might pu
them out of a job, but this is most unlikely to happen becaus
classification cannot accommodate all the factors influencing th
detailed preferences of retailers. Many secondary factors are impor
tant in the bargaining situation and someone responsible has t
decide what to do when, at any given moment, the supply an
demand do not coincide exactly.

In drawing up the beef classification scheme it had to be designed t
fit in with existing trade practices since there would be little hop
of maintaining a voluntary scheme, however desirable in theory, if
demanded much change in current practices. For example, i

ontrast to the American situation, it is not general trade practice to quarter, or even partially quarter, sides of beef at a fixed time post-slaughter. In fact, it is usual to quarter sides at a variety of positions only immediately before despatch and after personal inspection by buyers should they visit the abattoir. In addition, there is no guarantee that sides would always be available for inspection 24 hours after slaughter and, where they were, they could be packed tightly in chill-rooms. It is recognised that restricting classification to hot carcases as they hang in the abattoir, in the way described below, inevitably imposes some limitations on the value of the scheme, but in practice these are not thought to be great.

The scheme provides a standard definition of the following five primary factors which are applied accurately and independently by trained MLC staff:

CARCASE WEIGHT: determined according to one of two dressing specifications. The first specification includes the kidney knob and channel fat (KKCF) in the carcase weight, and the second excludes the KKCF. This weight is, of course, used primarily as the basis of payment to producers in deadweight transactions. Carcases, sides or quarters will normally be re-weighed when purchased by retailers from wholesalers to take account of carcase weight losses by evaporation which will have occurred since the original weighing of the sides.

SEX: S–steer, H–heifer, C–cow, B–bull. The main purpose of this is to distinguish the steer from the young bull, and the heifer from the cow. A cow is defined as any female which has conceived a calf except for pregnant animals where the quality of the carcase has not been seriously affected.

FATNESS: based on one of five fat classes according to the level of external fat cover, with 1 indicating the leanest class and 5 the fattest class. It was found that a large proportion of carcases fell in fat class 3 and so, if a participating wholesaler wishes it, this class is divided into two—3L (low or lean 3) and 3H (high 3). In addition, carcases in class 5 which are excessively fat are

175

classed Z. Carcases where the distribution of fat is unusual are identified by: P—patchy fat, U—excessive udder/cod fat and, where the KKCF is included in the carcase, K for excessive KKCF.

CONFORMATION: based on one of five classes with 5 indicating very good and 1 indicating poor conformation. In addition carcases with very poor conformation are classed Z. The class is determined by a visual appraisal of overall shape, taking into account carcase thickness and blockiness and fullness of the round. Fatness is allowed to play its part in influencing overall shape and no attempt is made to adjust for fatness. The conformation scale is illustrated in the form of silhouettes in Fig. 7. It emphasises the point that it is the overall shape of the carcase which is being considered in assessing conformation.

AGE GROUP: (at the wholesaler's option) identifies carcases from animals with calf teeth or two permanent teeth as Y (young) and those with four or more permanent teeth as T (teeth). The state of teeth development rather than the state of ossification was chosen since it was considered to provide a more precise assessment of the degree of maturity.

These foregoing factors describe the characteristics which are of importance to retailers preparing beef for sale. Amongst carcases of similar weight, the factor of most significance is the level of fatness since this has the greatest impact on the yield of saleable meat and on retail realisation values. Although the meat trade tends to think first of boinng loss, in fact cutting tests show that fat trimmings vary much more than bone even within batches of similar type.

Fat Trimming

The fat that is trimmed from carcases may come from one of three depots in the carcase:

    (a) external (or subcutaneous)—visible on the surface of the carcase and including cod fat, which may be used for covering joints with a layer of fat, or udder fat.
    (b) seam (intermuscular)—generally invisible on the uncut carcase
    (c) KKCF.

176

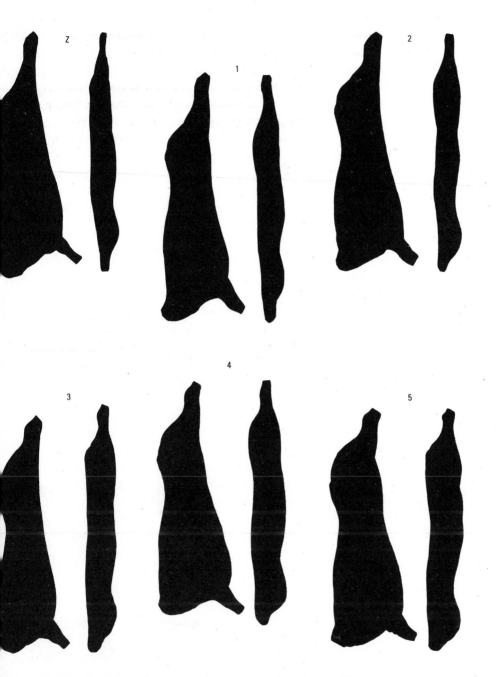

Fig. 7. Conformation scale based on five classes from 1 (poor) to 5 (very good). Carcases with poor conformation are classed Z.

KKCF

As noted earlier, the carcase weight specification may either include or exclude the KKCF, but there is now a trend for carcases to be dressed exclusive of KKCF. This simplifies the classification of carcases for fatness, as well as, for example, allowing better utilisation of the KKCF at the abattoir than would normally be possible in individual retail shops. It seems that if the KKCF is removed carefully from the hot carcase and the carcase is then properly cooled retailers need not fear any deterioration in the quality of the fillet. Of course, retailers need to ensure that when they are quoted prices for carcases they know whether they relate to carcases inclusive or exclusive of KKCF, because the cost/kg (lb) inclusive must be less than the value exclusive.

Given that the KKCF is excluded from the carcase, then the level of fatness in the carcase, which is closely related to the amount of fat which may need to be trimmed, is adequately described by the subjective assessment of external fat development, together with the application of the latters P and U to denote any carcases with unusual fat distribution. If trade practice permitted classification of the cold quartered carcase then the assessment of fatness could be improved to a degree.

It has not yet been found possible to replace the subjective assessment of fatness with an objective procedure, such as used for pigs, because a few simple fat depths in beef are not as good at assessing the level of carcase fatness as a subjective judgement made by a trained judge. This is partly because the external fat is laid down in cattle less evenly than in pigs and less of the total fat is, in fact, deposited externally compared to pigs. This means that it is more important to obtain an estimate of fatness over a wider area of the carcase than can be readily obtained by a few simple fat measurements.

So far as providing an objective measurement of conformation is concerned, no procedure has yet been found adequate for use on individual carcases. The best technique currently available for use under abattoir conditions is the carcase weight/length ratio. This is based on the fact that carcases which are short for their length will be more compact and have better conformation. While this ratio appears satisfactory for standardisation purposes it is not sufficiently

178

reliable to use as a replacement for the subjective conformation classification on individual carcases.

It is vital with classification schemes of this nature to ensure that standards are rigidly maintained, so that, for example, if the same carcase were classified in the north of Scotland and in the south of England it would be given the same classification in both places.

The combination of the factors used in classification can be used to provide guidance for those retailers concerned to try to identify breed type.

Amongst carcases of similar weight, the classification factors of most significance to retailers are fatness and conformation and the grid diagram shown below illustrates the way in which carcases are described separately for these factors. For example, ignoring the figures in the cells, a carcase in fat class 2 may have a conformation class ranging from 5 (very good) to Z (very poor). Similarly, it is possible to have a carcase in conformation class 3 with a fat class ranging from 1 (leanest) to Z (fattest).

Conformation Class	Fat Class						
	1	2	3L	3H	4	5	Z
5	★	0·3	1·3	1·9	1·4	0·4	★
4	★	1·8	7·0	7·1	2·7	0·4	★
3	0·1	7·3	21·3	13·5	3·3	0·3	★
2	0·3	6·2	10·3	4·3	0·7	0·1	★
1	0·5	3·3	2·6	0·8	0·1	★	★
Z	0·1	0·1	0·1	★	★	★	★

★ Less than 0·1%

The figures contained in the grid illustrate the proportion of classified steer and heifer carcases falling into each fat and conformation class combination in 1978. They illustrate the value of dividing fat class 3 into two. In using the scheme to define their buying specifications, it is important that retailers are not too demanding in their specification or wholesalers may not always be able to meet their requirements from supplies available at the time. Retailers should indicate their preference and, in addition, indicate acceptable alternatives for negotiation at different prices where the first choice cannot be met in full. Retailers wishing to evaluate the scheme and use it as an aid in purchase should ensure that the carcases bear the classification information.

Sheep Carcase Classification

Whenever the subject of sheep classification or grading is considered, the New Zealand Sheep Grading Scheme is often referred to. It is familiar to many retailers in Britain who have come to rely upon it for specifying their NZ lamb requirements, but it is not as simple as it is sometimes thought to be. There are, in fact, nearly forty possible New Zealand grades of sheep meat. Within weight ranges, the grades of lamb are based on different fat levels and, hence, conformation, since within New Zealand lamb there is generally a good relationship between fatness and conformation. The main features of the scheme for lambs as from October 1975 are illustrated in Fig. 8.

Modifications have had to be made to the scheme over the years to cope with diversification of lamb types in New Zealand. For example, in 1966, the Prime grade was sub-divided to include an Omega grade designed to pick out amongst lambs of similar fatness to the Prime grade those lambs with poorer conformation. The Omega (O) grade has now been distinguished as a separate grade.

While the New Zealand system may work adequately for a fairly narrow range of lamb types (although the introduction of the Omega grade indicates the beginning of modifications which will become necessary if lamb types continue to diversify), it would be inadequate in describing lamb carcase variations found in this country, where there is so much more variation in conformation at each level of fatness and vice versa.

The inappropriateness of the New Zealand type of grading system for use in this country led to the development of a sheep carcase classification scheme—the last of the classification schemes to be introduced. After an 18-month experimental phase, a formal classification scheme became operational during 1975.

The scheme provides a standard definition of the following four primary factors affecting sheep carcase value.

kg.	GRADE				
	A	Y	P	0	F
8.0 - 12.5	A	YL	PL	OL	
13.0 - 16.0	/////	YM	PM	OM	F
16.5 - 25.5	/////	YH	PH	/////	

Increasing fatness
and conformation →

Fig. 8. *New Zealand lamb grading. Increasing fatness and conformation.*
(The letters L, M and H are used as a quick guide to the weight range.)

CARCASE WEIGHT

CATEGORY: or age/sex group. L is used to identify lamb, Hgt—hogget and Shp—sheep. These groups are defined as in the Fat Sheep Guarantee Scheme.

FATNESS: Determined by a visual appraisal of external fat development. There are five classes, ranging from 1 (very lean) to 5 (very fat). In addition, the letter K is used to denote carcases with excessive KKCF development.

CONFORMATION: Based on four classes: extra, average, poor and very good. The definition of conformation is the same as that used for beef, i.e. it is essentially an assessment of overall shape, regardless of how much of that shape may be attributed to fat or lean development.

The sheep scheme has been simplified compared to the beef scheme in that provision is not made for every combination of fat and conformation. Since most carcases fall in fat classes 2, 3 and 4 and have average conformation they are identified by the fat class number only. Carcases of very good conformation in these fat classes are identified by the letter E (Extra) following the fat class number. Any carcase of poor or very poor conformation, regardless of fat level, is classed as C or Z respectively and there is no subdivision between average and extra in fat classes 1 and 5. The scheme is,

181

therefore, based on the class combinations shown in the following grid. The figures shown in brackets give the percentage of all classified carcases in each class during 1978 when approximately 24% of total clean sheep slaughtered were classified.

Conformation class	Fat class					
	1	2	3		4	5
	(very lean)		L	H		(very fat)
Extra		2E (0·8)	3LE (4·6)	3HE (2·5)	4E (2·1)	
Average	1 (1·0)	2 (23·2)	3L (32·8)	3H (17·6)	4 (7·2)	5 (1·5)
Poor				(5·8)		
Very poor				(0·9)		

The value to the retailer of certain aspects of the foregoing classification schemes will be considered further in the section entitled "Yield Variations".

LIVE ANIMAL ASSESSMENT

The more extensive use of carcase classification schemes will put increasing pressure on producers to try to meet particular wholesalers' grades based on classification. This highlights the need for them to become more expert at assessing fatness and conformation in the live animal. Since the penalties are likely to be greatest for cattle, consideration will be given here to how these characteristics may be assessed in the live beast.

The general appearance of a beast can be a guide to fatness. At the lowest level of fat cover, muscles can be seen rippling as the beast walks. At higher levels of fat cover, muscles are increasingly obscured, giving the beast a smoother appearance.

There used to be a preference for cattle which were fleshed well down to the hock and full in the flank. In fact, both characteristics indicate fatness. Today, most traders prefer a well-fleshed beast with a round which curves in well above the hock, and is "cut-up" in the flank. Fatness is also indicated by a brisket which is wide, deep and projects well forward.

182

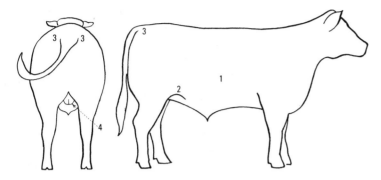

Fig. 9.   The four key handling points to feel depth of fat.   1. Over the ribs nearest to the hindquarters.   2. The curve of the flank close to the hind leg.   3. Over the pin bones and on either side of the tailhead.   4. The cod in a steer or udder in a heifer.

But these are only approximate guides and, to distinguish more precisely between cattle, it is necessary to handle them to feel the depth of fat. Furthermore, because fat is laid down at different rates in different parts of the body, handling at several key points is required. The following handling points, illustrated in Fig. 9, have been selected in parts of the beast where any confusion between lean and fat is least likely to arise:

1. Over the ribs nearest to the hindquarters.
2. The curve of the flank close to the hind leg.
3. Over the pin bones and on either side of the tailhead.
4. The cod in a steer or udder in a heifer.

The brisket is not included as a handling point because of the impracticality of using it with commercial cattle on the farm. Fat cover over the hook (hip) bone and shoulder appears useful only in indicating very fat or very lean animals.

So far as conformation is concerned, it will be recalled that this characteristic is defined as the thickness of lean meat plus fat in relation to the size of the skeleton and that good conformation is applied to animals which are thickly fleshed. In the live animal, the best types for thickness of lean have good length from hooks to pins, and good depth through the round. Viewed from the rear, such beasts stand wide and show a plump round which is wider than the back. In other words, the rear body profile is O-shaped, as shown in Fig. 9. From the front, they are wide between the legs, thick through the shoulder and have a trim brisket.

Long, shallow, leggy types have a poor thickness of lean meat and tend to have a poor lean to bone ratio. Short, deep, low-set types tend to be overfat.

Lean in the right places—the higher priced cuts—is obviously desirable, and with this in mind a good deal of attention is traditionally paid to shape. In fact, the differences in the proportion of high-priced cuts from one beast to another are much less than commonly supposed, and extremely difficult to detect in the live animal. Consequently, among animals having similar overall leanness, it is very unlikely that anyone could visually detect the beast with "a wealth of lean in the right places".

There are obvious limitations to any technique designed to predict from the living animal the composition and shape of the resulting carcase. Problems arise with batches of mixed breeds and crosses, steers and heifers. For example, some of the large, muscular so-called "exotic" breeds are often thought to be fatter when judged live than when seen as carcases. Nevertheless, the four key handling points and visual characteristics described above provide a useful guide both to carcase shape and, most important of all, to carcase fatness. The best way in which to gain experience in selecting cattle for slaughter is to follow cattle through to the abattoir. In this connection, the beef carcase classification scheme can be of considerable help since one can attempt in judging to place live cattle in carcase classes for fatness and conformation and compare these predicted classes with the actual carcase results.

In the case of sheep, use can again be made of the carcase classification scheme in gaining experience of predicting the carcase classes in the live animal. When assessing fatness in sheep it is particularly helpful to feel the development of fat over the spinous processse along the back and over the eye muscle.

Yield Variations

Although some young stock and store stock are still sold on a headage basis, virtually all trade in finished stock and in carcase meat is in terms of weight. But few buyers are able to purchase precisely what they want in the form they need; a few are, like the restaurateur who is supplied with individual steaks, within a narrow weight range and closely trimmed, the so-called portion control business. Most procure their raw material in one form and have to process this to a degree to obtain the product in which they subsequently trade. This means that yield and its variations is a vital concept to everyone in the meat trade.

First and foremost, the wholesaler or retailer buying livestock in the auction market is concerned with the yield of carcase meat and valuable by-products he obtains from the animal he buys on the basis of liveweight. The variation in the ratio of carcase to liveweight purchased (the killing out or dressing percentage) is considerable, but much of the variation is due to the gut fill, partly digested food retained in the digestive tract, and the buyer's skill is primarily to estimate the extent to which weight on the scale at the market is fill and, even, external dirt.

Apart from the effect of fill on the killing out percentage, it is commonly considered that as cattle become better "finished", i.e. heavier and fatter, they kill out better. However, the results of recent MLC trials on commercial cattle indicate that, among cattle giving similar shape, the effect of increasing fatness on killing out percentage was quite small. It was carcase shape at similar levels of fatness which was related to killing out percentage. Of course, as age and weight increase, fatness also increases and this generally improves carcase shape, so the conventional view is explained.

At one time, it was felt that the dressing percentage of cattle could be used as a numerical indication of "quality"—not nowadays, the sensitivity of this factor to external factors, such as time of last feed, conditions of transport, etc., and its high variability from animal to animal, have caused this approach to be played down. Furthermore,

185

traditional relationships have to a degree been upset by the introduction of new breeds. The lean, muscular "exotic" cattle breeds
have a high killing out percentage despite their relatively low fat
content. At similar levels of fatness, entire males kill out better than
heifers; however, most heifers are marketed at a higher fatness than
steers and so there is an advantage in killing out percentage.

**DEADWEIGHT
MARKETING**

Increasingly, there is a movement towards deadweight marketing
where animals by-pass the auction market and are sent straight to
the abattoir by the producer who is paid for the weight of carcase
obtained from individual animals.

The extent of deadweight marketing in pigs is large (approximately
85% of clean pigs slaughtered) and there has been a long history of
payment for carcase linked to the quality of the carcase. This has been
particularly evident in the bacon market sector, but in recent years
there has been an increasing degree of payment for quality in the
pork, cutter and heavy hog sectors of the market. So far as beef and
sheep are concerned, deadweight marketing is much less extensive
and there is a tendency for wholesalers to pay producers an average
price per kg (per lb) carcase weight for a batch of carcases unless there
are some exceptionally poor carcases in the batch. However, there
is now a movement towards payment for the quality of individual
carcases and this will be encouraged by the use of carcase classification
schemes which have already been referred to.

When the retailer comes to purchase carcases he could be faced
with an enormous range of quality which has a marked influence
on the yield of saleable meat he can highly obtain from a carcase.
The yield of saleable meat will depend firstly on the weight of
carcase purchased and, then, for carcases of similar weight on such
features as the fat content of the carcase, the fat requirements of
customers and the cutting method adopted. Given carcases of
similar weight which are jointed similarly and then cut to the same
standard of trim, the three factors which influence the yield and,
hence, carcase realisation value are, in decreasing order of importance, the amount of fat trim, amount of bone trim and inherent
differences between carcases in the proportion of high priced cuts.

186

To examine the effect of variation in fatness and conformation in pigs on the yield of saleable meat and some other carcase characteristics, the MLC conducted cutting tests on 255 commercial pigs in four weight groups (ranging from 41–77 kg (90–170 lb)) using a standard cutting technique and level of trim. Within each weight range, pigs were chosen to be in one of three backfat ranges (with low, medium and high optical probe fat measurements) and in one of three ranges of conformation (poor, medium and good). As an example of the fat levels chosen, the average $P_1 + P_3$ values for carcases in the lightest weight group 41–50 kg (90–110 lb) were 22 mm (0·86 in), 32 and 44 mm (1·26 and 1·7 in). The results averaged over the four weight groups are summarised in Table 2, p. 188.

These results confirm that fatness has a greater impact than does conformation on the yield of saleable meat in a carcase. Nevertheless,

TABLE 1

Weight of Selected Muscle Groups as % of Total Weight of
Carcase Muscle for Steers of Different Breed Types
(R. M. Butterfield, 1964)

| | Breed type | | | |
Muscle group	Hereford	Angus	Half Brahman	Unimproved Shorthorn
Upper hind leg	32·6	32·2	33·8	32·2
Lower hind leg	5·20	4·87	5·07	4·46
Spinal	11·9	12·2	12·1	12·3
Abdominal	9·69	10·5	8·85	8·65
Neck and thorax	9·69	9·66	9·84	10·6
"Expensive" muscles	55·6	55·6	56·8	56·1

a difference between yields from good and medium conformation carcases on the one hand and poor conformation carcases on the other, can be detected. This was due primarily to the rather higher yield of head and feet from the poor conformation group. The measurements of eye muscle size shown in the table indicate that selecting carcases with low probe fat measurements produced carcases with somewhat bigger eye muscles, and also that elimination of pigs with poor conformation weeded out the smaller eye muscles among pigs of similar weight and fatness. This was found to be particularly evident among the lighter pigs in the test.

TABLE 2

Average Yields of Retail Cuts as Percentage of Carcase Weight for Pigs
in each Fatness Range and of each Conformation Type

	Fatness			Conformation		
	Low	Medium	High	Poor	Medium	Good
Trimmed high priced cuts	44·8	43·2	41·5	42·5	43·5	43·4
Trimmed lower priced cuts	36·4	35·3	33·7	35·1	35·1	35·2
Fat trim	6·7	9·2	12·9	9·8	9·4	9·7
Head and feet	12·1	12·3	11·9	12·6	12·0	11·7
Total	100·0	100·0	100·0	100·0	100·0	100·0
Depth of eye muscle at last rib, mm (in)	45 (1·8)	44 (1·7)	44 (1·7)	43 (1·7)	44 (1·7)	45 (1·8)
Width of eye muscle at last rib, mm (in)	87 (3·4)	84 (3·3)	83 (3·3)	83 (3·3)	85 (3·3)	85 (3·3)

TABLE 3

Average Composition of Beef Carcases in Each Fat Class of the MLC Classification
Scheme (Percentage of Carcase Weight)

	Fat class					
	1	2	3L	3H	4	5+Z
Saleable meat	75	73	71	70	69	66
Fat trim	7	9	12	14	16	20
Bone and waste trim	18	18	17	16	15	14
Dissectable lean in saleable meat	68	64	61	59	56	52
Dissectable fat in saleable meat	7	9	10	11	13	14

An illustration of the relative significance of variations in fatness
and conformation on the yield of lean meat in a beef carcase is given
in Table 3, p. 188. This shows the results obtained from MLC studies
on many hundreds of cattle of mixed breed types which have been
cut up into lean meat, fat and bone in such a way that all the fat
covering the outside of muscles is removed. The lean meat produced
in this way obviously differs from saleable meat which has some fat
surrounding the muscles.

The effect of conformation on saleable meat yields is illustrated in
Table 4. Its smaller, but important, impact compared with fatness

Fat class

Conformation class		1	2	3L	3H	4	5
4+5	76	74	72	71	70	67
	78.4	75.6	72.8	71.4	70.0	65.8
3	75	73	71	70	69	66
	77.0	74.2	71.4	70.0	68.6	64.4
1+2	74	72	70	69	68	65
	75.6	72.8	70.0	68.6	67.2	63.0

TABLE 4

The value shown in the box in each cell of the grid illustrates the effect of yield on the carcase price a retailer could afford to pay to break even in terms of cost/lb saleable meat (assumes a carcase price of 70p/lb for an average carcase classifying 3H3)

can be seen. Among carcases of similar external fatness, therefore, those with better conformation will tend to have greater meat yields. This is largely due to the influence of breed, some breeds typically having better muscle to bone ratios and fat distribution than others and also better muscle thickness. Within breeds, those carcases with better conformation are very unlikely to show the effect shown in Table 4.

Table 4 also illustrates the financial value of changing saleable meat yields using, for the example, a base carcase price of 70p/lb (ex KKCF). It can be seen that for every 1% extra meat yield a retailer could afford to pay an extra 1·4p/lb carcase weight and still break even. As carcase prices rise, the value of an extra 1% increases.

Slaughtering

Historical Introduction

Before the 1939–45 war slaughtering of livestock for edible purposes was mostly on a small scale in Britain. Thousands of retail or "Master" butchers owned small slaughterhouses—usually at the rear of their shops. Here they prepared meat from carefully selected livestock. Many of these butchers produced meat of excellent quality and appearance. A high standard of dressing hygiene was achieved by pride in the skill of the individual. The slaughter room and hanging areas were kept clean. As they were only supplying their retail needs the preparation was not hurried and time was taken in washing away blood and apparent contamination. Cloths used to wipe the carcase were thoroughly cleansed. As a result a carcase relatively low in surface bacteria count, because of the quality of dressing and care of the product, was kept in good condition in a well-ventilated hanging room.

MARKETING AND SLAUGHTERING

Steadily the more active butchers killed stock for other retailers and the number of "Wholesalers" began to grow. Local authorities and particularly those of the larger cities were concerned that not all the facilities were as well constructed or as able to maintain hygiene as they should be. Some small and inaccessible slaughterhouses were increasing throughputs beyond the capacity where control of hygiene was considered adequate. They provided slaughterhouses to which butchers were able to bring their stock for killing. This benefited the local livestock markets from where the livestock was purchased. These livestock markets were formed so that the farming community were able to find buyers for their livestock under competitive conditions. Usually the slaughterhouse and market formed a composite unit. Some of the large market/abattoirs were built to handle hundreds of Irish cattle shipped in "on the hoof" for consumption in Britain.

SLAUGHTERING FACILITIES

The slaughtering facilities were designed in comparatively simple form. A building of high pitched ventilated construction was provided and equipped with individual "booths" or partitioned areas. Little mechanical equipment was installed—usually, means for pulling beasts down to the floor, sometimes steel boxes for restraining and stunning them. A hand-operated winch and simple overhead rails were the basic provision. The lairage area was normally common to all users but separated off into pens for entering the "booths". Hooks on stands were available for edible or "red offals". Some authorities built central areas to which the green offals or abdominal mass were taken for treatment. Later, companies set themselves up to separate and treat the offals in bulk for sale or further processing.

Butchers were able to slaughter and prepare meat under conditions of controlled hygiene, judged to be high by the standards prevalent at the time. Supervision was under the control of a responsible local government officer experienced in the trade and able to govern the hygiene standards set within the known facts of the period. This same officer was frequently responsible for all the various public markets common to the time.

In the late 1930's government consideration was given to the full centralisation of slaughtering in relatively few "factory abattoirs" strategically located so as to take all the stock from the various catchment regions. Theoretically this approach to the breeding, producing and marketing of the largest source of natural protein would seem the most effective. Who knows what the present-day structure of the industry would be if the proposals had gone forward? War intervened and the conversion of meat from stock became the responsibility of the Ministry of Food. Slaughtering was concentrated into approximately 450 premises in England and Wales and 150 in Scotland compared with a total of about 13,000 immediately before. The Ministry of Food was made responsible for arranging slaughter and distribution and the Ministry of Works for structural alterations and additions.

After the war control of meat remained in force until July 1954, during which time the central authorities responsible for slaughtering and facilities installed pilot equipment in some existing slaughterhouses, introducing simple line working into "bed systems". In the early 1950's a government policy of "moderate concentration" brought about the construction of two experimental slaughterhouses of different capacities; the first at Guildford, designed for daily throughputs of 120 cattle units, and the second at Fareham, Hampshire, for 250 cattle units. These are briefly reviewed on pp. 218–241. Both were multi-species slaughterhouses, all stock being slaughtered and dressed on a single floor. As a result of the experimental work five more units were built, at Canterbury, Wimborne (near Bournemouth), Swindon, Salisbury and Grimsby, before the end of Ministry control of meat in 1954. These five were constructed on the basic design of the Guildford experiment. A feature of them at the time was that the handling of the by-products of slaughter was carried out by elevating the abdominal mass and the hides and skins up to the first floor. Stomachs and inedible offals were separated and opened and various materials, hides and skins despatched by gravity from this floor. In retrospect these plants remained relatively successful for many years. A number of the principles of design are retained to the present.

A feature which remained a problem is that all equipment was suspended from reinforced concrete ceilings. Inserts were cast in these to support the equipment. This made any alteration difficult to carry out as breaking of the concrete affects the structural load bearing characteristics of the remaining monolithic slab. The slab being low over the tracking rails and of heavy solid concrete brings about condensation where humid air streams, a factor always present when dressing warm carcases, come into contact with the cold slab.

MARKETING AND DISTRIBUTION

Meat marketing and distribution were de-controlled in July 1954. A number of pre-war slaughterhouses were re-opened, many local authorities embarked on the construction of new facilities and many private enterprise ones were designed and constructed, so that by 1956 there were a total of about 4600 licensed slaughterhouses in England, Scotland and Wales. Since that time, owing to many factors including logical commercial trends and statutory obligations, in particular the Hygiene and Cruelty Regulations of 1958, the total numbers have declined. Tables 1 and 2 indicate the number of slaughterhouses in relation to total annual throughputs by unit numbers from 1968 to 1973.

Now, in the mid 1970's, the general size and scope of operation has increased and with an increase of refrigerated storage, development of the preparation and specialised marketing of different primal cuts, vacuum packing of meat and the supply to multiple retailers on an increasing scale, the industry now markets wholesale meat from about 1700 meat plants. Some large cities no longer provide slaughtering facilities for butchers and wholesalers. A number retain or have constructed new meat market centres where retailers may select and buy carcase meat or cuts.

MODERN SYSTEMS

Modern systems of line dressing of all species have been introduced.

Some of the all embracing skill has been removed by these systems which resemble on a smaller scale the production lines used in engineering and manufacturing processes.

TABLE 1

Slaughterhouse Numbers 1972–73. Analysed by Throughputs (cattle units)

Slaughterhouse throughputs Cattle units per year	England		Wales		England and Wales		Scotland		Great Britain	
	No.	%	No.	%	No.	%	No.	%	No.	%
1 to 99	184	12·1	12	9·1	196	11·9	24	20·5	220	12·5
100 to 499	569	37·5	44	33·6	613	37·2	13	11·1	626	35·5
500 to 999	141	9·3	19	14·5	160	9·7	2	1·7	162	9·2
1000 to 4999	253	16·7	33	25·2	286	17·4	28	24·0	314	17·8
5000 to 9999	123	8·1	14	10·7	137	8·3	17	14·5	154	8·7
10000 to 19999	96	6·3	8	6·1	104	6·3	13	11·1	117	6·6
20000 to 29999	59	3·9	—	—	59	3·6	7	6·0	66	3·7
30000 to 49999	58	3·8	—	—	58	3·5	6	5·1	64	3·6
50000 to 99999	24	1·6	1	0·8	25	1·5	5	4·3	30	1·7
100000 and over		0·7	—	—	11	0·6	2	1·7	13	0·7
Total	1518	100	131	100	1649	100	117	100	1766	100

cattle unit = 1 beast or 2 pigs or 5 lambs or 3 calves.

TABLE 2

Slaughterhouse Numbers 1978–79. Analysed by Throughputs (cattle units)

Slaughterhouse throughputs Cattle units per year	England		Wales		England and Wales		Scotland		Great Britain	
	No.	%	No.	%	No.	%	No.	%	No.	%
1 to 100	133	11·2	11	10·7	144	11·2	18	17·8	162	11·6
101 to 500	382	32·2	32	31·0	414	32·1	13	12·9	427	30·6
501 to 1000	122	10·3	14	13·6	136	10·5	3	3·0	139	10·0
1001 to 5000	206	17·3	24	23·3	230	17·8	18	17·8	248	17·8
5001 to 10000	94	7·9	8	7·8	102	7·9	13	12·9	115	8·3
10001 to 20000	94	7·9	10	9·7	104	8·0	16	15·8	120	8·6
20001 to 30000	53	4·5	4	3·9	57	4·4	7	6·9	64	4·6
30001 to 50000	50	4·2	—	—	50	3·9	4	4·0	54	3·9
50001 to 100000	43	3·6	—	—	43	3·3	8	7·9	51	3·7
100001 and over	11	0·9	—	—	11	0·9	1	1·0	12	0·9
Total	1188	100	103	100	1291	100	101	100	1392	100

After the Verdon Smith report of 1964 the Agriculture Act of 1967 brought into being the Meat and Livestock Commission. One statutory duty of the Commission was the setting up of a section whose activities must embrace the dissemination of advice and information to the industry on the efficient design and construction of slaughterhouses and anything contributing to the efficiency of them. Therefore the central area of technical information is the first since that provided by the Ministries of Food and of Works during control.

In the following pages an attempt will be made to provide information of a technical nature to assist the wholesale meat industry in selecting and analysing improvement of meat plant facilities and improving profitability and material utilisation.

Site Selection and Orientation

This section studies the technical implications involved in choosing a site relevant to: ease of operation, freedom from general criticism (of the public or of authorities concerned with nuisance), and ease of expansion. It also presents the manner of placing the building to follow the same criteria. It assumes that the commercial feasibility of a viable project has been satisfactorily determined, including the availability of the type of stock required within reasonable distance of the project.

Land is not easily obtained and it is hardly likely that every desirable feature can be achieved. Probably the intending owner is faced with a compromise choice. It is hoped that the following will help to resolve what compromise will accomplish the best result.

SELECTION

There are three defined land areas to which consideration is likely to be given: urban, nominated industrial, and rural. The availability of urban sites for processing and marketing of meat is receding.

Location: Urban land.
Nominated industrial sites.
Rural land.

196

This is hardly likely to be acceptable to the relevant planning authorities for the construction of meat plants; however, should the case arise and a site be available, there are many factors leading to the conclusion that it is unsuitable for even a small plant. Housing or commercial properties will inevitably be adjacent and unavoidable livestock noise, odours and traffic congestion eventually make conditions of operation untenable. The nature of business requires freedom in the hours of operation. Stock can frequently only be delivered out of normal working hours with some consequent disturbance. By-product collection cannot be easily concealed, with resulting objection from neighbouring property. Restriction on expansion is strictly controlled and if permission is eventually given, conditions as to size must be dictated, usually more costly and less functional. A final limit of expansion is always arrived at, inhibiting commercial life.

Most of the stated advantages such as availability of services and labour are offset by competition for both, and the difficulty of disposing of the waste products of the type of industry. In the past it has been claimed that proximity to marketing outlets is an advantage. This is now wholly offset by the efficiency and availbility of modern refrigerated vehicles, whereby carcase meat or cuts can be transported any distance to these outlets.

These are subject to many of the limitations of those in urban areas. Restrictions on expansion owing to planning requirements usually overtake the enterprising wholesaler. As neighbouring industries expand they tend to get closer to the less salubrious areas of the plant and nuisance problems arise. Transport problems increase as each business necessarily expands and personnel from all industries obtain and use their own means of transport. Usually the boundaries of these sites are established at the conception of the over-all development with consequent growth problems owing to natural expansion. Competition for labour is highlighted after years of growth and eventually the conditions approximate to those of urban areas.

The apparent advantages gained at the outset seldom prove so as the business progresses. The advantages are generally considered to be:—

Urban Land

Nominated Industrial Sites

197

adequate and varied labour resources, availability of public an
commercial transport and absence of residential and town develop
ment. The sites of course usually have some form of commerci
restriction such as leaseholds.

Rural Land

This is the most acceptable location from most points of view
provided that the responsible authorities are convinced and plannin
permission is possible. Objectively the type of processing an
distribution is parallel with that of livestock agriculture, to whic
it is so closely linked. A strong case for this approach is advocated
There is a distinct possibility that a site in rural land can be selecte
large enough to provide all the expansion of the enterprise for th
foreseeable future and to choose the most advantageous manner c
orientating the plant.

The correct parameters for obtaining maximum separation of th
"clean from the polluted areas" are easier to achieve and maximur
separation of them can be obtained. Urban traffic congestion i
avoided and the possibility of locating the site within reasonabl
traffic-free distance from trunk roads and motorways is increased.

There is less likelihood of severe restrictions being placed on th
size and types of buildings. They may approximate closely to thos
agricultural structures of the neighbouring farms and by prope
design some types of pre-fabricated agricultural buildings may b
used in the construction of the plant, increasing the likely viability

There is likely to be less restriction on working hours, and labour
once obtained and trained, will be more static.

A major influencing feature is of course the greater availability of a
least one species of livestock.

The following points are offered to assist anyone considering th
purchase of a site for developing a meat plant.

1. Consult people, especially experienced technicians from county
district or local authorities, notably those concerned with environ

198

ment, with planning and with future road development. People in these official capacities will generally be only too willing to advise and be associated with a properly conceived project. Advice may be obtained concerning the availability of mains services and the possible cost, the need for and degree of effluent treatment, subsoil condition and foundation requirements or the location of subsidence areas. Help should be forthcoming on future road developments and projected town boundaries. Consultation with the Deputy Regional Veterinary Officer of the MAFF will establish helpful advice concerned with stock problems and the advisability of some sites in relation to outbreaks of notifiable diseases. The Meat Plant Advisory Service of the MLC will always provide technical advice and have the facilities to offer functional designs relative to the throughputs and operating factors. Both the latter will provide facts involved in complying with the EEC Directives.

2. Try to purchase sufficient land for all your future needs. This should be undertaken to the limit of practical and financial considerations. A modern multi-species plant with a capacity of 250 beasts, 1000 lambs and 1000 pigs per week can be placed on a 4 acre rectangular site with space for expansion (see Fig. 1). However, with increased facilities for vehicles, requirements of the EEC Directive for proper hard standing, vehicle wash aprons, manure bays, and taking account of the turning circles of the larger modern transportation and car parking requirements, it can be seen that to avoid problems arising eventually when other facilities may become desirable, the prospective site should be ample. A stock-proof fence should enclose the perimeter of the site.

3. Bearing in mind that stock, by-products, carcase and cut meat, packing and boxing materials and possibly meat imported from a number of sources require loading or off-loading to and from the plant, a level site is the most functional choice. It means that the building needs to be raised over-all to tailboard height but the expenditure is justified by the saving of labour and mechanisation. Occasionally a rising or falling site can be used to advantage, that is a site rising towards the lairage areas will enable stock to be taken off at a level suitable for first floor dressing; a site falling from the meat despatch area will enable a loading dock to be constructed more

199

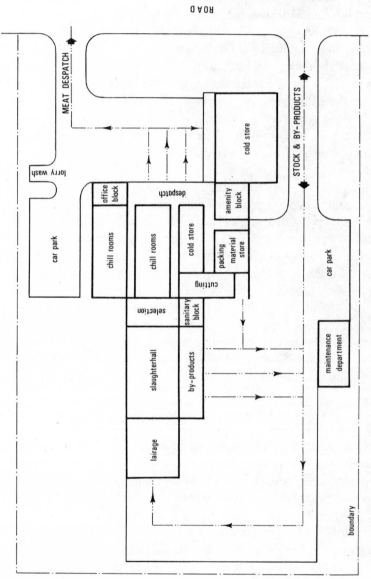

Fig. 1. A modern multi-species plant with ample expansion and illustrating "adequate separation".

200

cheaply. However, the advantages generally lie with a level area. Built-up ground should, wherever possible, be avoided as should an area with a high water table. Areas of disused mining often present problems but professional advice should naturally be sought when the prospective operator is uncertain of local conditions.

Wherever possible the meat plant should be located on the selected site to fulfil two basic requirements:—

ORIENTATION

1. It should present the most pleasant aspects of operation to the public at large, particularly relative to the angle from which it is approached. People whose occupations are remote from the agriculture and meat industry take a kinder view if the more obvious activities of the plant are the despatch of well presented carcase, cut meat and saleable edible products. Obviously the unloading and droving of stock and the process and despatch of the by-products of slaughter are essential. The public do not need and generally do not wish to see, hear or smell these areas, therefore they should be located at the rear, as far as possible from the main approaches to the plant. The edible despatch areas should conversely be located to confront the casual visitor. The operator is then encouraged to display the name of the enterprise to utilise the most publicity for his products.

2. The location described aids the functional purpose of the design in that operations of lairage, slaughter, dressing, refrigeration, processing and storage and the despatch of edible materials all progress systematically from the rear to the presentation point. Fig. 1 illustrates all the factors presented in this section and additionally shows how the requirements of the EEC Directive relative to 'adequate separation of the clean from the polluted areas" are complied with. Subject to road planning requirements two separate entrances are preferred but the alternative of separation by physical barriers and routes is indicated in Fig. 2. The directions of flow are clearly indicated.

This subject is one in which there is probably more controversy than any concerned with the conversion of livestock to meat. Individual

PRE-SLAUGHTER CARE

201

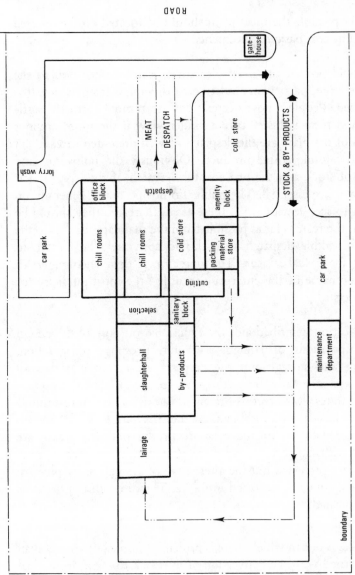

Fig. 2.   A modern multi-species plant with alternative physical barrier "adequate separation".

202

iews are widely separated; therefore it is not proposed here to take
n inflexible attitude, rather is it intended to put forward factors
vhich must be relevant to slaughterhouse activity.

dult bovines, calves, sheep and pigs are raised in an environment
f relative quietness and calm. From the time they are ready and
ntended for slaughter a cycle of activity commences, during which
hysical disturbances have some relevant effect on the animal.
nevitably they must travel in unusual surroundings, sometimes long
istances and either directly or via a livestock market to the point of
aughter. Stress in some form or other commences from the time
ne stock is loaded into transport at the farm.

Cattle

Many eminent wholesalers with reputations for producing high
uality beef carcase meat are adamant that cattle may be slaughtered
nmediately upon arrival at the point of slaughter without any
pecial precautions against stress. An example quoted is that of the
rrival of Irish cattle at the Merseyside ports where they were
laughtered almost immediately after disembarking from the ships,
he quality of meat showing no obvious signs of deterioration after
he whole chain of transportation and the numerous changes of
nvironment, droving and assembly.

Others with the same claims to quality insist that every beast subject
o even the minimum disturbance arising from transport and changes
f environment should on arrival at the slaughter point be given at
east a moderate feed and unlimited water. The feed is discontinued
 few hours before slaughter but water provided in abundance. It
 claimed that the resulting carcase shows a distinctive quality after
his treatment. Usually the same claimants maintain that a cold
airage floor without any straw bedding and washed down to the
equirements of veterinarians concerned with disease factors brings
bout dark cutting beef.

Calves

few will argue that careful handling of young calves can be anything
ut beneficial. Transport over long distances is detrimental and is
nown to multiply the chances of the incidence of Salmonellas.

Veterinarians concerned with animal welfare advocate strict contro of the manner in which calves are transported for slaughter. Poo handling of young animals transported long distances and laire under cold and uncomfortable conditions is commercially unre warding.

Sheep

The handling of lambs is adequately covered in MLC Technica Bulletin No. 5. However, in relation to immediate pre-slaughte care, the over-crowding of any sheep in transport is detrimental t meat quality. Pulling and picking up sheep by the fleeces spoils th appearance of the carcase and produces bruising and unsightly mark on the finished product. Young lambs which are handled roughl into the slaughterhouse lairage and stunning pen, frequently shov signs of "blood splash" in the finished product.

Pigs

Transport, holding and handling of pigs for slaughter is the mos controversial aspect of pre-slaughter care. Direct and contract buyin may reduce the number of environmental changes which affect th final carcase quality. Usually on arrival at the point of slaughte pigs which have grown up in each other's company and learned t live with each other will become separated. Stress factors emerg when they are penned before slaughter and natural aggressiv instincts are amplified in slaughterhouse conditions. Most peopl concerned with pig slaughter on a large scale are agreed that the should be slaughtered as soon as possible after arrival. Over crowding in the lairage should be avoided; insulated floors help t reduce stress and if they are to be kept overnight, clean, dry strav helps them to settle down. The straw must be removed and th pens cleared out before another batch are held in the same pen Every effort should be made to handle them without excitement Water should be provided.

Summary

The pre-slaughter care of stock requires as much skill and care as a any time in the animal's life. Breeding and stock-raising knowledg and skill have prepared the animal for conversion to meat. It i pointless if the quality is reduced by clumsy and careless treatmen hortly before conversion. Whatever the claims against this thesis, the foundation for stating it has many practical factors. Ruminants have active bacteria in the digestive system. If starved and stressed the bacteria enter the blood stream. The end result affects the keeping quality of the meat, if not the appearance. Bruising not only spoils the apparent quality but its commercial returns.

Pigs which are stressed produce some level of pale, watery muscle which is low in quality. Stress increases the possibility of "blood splash". If they are excited in holding pens they will fight, if they are driven through "races" which are wide enough to allow the stronger ones to overcome the weaker, they will scramble and scratch, resulting in unsightly and obvious scratch marks on the de-haired carcase. Either plenty of room for movement or narrow and controlled restraint will help to reduce the problem. Tepid water sprinkled from overhead in the restraining passages helps to calm pigs down.

There is a scientific basis for accepting that, to a degree, animals are similar in characteristics to humans, that behaviour, timid or aggressive, highly strung or docile, is dependent on the metabolic rate stimulated by thyroid activity. Some animals are naturally subject to stress and react quickly to excitement and some will be quiet and naturally unexcitable. This opinion is substantiated by the research on the Pietrain breed of pig originating in Belgium. The breed is generally excitable and highly susceptible to stress, producing pale muscle subject to excessive drip. Rigor sets in very quickly after slaughter with a lower than normal pH. It has been known for these pigs to be so stiff at the scalding and de-hairing stage immediately after slaughter that the legs are frequently broken in a de-hairing machine. Conversely, de-hairing is incomplete as the leg stiffness prevents the desired rotation in the machine.

Conclusions

Whatever extreme views are held relative to the care of stock destined for slaughter, moderate treatment in its transfer from its normal environment until slaughter is worthwhile. Stress and poor carcase quality may result from careless handling. Stockmen at all stages of this transfer should be instructed to treat animals in their care thoughtfully. Impatience vented on them during transfer to and from vehicles should be penalised and the stock should at least be allowed the dignity of slaughter with patient handling.

Cleanliness

This section of the chapter cannot be concluded without reference to cleanliness before slaughter. The fact that excreta and filth on any livestock destined for slaughter increases tenfold the probability that it will be contaminated in the carcase state is undeniable. It may be that soon there will be an obligation to present stock for slaughter in a reasonable state of cleanliness. If this becomes so, carcases will have far greater appearance value, but, more important, the hygiene of meat resulting in better keeping quality and reduced health hazards will be the end result. Pre-slaughter washing is sometimes advocated but has so far been impractical in the less warm climates as, particularly with regard to cattle, calves and sheep, intensive feeding of cattle and root crop feeding of sheep, combined with long-distance transport, cakes the dung so heavily into the hair and wool that the result of pre-slaughter wash has no advantage. More usually it will make the carcase dressing more difficult with higher contamination than otherwise, as the liquified excreta spreads easily to the carcase. In dry, dusty climates water sprays remove the travel dust and the animals are able to dry out before slaughter. We may one day resolve this particular problem by both washing and drying the stock, the problems being sufficient time and the amount of water and hot, drying air necessary at relatively high cost.

Pigs can be spray washed while closely restrained in a narrow controlled "race" where they are unable to lie down. They enjoy the water sprays, during which time they quieten down for slaughter. However, the technical problems associated with this apparently ideal treatment are numerous. There are in existence systems of brush washing pig carcases immediately before scalding for hair removal, this method proving a successful compromise as it lowers the bacteria content of the scald water at the same time as it prepares the hair and skin for scalding.

There are two broad principles which govern a good design for a slaughterhouse lairage:

CONSTRUCTION of a simple, well ventilated building of practical size and easy to maintain in a hygienic condition.

STOCK FLOW AND RETENTION under conditions of reasonable comfort preparatory for slaughter within the statutory (and EEC) conditions; allowing for free movement into and out of lairage pens and to the stunning areas; avoiding undue stress and damage and preventing injury to the operators.

Simple portal frames either of pre-cast concrete or galvanised steel provide the most satisfactory basis for construction. These frames, with cleanly shaped rafters, allow for a full high ridge to obtain rising ventilation and are easily cleaned. They minimise the collection of dust and cobwebs, and adjustable louvres set in the walls, protected by mesh to keep out birds and vermin, will control the ventilation.

An air change rate of some 40 changes per hour should be aimed at to prevent an accumulation of ammonia fumes and humidity. A double sheeted asbestos roof having an air space between the sheets improves roof insulation.

The infill walls between the supporting columns may be constructed of locally manufactured blocks or bricks of local authority requirements. These walls and the low separating partition walls, providing the stock pens, should be rendered in smooth cement to full height. The rendering should be steel trowelled and polished and made as impervious as the material allows. Coating in polymer high-build paints is more acceptable, provided it is so applied as to remain impervious and not flake or fall away. Rendering should extend to cover all internal block and brick work. All walls or piers should be rounded or bullnosed to prevent damage to hides and skins and bruising. Every effort should be made to obviate acute corners or sharp edges and provide an easily cleaned construction.

207

The floors may be finished in concrete screed; granolithic finish is preferable and will last longer. They should be tamped so as to give a non-slip finish. However, deep ridges made by coarse tamping are not only difficult to clean but cause discomfort to stock. The screed should be coved to all walls, simplifying cleaning. If possible the addition of insulation between slab and screed will improve conditions for stock, especially for pigs.

All the pens, with the exception of one which may be reserved for isolation, may drain from the back to the passageway. As all floors must slope at 50 mm in 3 m (2″ in 10 feet) this suits the location of drains in the passage. There is available a pre-cast heavy duty gully manufactured in cement products, with a narrow slotted inlet extending throughout its length. This may be installed down the full length of the passage.

Alternatively, a drain inlet may be located each side of the passage, enabling say two pens to drain away into these. In the latter instance the passage itself must be so laid that liquids run directly to drain. In the case of an isolation pen the drain should be placed centrally with all surfaces draining directly into it.

Windows are a doubtful advantage in the construction of lairage. They are frequently broken, the insulation factor being low they subject the lairage to temperature differentials and they are seldom cleaned. Their most useful function is the provision of light but this again is subject to fluctuation. The light intensity demanded by veterinarians is 3 m (10 ft) candles (110 lux) except in isolation when double this is required. A constant level is obtained by fluorescent lighting. If windows must be provided there should be no sills, the wall being sloped steeply from the frames for cleanliness and any openings protected by fly and vermin-proof mesh. No timber should be used so that any frame must be of non-corrodable material.

Stock Flow and Retention

It is a distinct advantage to offload stock on to a loading dock. It assists the unloading of animals from all transporter floor levels. The dock should be walled at each end and gates arranged so that stock have direct access to the desired passageway. Lairage for multi

IWEL sheep dressing conveyor

Courtesy Scottish Border Abattoir Ltd.

IWEL sticking and leg skinning operations on a sheep line

Courtesy Messrs Barretts and Baird Ltd.

Moving sheep cratch

Courtesy Meat and Livestock Commission

53lb plus   44–52lb   35–43lb   25–34lb

Champion lamb carcases

Large white gilts on test

Measuring back fat of a pig by ultrasonics

Courtesy Meat and Livestock Commission

Feeding pigs at MLC Evaluation Unit

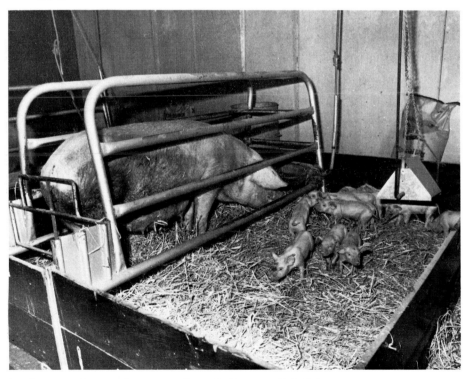

Confinement of sow, and warm condition for litter at MLC Unit

Courtesy Meat and Livestock Commission

IWEL Type H Pig Dehairer showing moving top, gambrel table, chute and operator's platform

Courtesy Messrs Barretts and Baird Ltd.

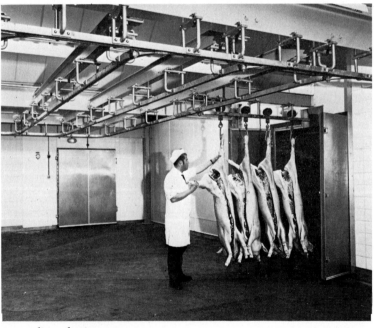

Pre-cooling selection area

Courtesy Prestcold

IWEL vertical beef dressing system

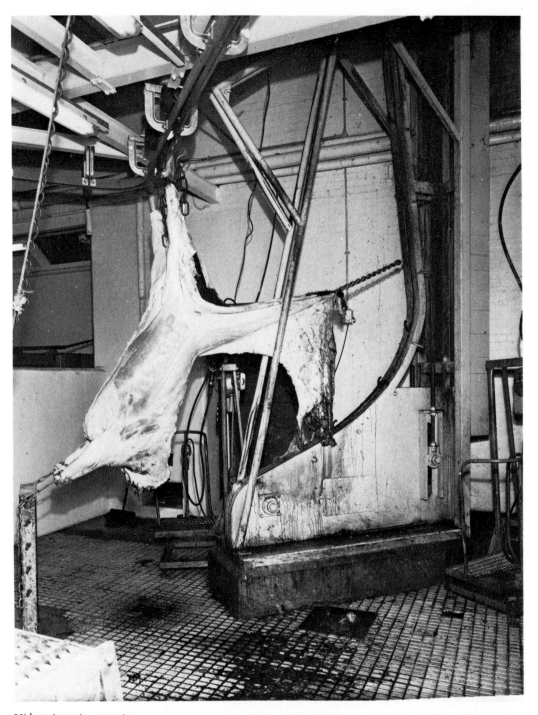

Hide stripper in operation

Courtesy Johnson, Bristol

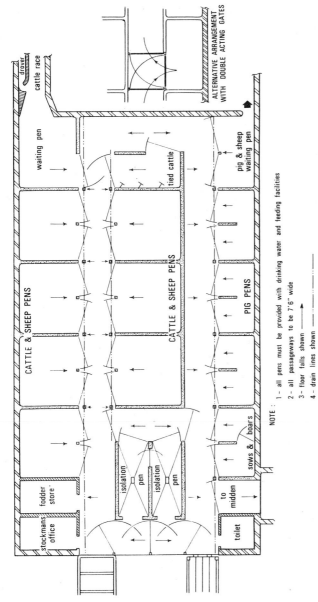

CATTLE & SHEEP PENS

CATTLE & SHEEP PENS

PIG PENS

waiting pen

tied cattle

drover

cattle race

pig & sheep
waiting pen

ALTERNATIVE ARRANGEMENT
WITH DOUBLE ACTING GATES

stockmans
office

fodder
store

isolation
pen

isolation
pen

to
midden

toilet

sows & boars

NOTE : 1 - all pens must be provided with drinking water and feeding facilities

2 - all passageways to be 7'6" wide

3 - floor falls shown ⎯⎯→

4 - drain lines shown ⎯·⎯·⎯·⎯

Fig. 3. Typical lairage arrangement.

species abattoirs are designed so that sheep may be held in cattle c
pig pens. The former may be about 4·57 m (15 ft) square. The latte
about 2·3m (7′ 6″) square. These sizes serve a number of purpose
Figure 3 indicates a typical lairage arrangement suitable for EE(
purposes, whereby pig pens are isolated from the remainder by
clearly defined barrier. The passageways are approximately 2·39 r
(7′ 6″) wide allowing plenty of access for a small tractor. A dun
trailer or a scraper can be used to muck out the pens. The gates, bein
of similar length to the width of the passage, will close it of
allowing simple loading and unloading of the pens. These gate
being of single swing are uncomplicated mechanically and requir
less maintenance than the double acting type frequently specifiec
The double acting gate by nature has to be of precision constructio1
its locking point has to become its hinge point (see inset Fig. 3). A
certain times the single swing gates may be arranged to open ou
and increase the area of two pens by 50%. Advisedly the separatio
of pens should be of solid construction which may be smoot
finished cast concrete, pre-cast blocks or be brick built. They can b
rendered smooth and impervious and thereby raise hygiene standarc
and make for draught-free and more comfortable conditions.

The gates may comprise tubular frames plated with steel. All stee
fabrications should be hot dip galvanised. Many are zinc sprayed an
therefore not very easy to keep clean. The use of tubular lairage per
has been common for many years. It may have originated i
livestock markets, probably in the first place because stock can b
viewed from further away than the actual pen within which anima.
are held. It was also easy for the local millwright or smith to con
struct. However, in lairages at abattoirs it has not proved wholl
satisfactory. This is accentuated with the advent of EEC where th
veterinarians require the lairage to be thoroughly clean and n
compromise is acceptable. The horizontal tubes become caked i
dung and thorough cleaning is labour intensive. If the tubes ar
spaced badly, animals get their heads caught between them. Ofte:
the stock situation at an abattoir is such that the lower bars have t
be plated. In the first place one is well advised to use rendered wal
together with tubular framed plated gates. The animals generat
sufficient heat for air to rise to the building ridge and be replaced b
circulating fresh air.

Horned cattle get scarcer but some ties must be available for the eventuality. There must be some pens available for completely separating large sows and also boars.

Water has to be provided in all lairage pens. Food should be given to animals when they are held for more than twelve hours. Open troughs and gullies are fouled by cattle and occasionally pigs climb into the gullies. If they are mounted high enough to prevent this, lambs cannot reach them. Therefore there are a number of problems to overcome. Most veterinarians advocate the provision of one of the various mechanical devices available on the market. They usually comprise a small metal bowl which is situated in a corner of the lairage. A valve is actuated by a lightly spring-loaded flap. Stock sense the water and muzzle the lever. This produces water for drinking. Others which are most suitable for pigs are shaped somewhat like a teat and are actuated by the animal itself as well. The types used in cattle/sheep lairages do not necessarily overcome fouling. There are various means to protect troughs and gullies

TABLE 3

Areas for Pens

Stock	Area sq. ft	Area sq. m
Cattle (collected)	25	2·3
Cattle tied singly	35	3·25
Sheep	6	0·56
Pigs—pork or bacon	6	0·56
Heavy pigs (collected)	8	0·74
Sows and boars, singly	25	2·3
Calves	8	0·74

TABLE 4

Volume per Ton (1·016 tonne) Bedding and Feed

Material		
Bedding straw	120	3·3
Hay in bales	100	2·9

which after all most closely represent the natural environment of the ruminant. Good hygiene practices can enable them to be kept clean. Fodder racks or feeding troughs should be provided. It is possible to construct the racks to protect the troughs below them, serving the double purpose of protecting the troughs from fouling. Provision is necessary to stop excessive feed falling into the water. A narrow passageway between two pens enables feed to be placed in the racks without disturbing the stock.

Tables 3 and 4 indicate the recommended areas which should be reserved for stock and indicate feed and bedding volumes.

Stunning and Slaughter

United Kingdom legislation enforces consideration for the welfare of animals to be slaughtered. All animals must be stunned by humane methods prior to slaughtering. No animal may be slaughtered within the sight of others awaiting slaughter. No animals awaiting slaughter should be able to see or smell blood or the by-products of slaughter while awaiting slaughter. Cattle must be restrained within a specially constructed stunning pen enabling humane stunning to be accomplished without unnecessary suffering or pain. The stunning pen should be manufactured in non-corrosive metal. It restrains the beast to prevent it damaging itself. It also protects the operator from injury.

Only licensed slaughtermen are allowed to kill animals within the UK. The person must have worked in a slaughterhouse for at least 3 months, and must be over 18 years of age. He receives the licence from the Local Authority when they are satisfied that he has the proper degree of skill. The licence is renewable every year and states the specie of animal he may stun and/or slaughter and the type of instruments to be used by that person.

CATTLE

The most common humane and practical method of slaughtering cattle is to drive them from the lairage into a race leading into the stunning pen. The race takes the form of a narrow passageway about

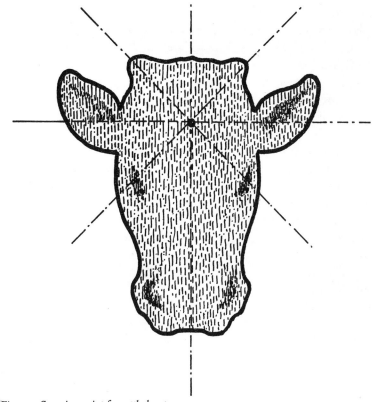

Fig. 4. Stunning point for cattle beast.

0·83 m (2′ 9″) wide. Once secured in the pen the beast is rendered unconscious by use of a captive bolt gun. The Temple Cox and the Accles and Shelvoke "Cash" are in common use. There are varying models but either a hand pistol or one mounted on a long handle to improve reach is used. The muzzle is placed very firmly at a point on the forehead at the intersection of lines theoretically drawn from just above the right ear or where the horn root is, across to the left eye, and from above the left ear to the right eye (see Fig. 4). It must be held at right angles to the skull. The gun is fired and a hardened steel bolt about 102 mm (4 in) long is driven into the brain.

Pithing of the beast is becoming less common. The lobby of experts **Pithing**

213

opposed to the process is increasing and includes eminent veterinarians. The process involves the insertion of a form of slightly flexible rod into the aperture resulting from entry of the captive bolt. The rod may take the form of a strong flexible cane or a tightly wound steel spring. The process of inserting the "pithing rod" destroys the Medulla Oblongata at the base of the brain. The intention is to accelerate the reflex muscular actions after which it is allegedly safer for the slaughterman to shackle the beast. It is claimed to save time by the accelerated reflex action. It is becoming less common for the following reasons: (a) The action of pithing takes time, especially as now the pithing rod must be sterilised between its use on each beast. Experienced men can attach and retain the shackle and get the beast up to the rail without trouble. (b) Over-zealous and careless pithing too deep into the spinal cord roots reduces efficiency of bleeding and produces blood saturation of parts of the viscera. (c) Veterinarians claim that lack of hygiene resulting in filth and contamination of the rod, more especially the cane type, results in poor carcase keeping quality and bone taint. Bearing in mind that heart action and circulation are necessary for efficient bleeding it is perfectly logical that bacteria are introduced into the system in this way.

The slaughtering procedures must take place quickly. The beast is ejected rapidly on to the floor in front of the pen and lifted to the bleeding rail (see p. 220). The beast is "stuck" by severing the carotid arteries and jugular veins. A special sticking knife is inserted just above the breastbone and central to the first ribs on either side. The knife approaches the backbone at an angle of about 45°. It must not touch the heart. To improve hygiene of the operation, especially when taking edible blood, the hide should be incised and "rimmed over" before sticking to reduce the possibility of contaminating the knife and the incision.

The operator should have two separate knives in use, and a knife steriliser containing water at 82°C (180°F) should be at hand. A knife is sterilised between each operation. Bacteria can enter the circulatory system even more easily from a contaminated knife than a pithing rod. Thorough bleeding is essential for carcase quality and appearance. It should take at least six minutes but animals should not

214

remain on the bleeding rail for more than about ten minutes. Dressing should be completed as quickly as possible. Good management ensures a continuous flow of stock throughout the process of bleeding and dressing. Total quantity of blood obtainable from bovines varies with the size. A beast of about 454 kg (1000 lb) liveweight will produce up to 16 kg (35 lb) of blood.

SHEEP

Normally sheep are stunned before slaughter by the electrolethaler process. They are driven from the lairage into a small room or "stunning pen". The corners of the room should have rounded corners at animal height and no recesses into which the animal may retreat. This ensures early capture and quick stunning, removing the need to chase and excite them unnecessarily. The rounded corners assist hygiene by simplifying cleaning. The common instrument used for electrical stunning is a pair of large tongs with electrode pads at one end and handles at the other. The operator quickly but carefully squeezes the electrodes below the eyes, a little in front of the ears. There are occasions when electrical stunning is unsuitable, as in the case of heavy rams and ewes or sheep with very close, thick wool. It is then preferable to use the captive bolt pistol. It has also been found that some very young lambs show signs of blood splash from electrolethaler stunning, in which case again the captive bolt is preferred. The muzzle of the gun should be placed directly and firmly against the crown of the head, aiming directly down towards the gullet.

Bleeding must take place as soon as possible after stunning, the animal being quickly elevated to a bleeding rail. The knife is drawn across the jugular furrow close to the head. Both carotid arteries must be severed. When the skin has to be pulled over the head the same arteries are severed by skilfully inserting a knife through the side of the neck. About five minutes bleeding time is required but on a mechanised conveyor rail six or seven minutes should be allowed before further knife work commences. From 1·81 to 2·28 kg (4 to 5 lb) of blood should be obtained.

PIGS

This species of stock is highly susceptible to stress, the more finely bred such as the Pietrain having an in-built stress characteristic, therefore many and varied methods of achieving restraint for stunning have been and are continually being developed. The scope of this chapter prohibits any exploration.

The most common facilities are as simple as those described for sheep. (The method used for administering CO_2 gas is touched on later.) Apparatus for administering electrical stunning is usually similar to that described for sheep. The electrodes should be placed on either side of the head at the intersection of a vertical line from the eyes and a horizontal line from the ears. The electrode pads should be immersed in a weak saline solution to improve conductivity through the skin. The surface skin over the head should not be wet as this can short circuit the voltage which must complete a path through the brain. Large boars and sows should be stunned by captive bolt. At times when electrical apparatus is faulty the gun may also need to be employed. The muzzle should be pressed firmly on the centre line of the forehead about 25·4 mm (1 in) above the line between the eyes. It should be aimed high up into the head.

It is most important that pigs should be shackled, elevated and stuck immediately after stunning. A time of ten seconds should be achieved to improve bleeding and minimise the possibility of "blood splash". The sticking knife should be inserted on the centre line of the throat in front of the breastbone. The anterior vena cava and the carotid artery are pierced. The knife should not go in so far that it penetrates the shoulder when blood will go back, spoiling any subsequent curing for bacon. The stick cut should be as small as possible to reduce the ingress of dirty scalding water. About 5 minutes is the proper bleeding time and it should not be placed in a scalding tank in less than 6 minutes.

Notes on Electrical Stunning and CO_2 Gas Anaesthesia

The established criteria for the electrical stunning of small animals are that a minimum voltage of 75 volts AC 50 cycles should be available at a current of not less than 50 milliamperes. All the usual meat plant equipment contractors market the apparatus and there is ample literature available. Briefly, it takes the form of "tongs" comprising a pair of electrodes at one end, with insulated handles at the other. Wire flex connects the tongs to a robust, well insulated transformer having an indicator to show it is functioning, together with overload protection against abuse. The supply ranges from 70 volts to 90 volts.

The common voltage used for small animals is 80 volts and the electrodes should be applied for a minimum period of 10 seconds. When these criteria are maintained the animal should stay completely unconscious for one minute. Its muscles are relaxed and bleeding takes place naturally and thoroughly. Unfortunately application is frequently incorrect and too short, resulting in stress and muscular paralysis, harmful to the finished carcase.

The equipment should be well maintained. The tongs and particularly the electrodes should be cleaned and inspected every day to prevent unnecessary suffering due to faulty stunning.

High frequency electrical stunning is emerging as a method which is claimed to be quicker and more effective. The mains voltage is converted by an electronic method into a high frequency. By further sophistication the wave form may be converted from the normal synosoidal form which gives peak amplitude according to frequency, to a square wave form. This form extends the period of maximum amplitude. It is claimed that conductivity is more positive and the time taken to render small animals quite insensible is down to two seconds. The Russians are claimed to have perfected the method but it is now becoming available in the UK.

CO_2 gas is used for anaesthetising pigs in some large plants of high throughput. It has been used experimentally in a number of small plants but owing to cost considerations is confined to the larger works. The pigs are induced by various means of conveyance into a chamber below slaughterhall level. They are exposed to a carbon dioxide concentration of about 85% for approximately 45 seconds. They remain unconscious for up to 3 minutes and it is claimed that bleeding is improved and the yield increased. Although also effective for sheep, carbon dioxide, being heavier than air, gathers and remains in the wool. During subsequent work the operators are affected by the gas. It is therefore impractical.

Multi and Single Species Carcase Dressing and Handling

Each section of this chapter concerned with slaughter and meat plants provides subject matter for a book devoted to that specific subject. It is problematical what material should be filtered out while maintaining a functional continuity. Much has already been said elsewhere of the relative merits of single floor, multi-storey or split-level arrangements. However, the chapter would be incomplete without some analysis. A complete description of *all* the systems employed in the slaughter and dressing of the three major species of stock would exceed the chapter limitations but again brief descriptions are essential. By these and with the aid of illustrative and flow diagrams it is proposed to pinpoint at least the fundamentals.

PRINCIPAL FLOW CONSIDERATIONS

Figure 5 is a flow diagram of a meat plant, either single or multi-species. Whether the plant is established entirely on one floor or several floors the desirable flow of materials through the various stages of conversion is the same.

It is likely that in the future almost the total national throughput of home killed livestock will be processed through about two hundred and fifty meat plants. Some will remain relatively exclusive and small, but generally the throughputs will be high. Many will be of split-level and multi-storey design. Those which continue to operate as multi-species plants on a single floor will be confronted to an ever increasing degree with the problems fundamental to this design. The hides, skins and various dirty by-products of slaughter which at present must cross at least one dressing line of another species, will need to be taken through, over or under these lines.

Plants which have been modified to comply with EEC requirements of separate pig killing facilities have, in some cases, installed pneumatic conveying equipment to transport the whole abdominal mass or the various components out of the dressing area. This and other

218

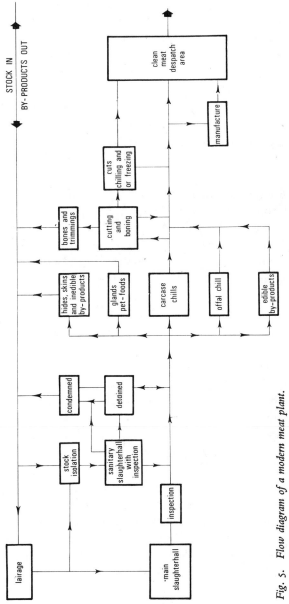

STOCK IN

BY-PRODUCTS OUT

lairage

stock isolation

condemned

detained

sanitary slaughterhall with inspection

main slaughterhall

inspection

hides, skins and inedible by-products

glands pet-foods

carcase chills

offal chill

edible by-products

bones and trimmings

cutting and boning

cuts chilling and or freezing

manufacture

clean meat despatch area

Fig. 5. Flow diagram of a modern meat plant.

219

forms of conveying equipment offer some temporary relief but where more than one species is handled, reference to Figs 6, 7, 8, 9 and 9a shows that the principles laid down in Fig. 5 are easier to attain with the multi-storey or split-level design.

Much has been said of the relative merits of single and multi-floor construction for meat plants. The author has elsewhere developed the case for the split-level construction which he designed as a result of observation of the manner in which the UK slaughtering trade was developing, together with experience of the disadvantages of single floor construction and the expense and complication of multi-storey. The split-level design is now used as a model by consultants and architects. Its full potential has yet to be fully exploited.

Cattle Dressing

Bovines are increasingly being slaughtered and dressed in higher throughputs which demand the adoption of modern vertical or "on rail" dressing. After being driven into the stunning pen or "knocking box" the beasts are rendered insensible and ejected on to a "dry landing" area in front of this pen. This area should be so constructed as to be kept relatively clean and dry. The beast is shackled by means of a short, heavy-duty shackle chain and is hoisted by (various) mechanical means to a "bleeding rail" above a properly constructed trough. This should have means of catching and retaining blood in hygienic conditions. The animal is "stuck" as described on p. 213.

After bleeding for 6 to 10 minutes, the head is "scalped" and removed; it should be placed so that the skinned head can be thoroughly washed outside and through the mouth and sinal passages. The tongue is dropped down through the lower jaw and transported in sequence with the carcase to a central inspection area where the head, "red offals", abdominal mass and split carcase may be inspected. The principle to be observed under modern conditions is that of bringing all these products of slaughter to the inspector properly correlated one to another. This aids proper inspection and relieves him of unnecessary and tiring movement. It ascertains the thorough examination of glands and sees that the relevant offals receive proper inspection.

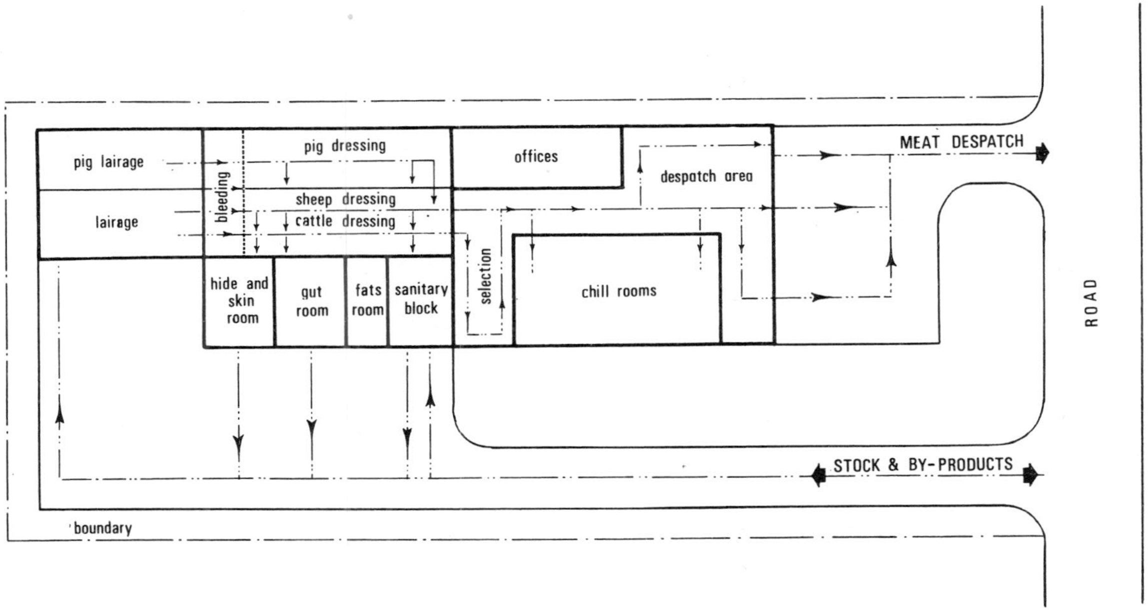

Fig. 6. Diagrammatic layout of a single floor multi-species abattoir.

221

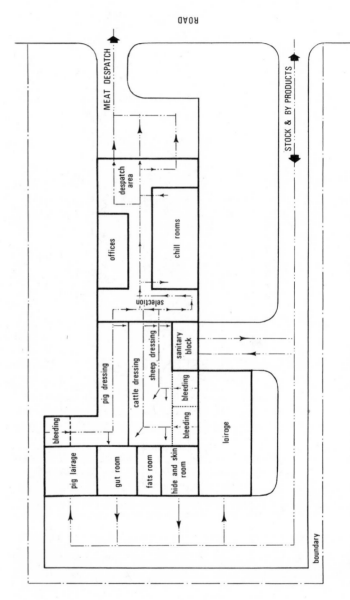

Fig. 7. *Diagrammatic layout of a single floor multi-species abattoir with improved separation of pigs.*

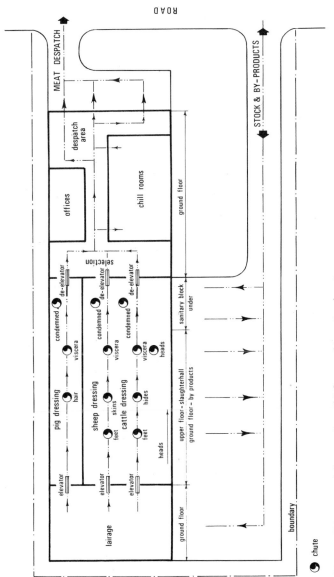

Fig. 8. *Diagrammatic layout of a multi-floor, multi-species abattoir.*

223

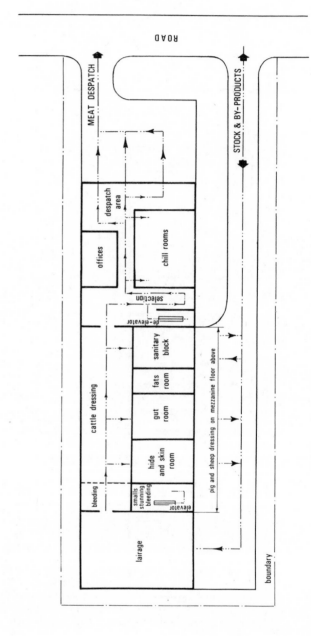

Fig. 9. Diagrammatic layout (ground floor) of split level construction.

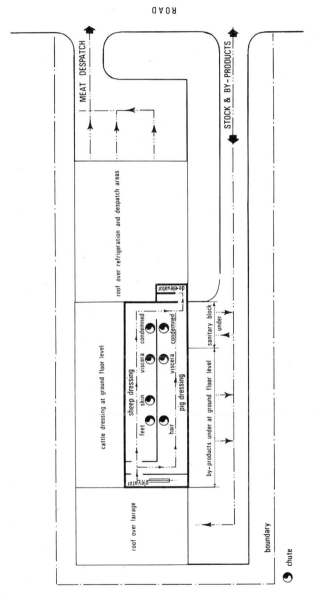

ROAD

MEAT DESPATCH

STOCK & BY-PRODUCTS

roof over refrigeration and despatch areas

cattle dressing at ground floor level

de-elevator

sheep dressing

feet   skin   viscera   condemned

hair   viscera   condemned

pig dressing

sanitary block

under

by-products under at ground floor level

elevator

roof over lairage

boundary

⊙ chute

Fig. 9a. Diagrammatic layout (mezzanine floor) of split-level construction.

225

The forelegs are removed at ground level and other operators "dress out" the hind legs while working on high level platforms. During this operation the "bung" should be cut and tied to prevent contamination of the carcase by excretal matter. In the case of cows the udder is removed and placed for inspection. The carcase is attached by the Achilles tendons to the two hind legs on to rollers which are mobile on the dressing rail. The hide is then rimmed over from the crotch to the neck and removed in stages by various operators. The complete hide may be removed by skilled operators with knives or mechanical hand tools. The more recent adoption of hide pulling or stripping machines is less fatiguing for the operators, more productive, and results in better quality hides. One type grips the flanks and pulls the hide back from the carcase, leaving a hand operation for complete removal off the back. Another strips the hide completely from the shoulders up and off the carcase, including the tail. Machines being developed strip the hide down from the rumps and completely over the head. After complete hide removal the brisket is opened and sawn. At this stage the weasand should be rodded and tied to further reduce the chance of contamination. The belly is opened and the entire abdominal mass removed to an inspection table to correlate with the head, carcase and offals.

The carcase is split down the entire length of the centre of the vertical spinous area from the aitch bone to the neck. In butchers' terms "the chine is sawn". In the two resulting sides the chine and neck are trimmed together with any bruising or loose external fat. After all the inspection processes the approved abdominal mass is passed to the Separating Room.

The edible (red) offals are trimmed and placed in the offal chill-room. The carcase is washed, weighed and classified and routed to the chill-rooms.

Many cattle dressing lines are conveyorised. Although it is usual for throughputs up to about twenty beasts per hour to be moved manually on a rail with a slight gradient, it is common to conveyorise above these rates. It has been claimed that over fifty beasts per hour have been dressed on the manual system. However, conveyors have a number of advantages. The carcases arrive at the work stations n sequence and without manual effort. The distance between carcases is fixed and ensures they do not touch, thus reducing the possibility of cross contamination. As there is less handling during their movement, the likelihood of contamination from the hands of the operators is reduced. This is particularly so at the areas where the hide leaves a clean carcase. There are numerous factors relative to this subject which involve complex detail. Insufficient room is available to argue the various opinions and technical involvement.

Dressing of lambs requiring throughputs from 100 per hour upwards is carried out on either of two partly mechanised systems: (a) mechanised on-rail, (b) moving cratch and rail. The on-rail system, with variations in detail, is common throughout the world. For instance, in New Zealand it is not uncommon to dress 500 lambs per hour on one "chain" or conveyor. The moving cratch originated in and to date is peculiar to the UK. It was conceived here and quickly gained popularity, probably because the operatives found that it required little change from the methods which they had been used to applying on fixed "cratches" or cradles.

Sheep/Lamb Dressing

MECHANISED ON-RAIL: The animals are stunned, usually in a small walled-in pen about 8 feet square. Electrical stunning apparatus is normally used and this is described elsewhere. Sometimes a captive bolt pistol is necessary. The lambs are shackled by means of a short, looped chain on one leg and elevated as quickly as possible to bleeding rail height. After stunning there is a rise in blood pressure and the sooner this is relieved by sticking the better for the resulting carcase quality. About 1·8–2·25 kg (4–5 lb) is the quantity to expect.

The area where the elevator is situated, that is the separating wall through which the animal passes as it is elevated, needs to be high enough to screen the sticking operation from other animals in the stunning pen. Usually heavy rubber double flap doors screen the opening above this. The Regulations in Britain governing animal welfare require that no animal shall be slaughtered in view of other livestock and the arrangements described fulfil this requirement. The rail is positioned over a trough so constructed that blood is contained within a restricted space and retained by draining into an outlet connected to a system of pumping or pneumatic transfer. Bleeding time is about 6 to 7 minutes, by which time a skinning knife may be used without danger to the operator from reflex movements. However, this is not normally the problem, rather is the reverse the case and the stock is left too long on the bleeding rail for good hygiene practice. If throughputs are above about 120 per hour conveyorisation of the bleeding rail should if possible be provided. By this means a regular flow of properly bled carcases is delivered to the first dressing station, in this case that of legging and transfer. Not only is bleeding improved but the risk of carcase contamination is also lessened. At this station the loose or "trailing" leg is skinned and the foot removed. A hook which is subsequently utilised as a "gambrel" for both legs is inserted in the dressed leg. The dressed leg, the flesh of which is exposed, is subject to contamination from the fleece of any other complete carcase coming into contact with it. The use of the conveyor governing the distance between carcases overcomes this.

After inserting the hook/gambrel in the first dressed leg it is transferred to a skid or runner on a dressing conveyor rail. The second or "trailing" leg is dressed out as the first and this leg hooked up on to the gambrel. The animal is thus suspended on the conveyor rail by both hind legs in a vertical position. In sequence the skin is opened down from the crotch to the belly and rimmed over the rump. The bung is cut and tied.

To simplify work on the neck, breast and flanks a specially formed spreader frame about twenty-four inches wide is used. It has at its extremities slots formed similar in manner to a shepherd's crook. It is now used to spread the forelegs. The toes of the animal are clipped into these slots and the frame with the fore-part of the animal slung up on to a separate travelling hook. This places the lamb in an upside down semi-horizontal position. In this position operators are able to open the skin and strip the forelegs to the breast. The skin at the neck is opened and where the two cuts up the legs meet, a "flap" is formed enabling the operator to punch a channel under the skin. This is done either with his fist or with a specially designed tool. This channel opens the skin of the breast to the position where the belly skin has been opened. Before cutting

228

he skin completely open the operator can clear it away from the reast to the shoulders, using leverage. After slitting the skin and learing the brisket this is sawn open. The weasand is stripped out nd tied and the forelegs released from the spreader and the front eet removed. The lamb now returns to a vertical position where the anks and shoulders are cleared. The skin is now completely pulled ff. Veterinarians advocate the complete removal of the skin from he whole carcase and off the head. This requires skilled knifing echniques but improves hygiene in that heads with skin attached end to be bloodied and contaminated. It is unhygienic if these dirty eads are retained for edible purposes. While still on the conveyor he carcase is opened so that removal of the complete abdominal nass is possible, followed by removal of the trachea, oesophagus, eart, liver and lungs, together being the "pluck". These are placed n inspection conveyors timed to correlate the carcase to the bdominal mass and plucks. Full inspection processes are possible vith the minimum of unnecessary movement by the inspector.

MOVING CRATCH & RAIL: The stunning and bleeding methods are lentical in this system to those for on-rail dressing. However, the led carcase is lowered on to a sturdy horizontal conveyor. The otal length varies with the rate of throughput. The appearance ears some resemblance to a long belt conveyor. In place of a onventional belt there are a series of horizontal steel plates about 5 mm (3″) wide set transversely across the conveyor (about 190 mm ′ 6″) wide). These plates are bowed slightly to cradle the animal. he plates are in sets which provide the equivalent of the traditional sheep cratch". Each set moves forward at a speed suitable to the ate of kill. Hence the description "moving cratch". The height is bout twenty inches and convenient for traditional legging. The nen work in pairs, both commencing on one lamb and between hem completing the opening and clearing of the skin, right to the pulling" stage. The feet are removed, the brisket is sawn and the arcase is elevated off the cratch on to the rail. As in the previously lescribed system, a gambrel is inserted in the two hind leg tendons. he gambrel travels on a skid or runner as before. The carcase should e spaced on a conveyor and the conveyorised evisceration inspec- ion routine followed out.

Pig Dressing

It is impossible within the scope of this book to give detailed descriptions of the numerous and varying systems in use for converting animals to pig carcase meat. A degree of sophistication has been reached beyond that of other species. Accuracy and consistency of pig production has made it possible to set up specialised plants for processing within limited weight ranges. Varying trading patterns influence the degree of sophistication incorporated in specific plants.

Some multi-species plants have very simple means for slaughtering comparatively few pigs of varying weights. Many specialised plants contain high levels of sophistication. They employ conveyorisation and machinery to slaughter and dress pigs at consistently high throughput under controlled conditions. Additionally, bacon factories vary in practices, size and diversification.

Figure 10 is a flow diagram showing the fundamentals of any modern pig slaughter unit. Within the functions (a) (b) (c) (d) and (e) conditions vary according to trading practices and hourly throughputs. Each may be dealt with simply or each separately may be highly mechanised. All the functions are sometimes included in a controlled mechanical sequence.

STUNNING: This is described adequately on p. 212.

BLEEDING: This takes place within a wide scope. In its simple form the animals are stuck and bled within a small area where they are suspended on an overhead track. A few butcher's slaughterhouses still slaughter the animals singly, bleeding taking place in an isolated floor area.

At the other extreme, after elevating quickly to an overhead bleeding rail they are bled in sequence, clearly spaced on a conveyor. The conveyor controls the time of bleeding accurately and delivers the carcases in orderly fashion at speeds up to 600 per hour to the scalding plant. A minimum bleeding time of 6 minutes is desirable and, depending on the weight of the animal, 2·25 kg (5 lb) to a maximum of 3·6 kg (8 lb) of blood may be obtained.

When pigs are slaughtered in batches and the bleeding is not

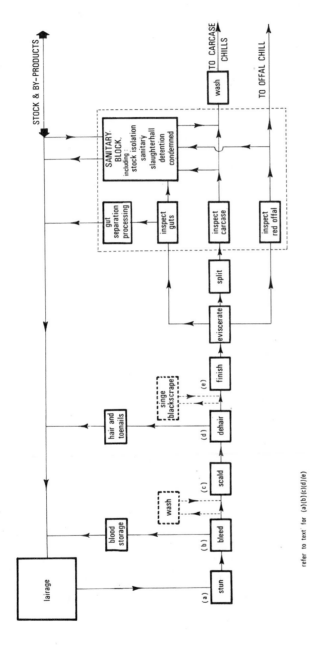

Fig. 10. Flow diagram of basic procedures in pig slaughter.

231

properly controlled not only is the meat quality reduced bu contamination in the scalding tank is increased. To the dirt from th carcase is added unnecessary quantities of blood. It results in heavy contamination throughout the scalding and de-hairing processes.

SCALDING: This is necessary to remove the pig hair from the follicle The time/temperature parameters are $5\frac{1}{2}$ to $6\frac{1}{2}$ minutes at from 61°C (142°F) to 59°C (138°F). The actual methods vary considerabl over a wide range of practices. The simplest system only involve lowering the carcase into a small tank of scalding water for th specified time. This is satisfactory if done properly. Unfortunately i is only too common for the tank to be too small, the water is heated by one open steam hose and the temperature erratic. The time o scald is decided subjectively, literally by rule of thumb. If the hai comes off when rubbed with the thumb, the pig is ready for scraping

Many variations of conveying through a controlled-temperatur tank are in use. They range from a form of moving beam to auto matic conveying from sticking point to de-hairing machine. The moving beam comprises long bars arranged longitudinally over the tank. The bars are braced by cross members. The beam moves to and fro by motorised cranks or compressed air rams. The forward movement gently propels the semi-floating carcases. At the end o the stroke the beam rises above the water level and moves back. The stroke is adjustable for varying throughputs and is said to be helpfu to the scald in that the massage movement and the bumping o carcase to carcase improves the penetration of scalding water.

Fully automatic conveying is installed in a number of high through put plants. The shackled pigs are carried by the conveyor from the sticking point through to the de-hairing position. Bleeding i controlled and the subsequent scald is accurate. The conveyor lower the carcase into the scalding water and pulls it horizontally through its scald for a precise time within the stated parameters. Speeds from 90 per hour upwards are within the scope of this system. The scalding tank is heated and the temperature regulated at a number of point and consistent temperature water is in circulation all the time during scalding. In conjunction with this system a method of pre-scald washing and brushing is used in the UK. Before being lowered into

ne tank the dirty carcase passes through a set of rotating brushes
tted with stiff nylon bristles. Warm water is liberally sprayed
while the brushes scrub the carcase. Blood and soil are removed
rom the pig, resulting in reduced contamination. It has been found
nat the scald is improved. The bristles and skin are prepared for
etter penetration.

or many years scientists and engineers have held the view that the
igh element of bacterial contamination in scalding tanks affects
ne keeping quality of pigs. Often the lungs are condemned, owing
o absorption of dirty water from this source; water enters through
ne mouth or the stick cut.

The logical solution is to achieve scalding by means of applying
eam vapour and/or hot water sprays. The cost of the method up
o now has been too high for commercial operation. Heat losses, to-
ether with capital equipment costs, have retarded adoption of an
bviously more hygienic means of scalding for hair removal. There
re however a number of these installations in operation. Sweden,
Holland and Germany use variations of the technique commercially.
There is evidence that the bacteria count is dramatically reduced and
ungs from this method are utilised for human consumption.

E-HAIRING: In its simplest form all that is required is a specially
ormed hand scraper. If the scald is effective a man can remove all
ne hair from a pig very easily. This of course is time consuming,
nerefore machines of varying capacity and sophistication are in use.

he transverse machine is common throughout Europe and the
straight through" or longitudinal machine more common in the
JSA. The transverse machine is set on the floor across the end of
ne scalding tank. Within the machine are one, two, or even three
rge rotating drums. One drum in all cases has heavy rubber beaters
nounted on it. The beaters are equipped with steel scraper blades.
Details of each form which this takes vary. The basis of operation
that the pig is rotated while at the same time the scrapers remove
ne hair from the carcase. This type of machine is normally equipped
vith a loading cradle or gate, the action of which is to lift each
arcase into the machine to a predetermined cycle. Pigs are ejected

233

by various means on completion of the cycle. It is also possible to convey carcases into the machine in sequence.

The "straight through" or longitudinal machine is sometime referred to as a "U" bar machine, owing to the construction of th bars inside. They form a "U" shaped cradle of separated bars. Drum with rubber mounted beaters are mounted on either side of thes bars. The scrapers project into and between the bars and are stag gered in their attachment to the drums. As the drums rotate, th scrapers not only remove hair but by a screw action propel the pi forward—straight through the machine. The time in the machine i directly related to the manner of attachment of the "staggered" beaters and controls the throughput. To gain additional scrapin time and improved de-hairing more than one machine may b arranged in tandem. The direction of rotation of the drums i opposite, resulting in a reverse of rotation one to the other. Thi further improves the finish of the pig.

FINISH: After removal of the hair, in particular by mechanised de hairing, some areas of the carcase need finishing. Hair still remain round the head, loins and under the legs. This may be removed b hand scraping and various degrees of singe. Modern plants with hig mechanisation usually include machinery for brushing and polishing The carcase is placed on a "gambrel" on leaving the de-hairin machine. The gambrel is attached to a skid runner or a roller and i moved manually or mechanically through the stages of finishing The high throughput line often incorporates conveyorisatio through a light singe machine, usually gas fired. It then include machinery for spraying the carcase with water while at the same tim the carcase is brushed and polished by rubber flayers and/or nylo bristles. The equipment can be arranged so that virtually any area o the carcase, including the head and the legs, can be finished.

If bacon is to be cured from the carcase or part of it a "heavy singe" is carried out. The carcase travels into and through a speciall constructed furnace. It comprises what is best described as two halves of a vertical steel cylinder about 2·7 m (9 ft) high and 0·91 n (3 ft) diameter. These half cylinders are braced and lined with furnac brick. An oil/steam fired burner generates approximately 935°C

,000°F) within the lined furnace. The two halves open to receive the ig carcase and close round it. A singe of up to 15 seconds is ommon. This heavy singe or burning of the skin achieves three hain objectives. It removes all remaining hair, sterilises the cuticles /hich improves the keeping quality and reduces slimy bacon, and "firms up" the rind. The resulting bacon, with its rind on, cuts etter. The appearance of the carcase, in spite of being hand or hachine scraped ("Blackscraped"), washed and polished, is less ttractive aesthetically than those produced from a light singe. The eavy singe is traditional, the advantages seem factual. However, here appears to be little controlled scientific test data available; for hstance, what the lowest time/temperature ratio is required to chieve the undoubted advantages. If these are established and they duce the unsightly burning and loss of usable protein, there may e additional commercial utilisation.

VISCERATION: After the intact whole carcase is washed, the bung is ut and dropped. The aitch bone is split with care to prevent con-hmination from the gut. The belly is accurately centred and opened, o avoid wasteful trimming later on. The breast is opened with a hall cleaver. The stomach and abdominal mass are removed and laced separately from the "pluck" or red offal for inspection orrelated to the carcase.

: must be noted that EEC inspection routines demand that corre-htion is maintained until after the carcase is split longitudinally own the centre of the spine from the aitch bone down and through he head.

fter inspection the red offals are trimmed and then chilled as soon as ossible in a separate offal chiller. They are passed into a separation oom for further processing.

he carcases are weighed and classified. They should be chilled as uickly as possible.

he 1966 amendment to the Slaughterhouse Regulations prohibited he use of wiping cloths from November 1968. Reaction from the

CARCASE SPRAYING

235

wholesale meat trade was emotive and continuous. There was
sincere belief that under most conditions use of wiping cloth
improved the "finish" of carcases. Under certain conditions, sub
jective visual judgement supports the belief. A cloth individuall
sterilised or thoroughly cleaned in hot water removes the obviou
staining. The subcutaneous tissues and fat, if then exposed to a rapi
flow of ambient air after being wiped, dry quickly. Surface drynes
retards bacterial growth. In isolation there appears to be a case fo
the use of cloths.

The more general manner and the conditions under which cloth
were used, together with results of controlled scientific tests presen
a clear case against their use. Frequently as operators came unde
great pressures of throughput a single bucket of lukewarm wate
was used, the cloth rinsed and squeezed served for a succession o
carcases. The soiling was spread evenly over the surface, apparentl
cleaning it. Many abattoirs do not have sufficient air movemen
Under *ideal* conditions a cleanly dressed beef carcase may carry fro
155–1550 bacteria per sq. cm (1000 to 10,000 per sq. in.). Where si
beef carcases have been dressed successively and wiped "clean" usin
the same wiping cloth, the surface may carry 155,000 per sq. c
(a million per square inch). Given conditions of high humidity an
poor air movement bacteria will multiply by simple cell divisio
one cell becomes two, two cells become four and so on. Doublin
can occur every 20 or 30 minutes. Slime forms on meat surfac
when the bacterial count reaches about 31 million per sq. cm (20
million per sq. in.).

The law does not permit the use of wiping cloths on meat. It do
not specifically require that the carcase shall be sprayed. If the carcas
is soiled it may be considered unfit for human consumption. Applie
and conscientious hygiene on the part of the operators is essential t
achieving the basic requirement of a clean carcase. If the carcase do
become soiled, proper spraying techniques, followed by earl
refrigeration in properly equipped chill-rooms, reduces the hazar
of bacteria growth. Soiled carcases should be sprayed as quickly a
possible after dressing. A well designed hand spray with a fan shape
jet should be used at about 30·5 cm (12 inches) from the carcas
surface. The impact temperature will be about 60°C (140°F)

water is delivered at not less than 88°C (190°F). A water pressure of 7 kg/sq. cm (100 lb per sq. in.) at a volume of about 8½ litres (2 gallons) per minute is necessary. The combination of warm carcase and hot water assists the surface drying of the carcase.

Chlorine of concentration 10 to 20 parts per million added to the spraying water increases the bactericidal effect of spraying without tainting the meat.

The main source of soil matter on the hands of the operatives is the hides and fleeces of the stock they dress. It is to be hoped that before long animals will be delivered to the slaughterhouse free of dung and dirt. To assist in keeping carcases as free as possible from soil and contamination during dressing there should be means for men to wash hands frequently. Stainless steel vessels without sink plugs should be placed strategically within the slaughter and dressing areas. Ideally they should take the form of protective funnels, they only need to prevent splash as the hands should be washed under running water. If it is not possible to pipe the water directly into a drain the outlet pipe should be taken to within two inches of the floor. The water should be piped at a temperature of approximately 43·5°C (110°F) to a single water tap on the unit which may not be operated by hand, owing to the possibility of cross contamination. The valve may be foot, knee or photo-electric cell operated. Knee operated is probably the best. A plentiful supply of liquid soap and paper towels should be ready to hand. To prevent blocking of drains and general untidiness a basket for dirty towels should be attached. The units must be placed so that any operator is able to move to and use them without inconvenience or they will not be used. The hand sprays which are placed strategically in high through-put lines for quickly removing dirt from carcases are often used as a subsidiary means of hand washing. Provided no carcase contamination results, this should not be discouraged. However, use of the main facilities, complete with soap and towels, should be encouraged.

In any meat plant there should be a means for properly *cleaning* cutting tools of any description. In simple terms that is the basic need

237

promoted by meat veterinarians and hygienists. Generally the term "implement steriliser" is used and many types are on the market. They are available as knife, cleaver and saw sterilisers. They usually take the form of stainless steel boxes in which boiling water is added so that the implements may be immersed. In practical terms many saws and particularly the now frequently used circular saws cannot be fully immersed in water. The following criteria of needs and facilities are submitted to comply with hygiene requirements in practical and labour-saving terms.

Water sterilising boxes are placed in positions where every operator who uses a knife of any description has immediate access from the station at which he is working. They should be of pot type construction, designed to hold six knives. The knife holders should be cut from temperature-stable plastic. These should have handles so that the complete holder may be immersed to briefly soak knife handles and withdrawn to stops to allow the handles to cool but the knife blade to continue soaking. They may be located for use by three operators for maximum utilisation, the essential motive being that one knife is in the unit while the other is in use. Each operator has two knives in use on any operation on which he is engaged. Cleaver units should be based on this same principle.

Units for immersing certain reciprocating saws are available. For obvious reasons they need accept only one saw.

For circular saws and other equipment which cannot be totally immersed, a stainless steel sink with draining board should be provided, together with detergent and suitable brushes so that saw blades and other equipment may be scrubbed clean.

All vessels should have a piped overflow enabling a slow but continuous overflow of water. Water for all the units should be centrally stored and heated so that it arrives at the unit at 82°C (180°F).

Hygienic Practices and Meat Inspection Facilities

The Meat Inspection Regulations of 1963 and 1966 established

238

standards of inspection governing the suitability of meat for human consumption, together with the power for local authorities to limit the number of hours of slaughtering. The wholesale industry moves steadily towards levels of throughput from a smaller total of plants. A pattern of slaughtering emerges where each man carries out a limited range of operations on separate carcases. The complete slaughter and dressing of single carcases by one man, carrying with it pride in individual skill, is now rare. To gain full advantage of good plant design and layout it is necessary for management to concentrate on improvement of facilities which encourage meat plant personnel to achieve better hygiene practices. Collective effort is required to instil an awareness of the need for them. Elementary bacteriology should be included in training programmes and advantage taken of courses which highlight meat hygiene.

Thorough meat inspection must be accepted as a routine necessity in plant activity and efforts made to improve the environment and facilities so that inspectors achieve it. The following notes highlight means for encouraging better hygiene practices and better meat inspection facilities:—

Provide good locker accommodation and allocate it individually. Construct showers that work effectively and which can be used in comparative privacy. Supply clean overalls and headgear daily and a well-defined storage bin for dirty ones. Inspect these and other amenities such as canteens frequently and establish a disciplined acceptance that they shall be kept clean and tidy. Establish a routine of entry into the premises directly to the locker room. Encourage work people to arrive and leave presentably dressed. Get them to shower and change back into this clothing after working and generally raise their status.

LOCKER ACCOMMODATION

Keep to a minimum any movement of personnel between the dirty and clean areas. The lairage staff should not have ready access to the slaughterhalls. The minimum transfer of dung from the lairage should be maintained; that which is brought in by stock is unavoidable but should be removed as quickly as possible. Blood

PERSONNEL MOVEMENT

troughs should be constructed so that personnel do not need to step into areas of heavy blood concentration. That which is deposited in other ways should be cleaned up quickly.

PROFITABILITY AND HYGIENE

Any blood, fat, trimming and sawdust which can be "squegee-ed" and lifted dry from the floor improves profitability as well as raising hygiene. Every gramme which is hosed down a drain raises effluent problems and costs; included in the by-products truck it raises this income. Floors should be washed down frequently but indiscriminate hosing not only washes unnecessary scraps into the effluent but is likely to cause splash-back on to clean carcase meat and contaminate it.

CONSTRUCTION AND REPAIR WORK

During construction or repair work round off every corner, and also cove the walls to floors generously. The structure is less susceptible to damage and far easier to clean. It encourages plant hygiene. Site overhead tracking far enough away from walls and doorways to reduce damage and contamination. If in doubt, during construction or repair, round off everything dictated by common sense and leave no ledges or flat surfaces which collect dust or upon which any random object may be placed for want of a home. Minimise the use of unwieldy trucks and carriers.

PROVIDING FACILITIES

Regard all movement of personnel as wasteful; therefore provide facilities for hand washing close to working stations. Encourage clean habits and the cleaning and disinfecting of all hand equipment. Provide paper towels and a receptacle for used ones.

CLEANING

Organise standard cleaning-down programmes at set intervals. Allocate management time once a week, to inspect the whole plant to ascertain that the cleaning programme is effective and complete.

WORKING LAYOUT

Provide a working layout which achieves the best possible conditions for proper meat inspection, free of congestion or hindrance.

Bring together in a common area the abdominal mass (stomach and guts etc.), the red offal (heart, liver, lungs etc.) and, in the case of cattle, the heads. They should all be correlated to each other and the carcase, enabling direct and speedy identification. If the plant is to be included on the EEC list, correlation is required relative to the *split* carcase.

All offals should be situated at a comfortable working height. In the case of cattle the head may be suspended from a rail or conveyor. It may be hooked by the lower jaw or held in a frame but it should be so placed as to enable inspectors to examine it thoroughly and incise where necessary. The red offals should be placed similarly, often on the same equipment. The abdominal mass must be placed on a work table identifiable with the carcase but far enough away to save the inspector making irritating contact with it. Generally the mass can be arranged to slide away from the carcase to a floor-level skip or table hoist against the separating room wall. It is lifted to table height and where there is room to hold enough sets to maintain identification.

WORKING HEIGHT

A good working relationship with the inspection staff should be established and maintained. Pleasant accommodation with locker and shower facilities is mandatory for the veterinarians. Inspection staff should be accorded this courtesy. Consult the inspector during equipment and structural alterations. Seek his advice before initiating new procedures.

INSPECTION STAFF

Modern abattoir practice demands the provision of a sanitary area. It should comprise facilities for isolation of suspected stock, a sanitary or emergency slaughter room, detention and condemned rooms. Bearing in mind that samples from detained carcases may need to be examined pathologically, insulation and refrigeration of the detention room should be considered. All isolation, detention and inspection areas should be lighted at least to statutory requirements.

SANITARY AREA

GENERAL

Possibly as a result of its early role as a predominantly meat import-ing country, Britain has not utilised the by-products of the slaughter of food animals to its reasonable potential. Market forces are un-stable due to the importation of a wide range of animal product from abroad. Imported hides and skins, frozen offals, processed edible fats and tallows affect home market prices extensively. Instability resulting from fluctuating imports discourage home investment in plant to fully utilise our own animal by-products particularly those suitable for edible and pharmaceutical use.

Our wholesale industry has for so long been based on fragmented units of low throughput, concentrating almost entirely on convert-ing stock in the simplest manner. Carcase meat, offals and those by-products requiring the least processing dominate our marketing. This may be the result of high levels of employment. Management may not be prepared to do its sums properly or the public may be unreceptive to the form that other materials take. The fact remains that "bone-in" carcase meat represents only a part of the edible protein total available.

Our meat plants are becoming larger single units and it is to be hoped that this will encourage the wider use of materials. As fresh meat becomes relatively more expensive the remaining edible pro-teins may be utilised to improve conversion of stock.

Hides and Skins

Hides and skins are sold but owing to wide price fluctuations are seldom treated properly at the meat plant, seldom are they even primarily cooled. This may be the fault of our marketing system derived from fragmentation.

Blood

Blood is almost never processed into edible plasma, seldom properly stored for pet food utilisation but frequently disposed of into the drains where it raises acute effluent pollution. As pet food the whole blood from one bovine animal will sell at about 18p, the only equipment required being clean storage tanks and means of collect-ng and pumping to these. Dried blood plasma is imported at about £32p per kg (60p per lb) and nearly 0·68 kg (1½ lb) is obtainable from a bovine animal. Equipment costs are relatively higher to separate and store edible plasma.

The greater part of the by-products of slaughter are processed by established companies who originated as collectors. In this manner they give service to the industry by removing materials which quickly deteriorate. Early collection reduces the problems of fly and blowfly's infestation. When the industry was made up of many more small slaughterhouses than now and all meat was sold from small retailers there were many traders who collected bones and waste fat. Some only bought the materials and sold them again to the processors.

Bones and offals are rendered. The fact that materials are not processed for varying periods of time usually results in the production of a moderate general industrial tallow, together with a meat and bone meal. The blending of the meal for animal feeds and fertilisers is carried out by large processor/merchants.

Bones and Offals

Many animal glands are used in the production of pharmaceutical extracts from which medicines for humans are manufactured. The pharmaceutical industry would utilise almost all the glands other than the small lymphatics, from all slaughtered animals. Unfortunately the labour required to remove and preserve some of them is too high for economy. As the numbers of animals killed in any one plant increase, the removal and preservation of a wider range of glands becomes economically viable. Many of those which produce the larger quantities of hormones or enzymes can be removed either at the offal trimming stage or during separation of the abdominal mass. Our meat plants are attaining a higher degree of sophistication as well as concentration owing to national economic forces. Progressive management will direct its attention to the items of relatively high return.

Glands

The viscera has traditionally been dealt with as a means of gaining

243

TABLE 5

Approximate Yields from a Commercial Steer, 454 kg (1000 lb) at Liveweight
(All weights taken at slaughter) (Source: MLC)

Liveweight	454 kg (1000 lb)		
	%	Weight	
		kg	lb
Carcase weight with kidney knob			
and channel fat	55·0	249·47	550·0
Intestinal fat	2·4	10·88	24·0
Caul fat	3·0	13·60	30·0
Intestines empty	2·1	9·52	21·0
Stomachs empty	2·5	11·33	25·0
Gut fill	14·8	67·12	148·0
Heart, lungs and trachea	1·9	8·61	19·0
Liver and gall bladder	1·4	6·35	14·0
Pancreas	0·1	0·45	1·0
Spleen	0·15	0·68	1·5
Hide*	7·0	31·75	70·0
Feet*	1·9	8·61	19·0
Head (inc. tongue)	2·9	13·15	29·0
Blood	3·5	15·87	35·0
Skirt	0·3	1·36	3·0
Tail	0·25	1·13	2·5
Waste	0·8	3·62	8·0
Totals	100·0	453·5	1000·0

* Includes adhering dung, mud and water

TABLE 5(a)

Approximate Yields from the Carcase of a Commercial Steer
of Average Fatness and of 454 kg (1000 lb) Liveweight
(Source: MLC)

	%	Weight	
		kg	lb
Kidney	0·2	0·907	2·0
Kidney knob and channel fat	2·0	9·071	20·0
Carcase, exc. KKCF	52·8	239·496	528·0
Total carcase	55·0	249·474	550·0

continued on facing page

TABLE 5(a) — *continued*

Carcase (exc. KKCF) breakdown	%	Weight	
		kg	lb
External fat	8·9	21·318	47·0
Seam fat	13·6	32·205	71·0
Cod/udder fat	1·4	3·628	8·0
Lean	60·0	143·789	317·0
Bone	14·7	34·926	77·0
Waste	1·4	3·628	8·0
Totals	100·0	239·494	528·0

TABLE 6

Approximate Yields from a Commercial Lamb 45·4 kg (100 lb) Liveweight
(all weights taken at slaughter)

Liveweight	% of liveweight	45·4 kg (100 lb) Weight	
		kg	lb
Carcase weight	50·0	22·679	50·0
Intestinal fat	1·3	0·587	1·3
Caul fat	1·7	0·770	1·7
Intestines empty	3·3	1·495	3·3
Stomachs empty	2·9	1·315	2·9
Gut fill	13·0	5·896	13·0
Heart, lungs, trachea and oesophagus	1·4	0·635	1·4
Liver and gall bladder	1·5	0·680	1·5
Pancreas	0·1	0·045	0·1
Spleen	0·1	0·045	0·1
Fleece and pelt★	11·7	5·307	11·7
Feet★	2·0	0·907	2·0
Head, inc. tongue	4·7	2·131	4·7
Blood	4·5	2·041	4·5
Skirt	0·5	0·226	0·5
Waste (inc. tail)	1·3	0·587	1·3
Total	100·0	45·335	100·0

★ Allow for time of year, adhering dung, mud and water

TABLE 6(a)

Breakdown and Dissection Yields of Commercial Lamb Carcase
45·5 kg (100 lb) Liveweight

Carcase weight	% of deadweight	23 kg (50 lb) Weight	
		kg	lb
Kidney knob and channel fat	3·6	0·815	1·8
Kidneys	0·6	0·136	0·3
External fat	12·6	2·857	6·3
Seam fat	11·4	2·585	5·7
Lean	55·4	12·564	27·7
Bone	15·6	3·537	7·8
Waste	0·8	0·181	0·4
Totals	100·0	22·675	50·0

TABLE 7

Approximate Yields from a Commercial Pig 91 kg (200 lb) Liveweight
(all weights taken at slaughter)

Liveweight	% of liveweight	91 kg (200 lb) Weight	
		kg	lb
Carcase weight with kidney, fat, head and feet	75·0	68·039	150·0
Intestinal fat	1·0	0·907	2·0
Caul fat	0·2	0·181	0·4
Intestines empty	2·7	2·449	5·4
Stomach empty	0·65	0·589	1·3
Gut fill	10·0	9·071	20·0
Heart, lungs, trachea and oesophagus	1·4	1·270	2·8
Liver and gall bladder	1·9	1·723	3·8
Pancreas	0·1	0·090	0·2
Spleen	0·1	0·090	0·2
Blood	4·5	4·082	9·0
Skirt	0·5	0·453	1·0
Hair scrapings and hooves	1·0	0·907	2·0
Bladder	0·05	0·045	0·1
Waste	0·9	0·815	1·8
Totals	100·0	90·712	200·0

TABLE 7(a)

Breakdown and Dissection Yield of Commercial Pig Carcase
91 kg (200 lb) Liveweight

Carcase weight	%	68 kg (150 lb) Weight	
		kg	lb
Head (inc. bones)	8·3	5·669	12·5
Feet (inc. bones)	3·0	2·040	4·5
Kidneys	0·4	0·272	0·6
Flare fat	2·4	1·632	3·6
External fat	21·5	14·605	32·2
Seam fat	5·2	3·537	7·8
Lean	47·0	31·977	70·5
Bone	8·1	5·532	12·2
Waste (inc. skin)	4·1	2·766	6·1
Totals	100·0	67·030	150·0

the best available "by-products price". Tables 5–7a, p. 244–247, set out various yields of carcase and viscera. They include glands but the more notable facts are not out of place here. Medicine needs every pancreas available, yet many factories do not consider the possibility of saving them. At least one pound of pancreas glands is available from two bovine animals or six to ten pigs. The gland is easily removed and stored. The extract is of course insulin, the means of keeping diabetics alive. The pancreas yield is higher than most glands; for example, the pituitary which is located in the base of the brain is utilised for several glandular extracts associated with growth stimulation and treatment for human thyroid malfunction. One hundred and eight-odd beasts or eighteen hundred pigs only produce one pound of this gland.

Intestines

Many plants process intestines to the extent of stripping and storing in preservative. The final production may be sausage casing of varying sizes or ligatures for medical use. Usually the mucus is washed away. Heparin, an extract of this mucus is necessary in the treatment of blood disorders.

Many of the glands and biological materials can be removed or set aside during the otherwise normal separation or disposal at the meat plant. An intelligent and perhaps more practical approach is required from the floor or direct slaughter management to find out how to identify, remove and preserve the components. Gut room staff hitherto chosen more for tolerance and endurance than skill will need to be taught routines and encouraged to develop skills. The old "Gut Room" conception can be improved immensely by better layout and design. Instead of a dirty conglomeration, an area conceived to attain a hygienic separation of viscera progressively disposing of manural content and preserving clean edible and pharmaceutical materials may emerge.

Glands when identified should be removed as soon as possible from the viscera whole. Fat, veins and ligaments should be trimmed off immediately. They must not be immersed in water as much of the valuable extract is soluble, neither should any gland have direct contact with ice. Soil or blood may be rinsed off quickly by clean cold water. They should be kept cool at all times and it is best to place them in separation trays, individually identifiable and frozen.

The present marketing structure is such that a few specialised companies organise the collection of glands and offals suitable for pharmaceutical production. These supply many of the principal manufacturers of drugs and extracts with collected materials.

Refrigeration Design Considerations

The scientific approach to meat refrigeration is covered elsewhere in this book. This part, concerned with slaughterhouse and meat plant basic principles, techniques and practices, requires some practical reference to the necessities of chilling and freezing. There is considerable research and development of meat refrigeration proceeding all the time. Research entails the investigation and detailed consideration of many aspects. This very detailed consideration introduces conflicting opinions as to the effects of temperature, cooling rates, air velocities and humidities. Meanwhile, in spite of many imperfections it is necessary to apply fundamentals, go ahead and refrigerate. British slaughtering legislation may soon fix statutory obligation to refrigerate freshly slaughtered meat. The EEC Directive specifically requires the carcase to be reduced to 7°C (45°F) as soon as possible after slaughter. Reduction of growth of harmful bacteria is the primary aim of refrigeration. There are many issues which the well informed may care to bear in mind or speculate about. These include reduction of weight loss or the dangers of toughening by quick temperature reduction ("cold shortening"). If one has the means to bring about this phenomenon it is a comparatively simple matter to slow the rate of cooling. The most serious problem in a hitherto fragmented industry is basic inadequacy of meat refrigeration capacity. Therefore no apology is made for adding the following it is hoped, helpful principles.

Refrigerate any carcase as soon as possible after slaughter. Try to dry off excessive spray moisture while the carcase awaits weighing and classification. Use the following times as a basis for reducing the *deep muscles* to 6°C to 10°C (42°F to 45°F):—

TEMPERATURE

Beef	:	28–36 hours
Pigs	:	12–16 hours
Lambs	:	30 hours

Remember that the temperature of the meat is the criterion, not necessarily the room temperature. Carcase temperature is measured by inserting a probe thermometer into the muscle.

REFRIGERATION EQUIPMENT

Only purchase refrigeration equipment to a detailed specification. This should state the types of carcase to be chilled, the specific reduction of that carcase temperature and a fairly accurate estimate of weight loss under certain humidity conditions, combined with controlled air flow and velocity. Compare specification and bear in mind cost difference in relation to other constructional or preparatory work for a particular installation. Ascertain that the entire specification is fulfilled after installation. Be sure that all equipment to be installed inside a chill room is fully protected against corrosion for its useful life. Evaporators, trunking and steel runways are constantly subject to saturation.

FUNCTIONAL NEEDS

Give full consideration to the functional needs *before* designing a chill-room building. Decide the carcase holding capacity first. Establish if the overhead tracking shall be low level for "smalls" and beef quarters or high level for beef sides and smalls on carriers. Consult and decide the refrigeration equipment needs. Evaporator cooler units and air flow trunking need space. Floor standing units require additional area and overhead units and trunking need extra height over the rails. Overhead rails may be suspended directly from ceiling inserts or from galvanised steel. The former need latitude for rail adjustment, the latter are more flexible but the steelwork is subject at times to condensation and drip.

ENSURING GOOD AIRFLOW

Leave sufficient room between carcases for the equipment to be effective. Between carcase surfaces there should be at least 38 mm (1½ inches) to ensure good airflow. Beef rails should be at least 0·9 m (2′ 6″) apart, not allowing for inspection alleyways. "Smalls" rails should be at least 0·47 m (1′ 6″). Where it is necessary to allow walking access between carcases and a wall the first rail should be about 0·84 m (2′ 9″) from the wall face. Give individual protection to those wall surfaces at the loading and unloading positions of the chill rooms. Where carcases are moved quickly through bends and curves they swing violently and damage almost any surface. Heavy gauge non–corrodable steel is the only sure protection.

ROOM SIZE

Size the rooms to suit the throughputs and time/temperature ratio of cooling. Large high throughput multi-species works are likely to employ low level and high level railed rooms. It should be possible to size the rooms for approximately two hours' killing time. Other plants may have to arrange to size the rooms to receive four hours of kill. The basic principle is to close down the chill room as soon as possible to enable the equipment to fulfil its intention. If possible the killing programme and transport arrangements should also be geared to refrigeration times. Working capital involvement

250

justifies consideration of a 24-hour cycle to enable meat to be despatched rather than held unnecessarily.

Always consider the use of relatively small chill rooms holding specific numbers of carcases, rather than large unmanageable rooms. Keep cold meat cold and do not put hot carcases into a chill room containing cooled meat.

Boning and Cutting at the Meat Plant

Import restrictions on carcase meat after the 1968/9 Foot and Mouth Disease epidemic stimulated centralised boning and cutting. Acceptance by retailers of this form of imported meat preparation encouraged many wholesalers to purchase cutting and packing equipment and follow suit. For several reasons some of the early attempts were discouraging. The wholesalers experienced fluctuating profitability. Where accountancy systems are lacking it is difficult accurately to cost additional labour and plant capitalisation and the trim losses. Retailers accustomed to utilising skilled cutters in a dual role as salesmen resisted film packed boned home killed meat. In addition, lack of experience in vaccum packing brought about some spoilage due to faulty sealing of the packs.

This, though an incomplete picture, briefly indicates why the adoption of such a logical progression for slaughtering and chilling is only now gaining momentum. Multiple retailers controlling many outlets are increasing the opportunities for such preparation and the wholesale trade in prepared beef and pork cuts increases steadily.

Many existing meat plants originally designed exclusively for slaughtering and despatch of carcase meat cannot easily be converted for large throughput of cuts. Some companies, and in particular groups who have several slaughtering units, construct a separate cutting plant to which carcase meat is transported. New, composite plants require careful planning to include them or leave the options open for their addition. Fig. 11 is a block flow diagram indicating the logical and practical flow of full-scale boning and cutting, either as an individual unit or a composite slaughtering and cutting plant.

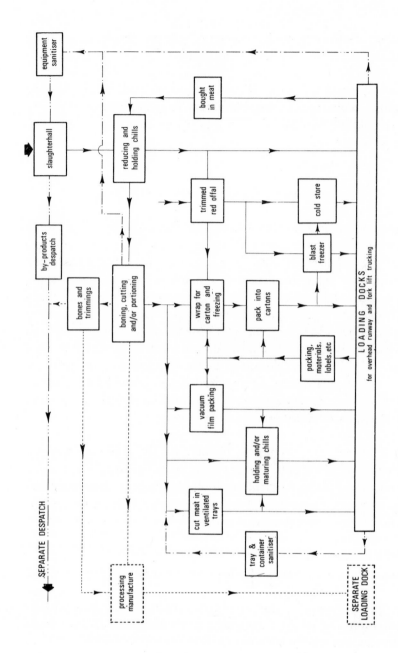

Fig. 11.   *Block flow diagram of boning and cutting.*

Carcases are chilled down after slaughter in the carcase reducing and holding chillers (7°C (45°F) in the deep muscles is the minimum requirement for EEC inclusion). Carcases arriving from other sources should also be held at this temperature for cutting and a separate chiller for this purpose may be provided if advantageous to the particular business. Meat sold in carcase form is despatched directly to the loading dock. That for boning and cutting proceeds to an air-conditioned room which should be maintained by moderate air velocities at 10°C (50°F).

Boning and cutting now commonly take place by conveyorised methods where fairly high throughputs justify it. Primals may be removed while the carcase is on overhead rails. These are then placed on an impervious conveyor transporting them to work stations placed strategically. Each operator bones and trims specifically. Bones are placed on a transporter conveyor, as are fat and sinewy tissues within stated specifications. Materials for inedible sale should be despatched via a widely separated exit.

It should be noted that to comply with hygiene requirements and in particular those of EEC there must be knife and implement sterilisers available for all operative staff, together with hand-washing facilities as set out in MLC Bulletin No. 13.

Vacuum-packing may take place in the boning room but EEC requirements so far are that packing into cartons must be carried out in a separate room. Wrapping and packaging materials should be stored so that they are convenient to all areas for use and do not interfere with carcase and cut meat loading and despatching. All meat trays and containers and hooks and ironmongery used for conveying and transporting carcase meat and cuts should be thoroughly cleansed before re-use.

Figure 12 indicates diagrammatically a manner of achieving the flow principles set out in the block flow of Fig. 11. As the future handling of meat is likely to proceed increasingly in the direction of packing and despatch in ventilated trays or caged pallets, note that the two

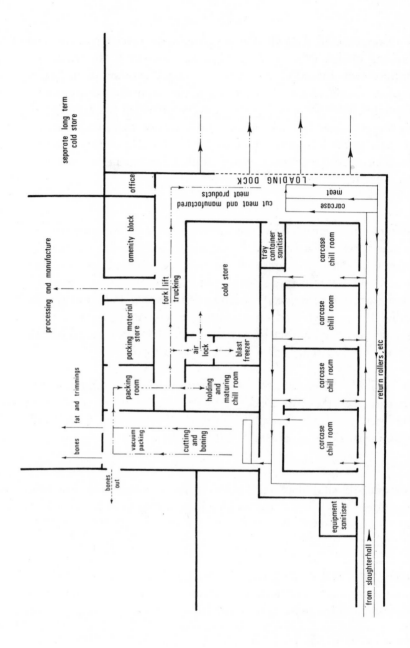

Fig. 12. *Diagrammatic interpretation of Fig.* 11 *to show means of attaining minimal movement with varying transport methods.*

254

areas of despatch, that is, carcase meat from overhead runways and cartons, trays etc. by means of pallet type trucks, are separated. Palletisation utilises height and where there is active trucking overhead runways are not compatible. For security, offices for despatch supervision should be placed so that all activity in this direction is observed. For many reasons, including hygiene and supervision, separate amenities in the form of lockers, showers and washing facilities should be provided for cutting staff.

Construction and Equipment Finishes
An Aid to Efficiency and Hygiene

The required properties and materials for floors and walls are described in detail in MLC Bulletin No. 1, available from the Meat and Livestock Commission. This section reviews the problems throughout the slaughtering and holding areas where a little forethought can eliminate dirt traps and collected bacteria. Good finishing simplifies maintenance and raises hygiene levels. Pages 207–212 offer recommendations for the specific lairage area which is less complex.

The choice of finishing or topping materials is logically reduced to three alternatives:— granolithic, polymer or ceramic tiles. There are other materials offered and by specialist advice and selection they are used. However, the three are most common and it is fair to say that in the order stated they offer a direct life/cost ratio.

THE HEAVY DUTY CARBORUNDUM IMPREGNATED CERAMIC TILE, laid to specification and properly maintained, has the longest life and carries most of the arduous properties demanded, that is, resistance to wear, impact damage, slipperyness from fats and proteins and resistance to chemical and bio-chemical attack from slaughtering, processing and heavy duty cleaning. Special coving tiles are available for jointing to walls. Tiles are by far the most expensive.

MODERN POLYMERS, usually polyester or epoxy or patent variations, have many of the qualities demanded with some reservations

CONSTRUCTION FINISHES
Floors

concerning resistance to impact. It now appears that the life expectancy is less than ceramics but possibly more than granolithic concrete. One of its notable assets is that it can be in operation immediately after laying. The latter is an excellent material in relation to its cost which is the lowest.

Assuming that the base floor construction is good, the life and quality of all three materials depends on the care and skill of the floor contractor. If the tiles are laid unevenly and to improper levels, the coves badly matched or the wide, impervious grouts badly mixed and placed, if chipped tiles are laid or voids left under them, the floors are a failure. Polymer floors lend themselves to the making of accurate coves and falls but skill is required in the preparation and laying. They rely on specific chemical action between accurate quantities of base material and hardener; therefore they must be accurately measured and mixed, and skilfully trowelled on to a well-prepared base and coved to defined edges. Foreign matter, oils and dirt will destroy their properties and bonding.

GRANOLITHIC FLOORS need the same care and preparation as the more expensive materials. Cementatious products are subject to attack from blood and protein materials but if a well-prepared grano floor is laid properly and with graded aggregates, and kept moist for the prescribed (seemingly unnecessary) length of time, so that the surface not only sets but hardens, it will give long life and fulfil most of the stated requirements. It should be tamped to provide a non-slip finish but the aggregate should not be exposed. Coving to the wall is simple and should be to a stepped face, making a flush joint with the wall material. A radius of 50 mm (approx 2″) is easy to construct. It simplifies cleaning. Joints should be impervious.

Walls

Heavy duty ceramic tiles or cement rendering are the choices open. The rendering can be covered with a large variety of materials. The heavy duty tile is extruded from clay which is obtained from only a few sources; they are extruded in heavy section, cut and glazed at very high temperatures. They are frost proofed, impervious and fairly resistant to impact. Specialist contractors must apply them

256

and all the precautions stated for floor tiles should be watched. The wide, impervious grout must be properly placed and smoothed.

Cement rendering may be used without any additional finish. It should however be placed by a craftsman. Washed fine sharp sand is used and final steel trowelling should produce a polished finish. For appearance and ease of cleaning, polymer paints can be applied to the trowelled rendering. A high-build epoxy material mixed and applied exactly to specification after the cement has dried slowly and to exact maximum moisture levels gives a durable finish. Sheets of materials such as aluminium, stainless steel and impervious plastic may be attached. In the slaughter and holding areas metal sheets are used and are bonded to level cement. In chill rooms it is usual to attach them to battens secured through the insulation to the walls. Extruded aluminium sections will in future replace the old timber battens.

The choice of sheeting material must depend on economic factors but the order of durability combined with hygiene and appearance is stainless steel, heavy gauge anodised aluminium and galvanised sheet. Sheets are manufactured in a number of standard sizes, therefore there must be many joints between sheets and at corners and floor and ceiling butments. At sheet joints an impervious seal must be made.

When choosing the type of material and the form of construction which either floors or walls are to take, there are numerous details which, if properly thought out, will improve life, maintenance and above all hygiene. The coving of floors to walls is now common. The main cause of failure is lack of foresight in producing a flush, impervious bond. Male corners or wall ends should be rounded off; this reduces damage. Right-angled wall joints or wall to ceiling joins should be radiussed; they clean more easily. Where possible piers or concrete stanchions should be built into the outer wall skin. If they must project within the wall lines they should be radiussed and rounded. Cills in wall apertures should be sloped at 30° from the vertical and all ledges avoided, including the tops of dividing walls which do not reach to roof level. Study the foundation details

257

offered for fixing machinery. A few simple modifications will eliminate corners and ledges which are difficult to clean.

Where sheeting is applied there are available aluminium extrusions and galvanised formers to form coves and corner radii and angled or rounded wall edges. Clean portal frames eliminate the ledges which are unavoidable with fabricated roof trusses. Doors and door frames are available in non-corrodable metals and factory built reinforced plastics.

EQUIPMENT FINISHES

These are no less important relative to efficiency, maintenance and keeping up standards of hygiene than the materials of construction. Flaking paint and dirty equipment oppose the ideals of cleanliness desired for producing clean meat. Maintenance costs are reduced if proper consideration is given to equipment investment.

High level Equipment

Overhead tracking systems vary considerably in the UK for historical reasons; there may eventually be standardisation. Some aluminium alloys are beginning to be adopted but generally they take the form of steel sections of varying form. These and the support steelwork grids should be galvanised after fabrication and the fixing bolts plated. All rollers and trolleys should be rustproofed and the hooks made in stainless steel. Rollers and trolleys need a cleaning and maintenance routine. After cleansing, the wheel bearings need light lubrication.

Machinery & Equipment

Choose carefully that which is protected by galvanising and plating. Well-fitting cover plates give access to simplify cleaning. In general additional cost for stainless steel incorporated practically into machinery is a good, long-term investment. Steam, air, refrigerant and hydraulic piping and fittings are only too frequently placed badly or exposed to collect dirt and animal waste.

Trucks & Carriers

Select well designed, tubular stainless steel or alloy components without ledges. Working platforms on single columns with removable non-slip flooring are easier to clean than spidery scaffolding construction, and working time reduced soon justifies the extra cost. Loose equipment should be cleaned in a separate sanitising area; this should be placed so that there is, after the despatch of meat and offals, a practical flow from the despatch back to the working areas.

Stainless steel is the best material, particularly for those tables needed for inspection and handling of offals and similar materials. Large galvanised steel tanks can be fabricated in a manner so that they can be kept clean with minimum labour. Welded structural sections used in some tank construction collect dirt and contamination. Remember that all areas need cleaning if they are not totally enclosed; therefore where dirt and protein can collect the equipment should be accessible or easily moved for cleaning.

Tanks, Tables & Floor-mounted Conveyors

Too frequently the numerous valves and sets of lagged and unlagged piping serving this type of equipment are placed so that they are unnecessarily unsightly and collect dirt. Refrigeration engineers can use foresight in this respect. The evaporation cooling units, trunking and supports must be corrosion resistant. Chill room and air conditioned cutting-room maintenance should be reduced.

Refrigeration Equipment

Principles of Costing

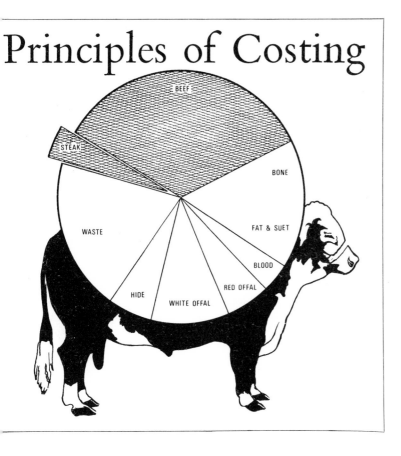

Costing and pricing are two of the most vital factors in a trading enterprise—especially the meat trade. They are fundamental factors in achieving those elements that are the prime essentials for success—competitiveness and profitability.

The definitions of costing and pricing are often confused by members of the trade because they are closely related to each other: indeed inter-related.

Costing is the process of calculating, and sometimes apportioning, the various costs that arise in running a business; e.g. performing a

COSTING AND PRICING

service, manufacturing a product or preparing and presenting a product for sale. The end result of a costing is to ascertain the costs of the enterprise or some particular part of that enterprise.

Pricing is the process of establishing a selling price or a set of selling prices after considering the various costs involved to sell the product i.e. overheads, labour, buying-in price, packaging, profit etc.

Costing is therefore an exercise which must generally precede pricing because the various costs of running a business must be covered by the prices charged for the goods sold. There is the exception when the market price limits may be in complete disregard of the costs of producing or handling and selling a particular product.

Example:
> If an importer imports frozen pigs' liver at a total cost of 75p per kg (34p per lb) (cost of liver from packers plus freight, insurance, landing charges, etc.) and the wholesale price of pigs' liver is only 66p per kg (30p per lb) then the importer will have to sell at a loss.

Meat wholesalers often find that the market price of meat has a greater influence over their wholesale prices than the overheads of their businesses.

The retailer has more control over his retail prices but he must also have regard for the competitiveness of his prices. He cannot charge prices way above those of his competitors otherwise he will soon find his customers drifting away.

COSTINGS

Costings for the meat trade can be divided into two categories:—

> Costing related to overheads such as wages, rent, rates, transport, packaging materials, etc.

> Costings related to the input material, e.g. live animals, carcases, offals, etc.

262

The importance of costing the overheads of a business is firstly to establish the cost of overheads as a foundation to the calculation of prices; that is to establish how much gross profit is needed and secondly to check these costs and to consider whether there is scope for economies. It is quite possible that costings can reveal that alternative methods of running an enterprise would result in monetary savings and improved profitability, for example centralised cutting, closing down a depot, use of different packaging techniques, etc. Costings have often revealed that some sections of an enterprise are running at a loss. This has, in many cases, caused managements to take drastic action by closing down or eliminating these uneconomic sections of their businesses. These actions are very often the sort of actions that many members of a company or business dislike taking, especially if that part of the business to be discontinued has been a significant part of a business for many years, but extreme situations, from an economic point of view, require drastic action.

Profitability

In carrying out costings to compare two systems or to evaluate the effect of a section of an enterprise on the whole enterprise, the criteria for judging must always be overall profitability, not only in the short term but also taking into consideration long term profitability.

Services

In the trading situation, some services given, or the supply of certain commodities, may not in themselves be profitable but there is often the argument that these goods or services help to attract and hold customers. The acid test would be if an assessment could be made of how much trade, therefore how much profit, would be lost by discontinuing these lines or services, compared with how much saving would result. If the saving would exceed the loss of profit, then the decision must favour the discontinuance of the lines or services.

Although on commercial grounds profitability must be the ultimate criteria for assessment and for taking decisions in respect of all aspects of a business, there are exceptions such as in the case of a small business proprietor whose business has become a way of life.

263

It is quite possible in such a case that making alterations in th
business to improve its overall profitability would virtually destro
the way of life enjoyed by the proprietor.

Overheads

When costings of overheads are complete, the results may b
adequate as a basis for pricing or they may satisfy the managemen
that the enterprise is profitable: nevertheless the question should b
asked "Is there room for even greater efficiency and even highe
profits?" In many sections of a business there is always room fo
some improvement of efficiency but to ascertain that room fo
improvement exists, requires some method of measuring efficienc
and also any improvement in efficiency that may be achieved. On
method of determining efficiency is to compare the level of eacl
overhead of a business with the level of that overhead in simila
businesses—if such figures are available. For comparison, the level o
each overhead can be expressed as a percentage of the output o
turnover; these percentages would constitute management ratios.

As there is almost no published data relating to percentages of over
heads for businesses in the meat trade, the only form of compariso
open to most persons in the meat trade is:—

Comparison of these overhead percentages between the separat
shops, depots, abattoirs, etc., owned by one company.

or Comparison (with an aim at improvement) of these overhea
percentages between one trading period and another.

This latter method of comparison can be used even by the one sho
business. The proprietor can aim at progressively lowering thes
overhead percentages by improved efficiency and thereby increasin
the profitability of the enterprise.

Labour Costs

Labour costs of wages constitute a major cost for any section of th
industry and there can be considerable variation between simila
types of business in the level of labour costs. Apart from inpu
materials, labour costs are an area where maximum economies ma

264

e effected through improved efficiency. To get costs to an absolute minimum often entails finding the optimum level between labour costs and either the value of meat or the required standard of the end product. The removal of the last scraps of meat from bones etc. would cost more in labour than the extra meat yielded would be worth, whereas very careless boning can result in a great deal of expensive meat being wasted. A pedantic concern for the smallest detail in the preparation of meat could lead to an excessive labour cost on the final product; on the other hand, badly prepared meat may have such a poor sales appeal that loss of profit would result from a lower turnover. Management must strive to find the right, that is the most profitable, balance between the extremes.

Efficiency

As output or turnover of a business increases, so invariably does the opportunity for greater efficiency of labour resulting in the reduction of the cost of labour per unit of throughput or turnover. Table 1 shows a comparison of wages expressed as a percentage of sales for two shops having different levels of sales.

TABLE 1

	Average weekly sales (4 week period)	Wages % of sales	Rent and rates % of sales
Shop "A"	£1025	12·4	2·0
Shop "B"	£2450	10·1	1·7

The wages figures above are exclusive of National Insurance.

This comparison demonstrates the considerable economy on labour costs that may be effected with higher level of sales per shop.

PRICING

This fact, coupled with the similar relationship of fixed overheads and sales, opens up the prospect of cutting back gross profit margins to reduce prices and thereby attracting a higher volume of sales, which in turn allows a sufficient saving on wages and fixed overheads to compensate for the reduced level of gross profit margin. Indeed, the saving could be such that the ultimate outcome could be a higher net profit, as demonstrated in Table 2.

TABLE 2

Effect of Gross Profit Reduction and Sales Increase on Overheads

	A Before sales increase		B After sales increase	
	£	%	£	%
Sales	1200		2400	
Fixed overheads	47	3·9	55	2·3
Semi-variable overheads	62	5·2	92	3·8
Variable overheads	40	3·3	79	3·3
Wages, Nat. Ins., etc.	170	14·2	266	11·1
Net profit	23	1·9	120	5·0
Gross profit	342	28·5	612	25·5

"A" above shows the overheads in total amount and expressed as percentage of sales for a shop with sales of £1200 per week. Assuming this shop has a very high sales potential, "B" shows the possible effect of lowering the gross profit margin which would enable an aggressive pricing policy to be pursued. The sales figure is raised to £2400 and because of the savings on overheads, especially the savings resulting from increased labour efficiency, the net profit is increased in both total cash and percentage of sales. The overriding restricting element in such an exercise is the sales potential of the shop concerned. If the shop is in a premier position, having a much higher sales potential than the existing sales level, then this should give adequate scope for the success of such a change in pricing policy. On the other hand, in a shop where the potential is limited to only a moderate increase in sales, the end result may be that the small increase in sales would not compensate for the reduction in gross profit margin; therefore the overall net profit would be less.

A premier position for a shop would be a good position in the main shopping centre of a town. It may be argued that the competition is stiffest in such a position but also the potential for high sales level exists in these positions and it gives the opportunity for the manage with the flair and ability at meat merchandising, salesmanship and skilful pricing, to get the highest rewards for his talents. Where competition is lacking it generally follows that potential for sales increases is also lacking.

When management consider the prospect of reducing prices and therefore reducing the gross profit margin as a change in pricing policy in an effort to increase sales, it is useful to calculate what percentage increase of sales is necessary to produce the same total gross profit at the reduced gross profit margin. This can be calculated as follows:—

$$\frac{\% \text{ gross profit margin reduction} \times 100}{\text{Proposed } \% \text{ gross profit margin}} = \begin{array}{l} \% \text{ sales increase} \\ \text{necessary to produce} \\ \text{same total gross profit} \end{array}$$

Example :

If the proposal for consideration by the management is for an overall reduction of 3% on prices which would mean a reduction of the gross profit margin from 30% to 27%, the required increase of sales necessary to produce the same gross profit would be:—

$$\frac{3\% \times 100}{27\%} = 11 \cdot 1\%$$

This means, therefore, that it would need an increase of sales in the order of 11·1% to produce the same gross profit so that if the sales could be increased well above 11·1% due to the adoption of a more aggressive pricing policy, then it could result in greater profitability. Knowing the required percentage sales increase to cover the reduction in gross profit margin is an important piece of information for the management to have when considering such a proposition. The management should be able to assess whether the shops concerned have a potential for increasing sales by more than 11·1% if they adopted the change in pricing policy.

TURNOVER

The relationship between labour efficiency and turnover or through-put also applies to wholesale depots and abattoirs so that saving labour costs is often best effected by increasing turnover or through-put. This will again rely on the potential for extra sales in the area and to some extent, the potential for procuring the extra meat or animals, of the required quality, to produce the extra sales or throughput. There may exist wage agreements which tie wages to

throughput so that there would be no saving on labour costs by th
management but there would be savings on fixed overheads. Mos
wage agreements include a bonus related to throughput; therefor
while the employees benefit by higher wages due to increase
throughput, there would also be an appreciable labour cost saving
for the management.

An important feature of labour efficiency related to turnover o
throughput is that while the improvement of efficiency can be quit
considerable with rises at the lower end of the scale, as the turnove
or throughput reaches a higher level, less improvement of efficienc
and labour cost savings can be expected from further turnover o
throughput increases (see Table 3). Each type of business establish
ment will, in most cases, have an optimum level of turnover o
throughput for overall efficiency and for increases above thi
optimum there will be no improvement in efficiency: indeed, ther
may even be a loss of efficiency with the throughputs abov
optimum.

This Wages % to Sales graph shows how the percentage of wages t
sales can be improved quite appreciably as the sales of a shop rise u

TABLE 3

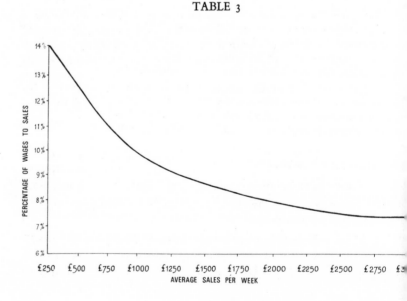

about £1500 per week but with sales increases above this figure the improvement becomes more marginal.

Fixed overheads, rents, rates etc. are largely outside the control of the management of a business and the level of these will be fixed in relation to property values in the area and not in relation to the economics of the business. As previously outlined, by increasing the level of sales, the percentage of fixed overheads to sales will fall (Tables 1 and 2).

To assess whether the rent and rates of a new shop are reasonable, the sales potential of that shop must also be considered. Providing the rent and rates figure is not more than $2\frac{1}{2}$–3% of sales, it should work out economical for the business. This would mean, therefore, that a shop with a high rent and rates figure, coupled with a high sales figure, would be more profitable than a shop where both these figures are low. For example, a shop in a premier shopping centre position with rent and rates of £50 per week would be a good proposition if the sales potential of the shop was £2000 per week (rent and rates = $2\frac{1}{2}$% of sales).

There are many other expenses that must be included in costing the running of a business, some of which may be overlooked in the daily running of a business but must at some time be met. All equipment has eventually to be replaced, even such seemingly permanent structures as cold stores. All equipment should be costed against the business at its replacement value, divided by the estimated number of years of serviceable life the equipment will have. In the estimation of the serviceable life of equipment, consideration must be given not just to its functional life but also to the equipment being superseded by more sophisticated and more efficient types of that equipment, for example new types of scales, better types of display equipment, more efficient types of processing equipment, etc. These replacement values will have to be adjusted upwards constantly to keep pace with the ever increasing rise in equipment prices. It is advisable to ensure that these replacement costs are within $1\frac{1}{2}$–2% of sales of a retail business.

269

Table 4 is a list of the various costs of a retail business showing an average percentage of each cost compared with sales for shops with different levels of sales.

TABLE 4

Retail Butchery Overheads

	"A"	"B"
	%	%
Rent and rates	2·8	2·4
Decorations and premises maintenance	0·7	0·6
Interest on capital	3·0	3·0
Replacement cost of equipment	1·8	1·8
Maintenance of equipment	0·6	0·5
Wages and National Insurance	14·2	11·1
Holiday relief	0·6	0·5
Electricity and gas	1·3	1·0
Paper, string, tickets, etc.	0·8	0·8
Carriage	2·5	2·5
Trade insurance	0·4	0·3
Book-keeping and administration expenses	0·5	0·4
Sundries (laundry, overalls, tools, cleaning materials, etc.)	0·5	0·4
	29·7	25·3

The percentages are of each overhead as a percentage of sales. The "A" list represents a shop with average sales of £600 per week whereas the "B" figures are for a shop with average sales of £150 per week.

Input Materials

Costing the input materials, carcases, live animals, etc. is often an exercise very similar to the process of pricing: the essential difference is the purpose. Costing carcases, live to deadweight etc. is to check the yields and profitability of carcases. It is an activity that should be carried out more frequently than is generally the case in the meat industry.

These costings can be used as a control factor related to the following:—

As a check on the evaluation (profitability) of carcases supplied on contract to specification or of carcases within a particular grade or classification, or a check on the efficiency of carcase selection by the company's buyers. In this way it can be determined whether the standards in respect of carcase values are being maintained.

Regular yield percentages or carcase evaluations for comparison with the standard percentage figures or charts used in pricing meat.

As a check on the methods of cutting and standards of preparation for retail sale, packaging or processing.

Carcase or other input costings should be carried out quite apart from a regular check on overall gross profit. The advantage of individual carcase costings is to confirm that an adequate level of profit is being maintained on beef carcases, lamb carcases and pork carcases, as well as on all other commodities. This will help to show up the type of situation where a good profit on one commodity is masking the fact that another commodity is being sold at a loss.

When costing carcases for the purpose of the first two above, it is necessary to have a tightly controlled method of cutting and preparation, otherwise variations in cutting, trimming etc. could result in a greater degree of difference in the evaluation than would result from the difference in carcase. If the same butcher does all the sample costings, this will help to reduce the variations due to cutting, but where the purpose is to obtain percentage yields as a basis for pricing, variation in the standard of cutting by other butchers in a company may result in some shops achieving less than the required profit on the basis of the company pricing method. One way which may help to reduce this problem is for one butcher to carry out costings to set standard evaluations or percentages for a company, and all other butchers to carry out costings from time to time; the results of the latter could be compared with the standard evaluation or percentages.

Trimming the various cuts from a carcase should be carried out to

an extent that warrants no further trimming before sale, otherwis
the carcase costed could show a higher profit margin than it woul
produce in normal sales.

The most common method of costing carcases for evaluation is t
use the current selling prices of a company and to calculate th
profit margin on the carcase that is to be costed. On the basis of th
standard percentages used for pricing carcases, the profit expecte
from current selling prices at a particular cost price could be fixe
and then compared with the carcase being costed.

Example:

Test costing for English lamb carcase
Total weight 16·59 kg (36½ lb)
Wholesale cost price 150p per kg (68p per lb)

	Weight		Price		Total value
	kg	lb oz	kg	lb	£
Legs	4·16	9 2	296p	135p	12·32
Shoulders	3·76	8 4	230p	105p	8·66
Chump chops	0·54	1 3	318p	145p	1·72
Loin chops	1·87	4 2	275p	125p	5·16
Chump ends	0·45	1 0	144p	65p	0·65
Best end chops	1·65	3 10	208p	95p	3·44
Middle neck	0·68	1 8	152p	69p	1·04
Scrag	0·79	1 12	144p	65p	1·14
Breasts	1·47	3 4	86p	39p	1·27
Kidneys	0·11	0 4	127p	55p	0·14
Trimmings	1·11	2 7			
	16·59	36 8			35·54
			Less cost price		24·82
			Gross Profit		£10·72

$$\text{Gross Profit Margin} = \frac{10\cdot72\times100}{35\cdot54} = \frac{1072}{35\cdot54} = 30\cdot16\%$$

If the profit margin expected at these prices on the basis of a com
pany's standard percentage yields is less than 30·16%, then the carcas
tested would have a good evaluation, but if the expected profit wer
more than 30·16%, the carcase tested would be a poor one. The sam
type of costing can be used to check the evaluation of carcases agains
pricing computation tables, such as those contained in "What Pric

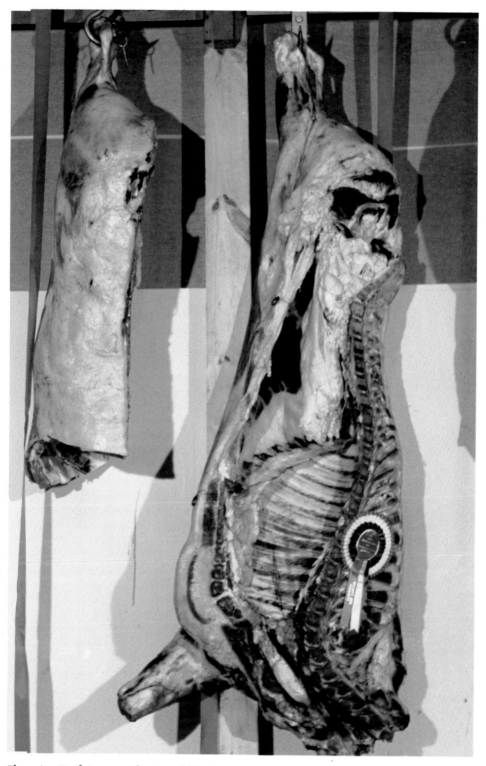

Champion Beef Carcase, side view of hind

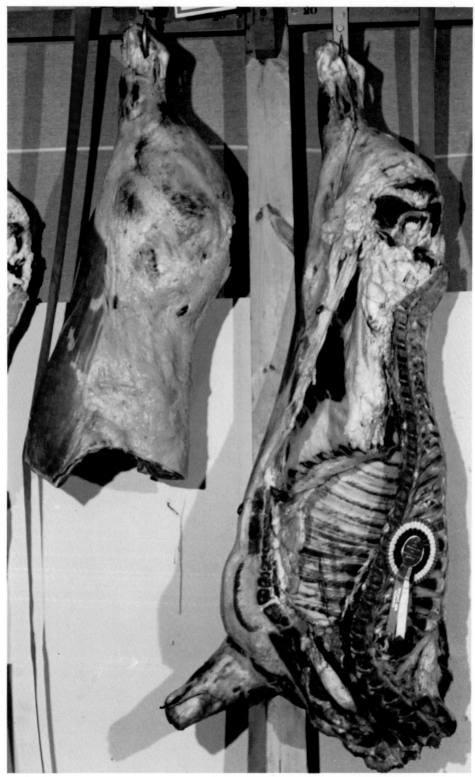

Champion Beef Carcase, back view of hind

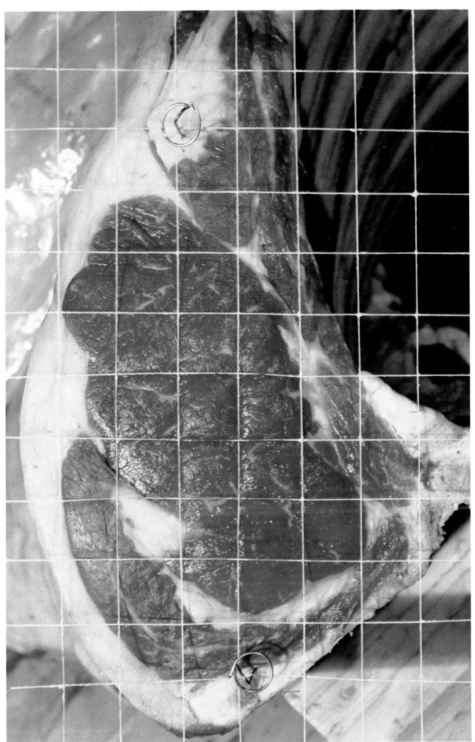

Champion Beef Carcase, eye muscle

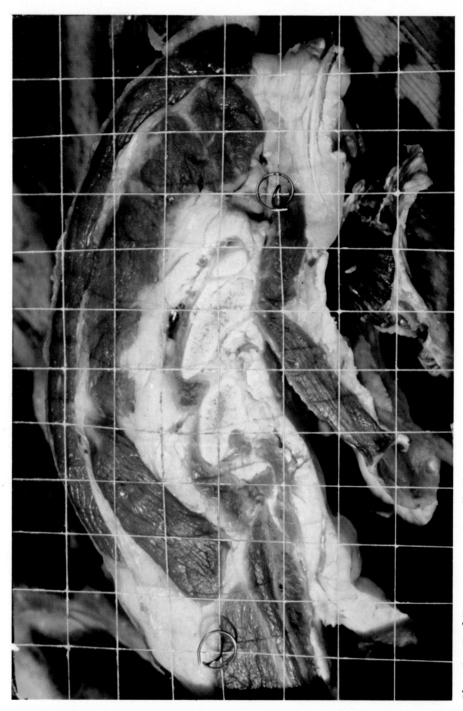

Champion Beef Carcase, cross section of brisket

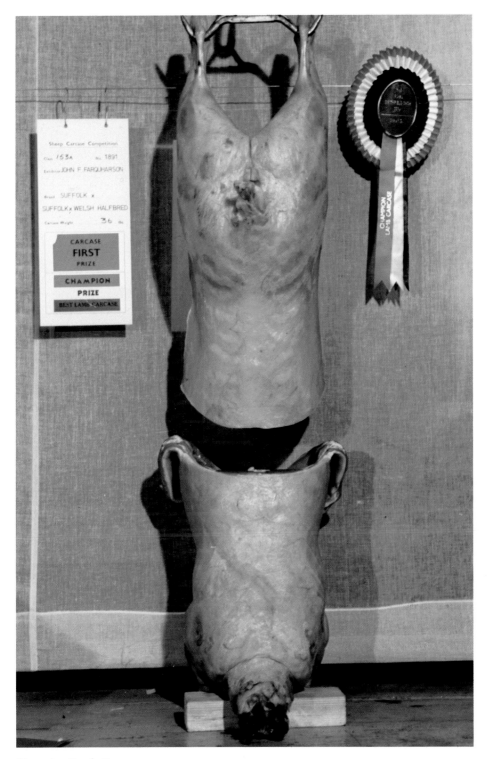

Champion Lamb Carcase

Courtesy Royal Smithfield Club

Champion Lamb Carcase, cross section

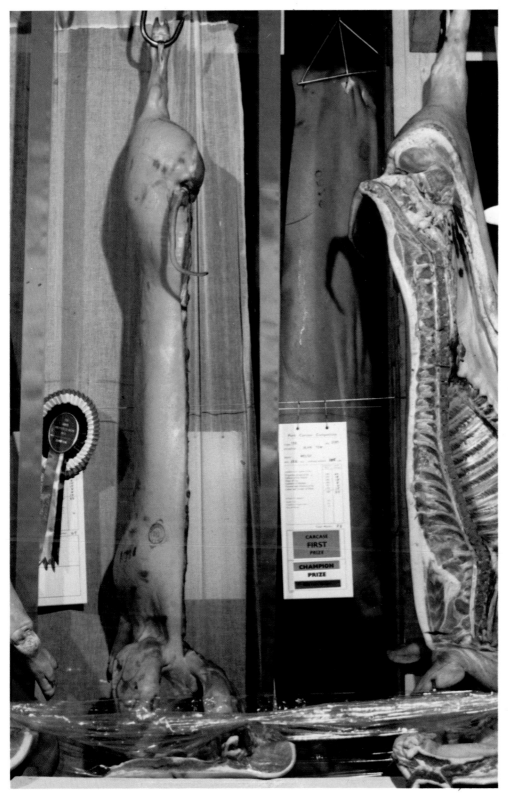

Champion Pork Carcase, side view

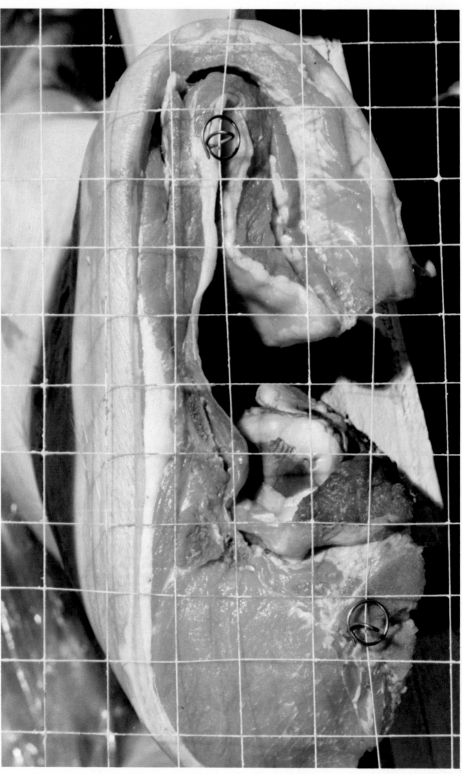

Champion Pork Carcase, cross section eye muscle and belly

Meat". In the same way as above, if the carcase costed shows a higher profit margin than the profit margin shown by the computation tables, it would indicate a good carcase, whilst a lower profit margin than that shown by the tables would indicate a poor carcase.

When cutting carcases to produce percentage figures which can be used as a basis for pricing, the same care must be taken to ensure that all necessary trimming for final sale takes place before weighing each cut.

To calculate the percentages of each cut the formula is:—

$$\frac{\text{Weight of cut} \times 100}{\text{Weight of carcase}}$$

Example:
Using metric weights shown in previous example—English Lamb 16·59 kg. The percentages of the cuts are as follows:—

Legs	$\frac{4\cdot175 \text{ kg} \times 100}{16\cdot59 \text{ kg}}$	$= \frac{415\cdot7}{16\cdot59}$	$= 25\%$
Shoulders	$\frac{3\cdot765 \text{ kg} \times 100}{16\cdot59 \text{ kg}}$	$= \frac{375\cdot6}{16\cdot59}$	$= 22\cdot6\%$
Chump Chops	$\frac{0\cdot540 \text{ kg} \times 100}{16\cdot59 \text{ kg}}$	$= \frac{54\cdot0}{16\cdot59}$	$= 3\cdot3\%$
Loin Chops	$\frac{1\cdot873 \text{ kg} \times 100}{16\cdot59 \text{ kg}}$	$= \frac{187\cdot3}{16\cdot59}$	$= 11\cdot3\%$
Chump Ends	$\frac{0\cdot454 \text{ kg} \times 100}{16\cdot59 \text{ kg}}$	$= \frac{45\cdot4}{16\cdot59}$	$= 2\cdot7\%$
Best End Chops	$\frac{1\cdot646 \text{ kg} \times 100}{16\cdot59 \text{ kg}}$	$= \frac{164\cdot6}{16\cdot59}$	$= 9\cdot9\%$
Middle Neck	$\frac{0\cdot681 \text{ kg} \times 100}{16\cdot59 \text{ kg}}$	$= \frac{68\cdot1}{16\cdot59}$	$= 4\cdot1\%$
Scrag	$\frac{0\cdot794 \text{ kg} \times 100}{16\cdot59 \text{ kg}}$	$= \frac{79\cdot4}{16\cdot59}$	$= 4\cdot8\%$
Breast	$\frac{1\cdot468 \text{ kg} \times 100}{16\cdot59 \text{ kg}}$	$= \frac{146\cdot8}{16\cdot59}$	$= 8\cdot9\%$
Kidneys	$\frac{0\cdot113 \text{ kg} \times 100}{16\cdot59 \text{ kg}}$	$= \frac{11\cdot3}{16\cdot59}$	$= 0\cdot7\%$

The easiest method of all for working out percentages is to use the "Meat Traders Calculator" or an electronic calculator.

K

Costing the trimmed saleable boneless meat from a primal cut or carcase is as follows:—

$$\frac{\text{Cost per kg (lb)} \times \text{Total weight of primal}}{\text{Weight of boneless trimmed meat}} = \text{Cost price per kg (lb) of boneless trimmed meat.}$$

Example:

Steakmeat cost 128p per kg (58p per lb) wholesale
Total weight of steakmeat 13·6 kg (30 lb)
Weight of boneless trimmed meat 9·5 kg (21 lb)

$$\text{Cost price of boneless trimmed meat} = \frac{128 \times 13 \cdot 6}{9 \cdot 5} = \frac{1741}{9 \cdot 5} = 183\text{p per kg}$$
$$= \frac{58 \times 30}{21} = \frac{1740}{21} = 82 \cdot 9\text{p per lb)}$$

Many meat manufacturing and processing concerns will use standard or average yield percentages of various primal cuts and carcases for calculating the cost price of boneless meat. These standard percentages are obtained by taking the gross weights and the trimmed yield weights of a number of primal cuts or carcases and calculating the average percentage yield. The larger the number of any cut or carcase used in this exercise the more realistic will be the standard percentage figure obtained. This standard percentage figure can be adjusted down to allow for subsequent evaporative or drip losses occurring before the final use or sale of the boneless trimmed meat. Standard yield percentages can be calculated as follows:—

$$\frac{(\text{Weight of boneless trimmed meat} - \text{Evaporative or drip loss}) \times 100}{\text{Total weight of primal cuts or carcase}} = \begin{array}{l}\text{Average or} \\ \text{Standard} \\ \text{yield \%}\end{array}$$

Example:

40 Steakmeats—total weight 544·8 kg (1200 lb)
Yield of boneless trimmed meat—390·4 kg (860 lb)
Evaporative and drip loss to use—8·2 kg (18 lb)

$$\text{Percentage Yield} = \frac{(390 \cdot 4 - 8 \cdot 2) \times 100}{544 \cdot 8} = \frac{382 \cdot 2 \times 100}{544 \cdot 8} = \frac{38220}{544 \cdot 8} = 70 \cdot 2\%$$
$$= \frac{((860 - 18) \times 100}{1200} = \frac{842 \times 100}{1200} = \frac{842}{12} = 70 \cdot 2\%)$$

In most cases companies would decide to use 70% as their standard yield percentage for steakmeats on the basis of the above test.

To calculate the cost price of boneless trimmed meat using standard percentages, the calculation is as follows:—

$$\frac{\text{Cost per kg (lb) of primal cut or carcase} \times 100}{\text{Standard yield percentage}} = \text{Cost price of boneless trimmed meat}$$

Example:

Steakmeats cost 128p per kg (58p per lb) wholesale
Standard yield percentage—70%

Cost of boneless trimmed meat $= \dfrac{128 \times 100}{70} = \dfrac{1280}{7} = 183\text{p per kg}$

$\left(\dfrac{58 \times 100}{70} = \dfrac{580}{7} = 82 \cdot 9\text{p per lb}\right)$

Costing cattle liveweight to deadweight is often carried out excluding the value of the offals ("sinking the offals"): the value of the offals can be offset against the cost of slaughter. The cost price of the carcase per kg can be calculated from the liveweight price per kg:

$$\text{Liveweight price in pence per kg} \times \frac{\text{Liveweight kg}}{\text{Carcase weight kg}} = \text{Carcase price per kg}$$

Example:

Beast cost   80p per kg
Liveweight   485 kg
Carcase Weight   272 kg

Carcase price $= \dfrac{80 \times 485}{272} = 142 \cdot 7\text{p per kg}$

To find the price per lb:—

    price per kg × 0·454   = price per lb
or   price per kg ÷ 2·2   = price per lb
Price per lb = 82·1 × 0·454 = 37·3p per lb
The same formula is used for pigs:

Example:

Pig cost	65p per kg
Liveweight	68 kg
Carcase weight	49 kg

$$\text{Carcase price} = \frac{65 \times 68}{49} = 90 \cdot 2\text{p per kg}$$

Price per lb = $81 \cdot 9 \times 0 \cdot 454 = 37 \cdot 2$p per lb

Calculating the killing out or dressing percentage can be a useful guide to the economic value of different types of cattle (e.g. plain cows, fat cows, intensive steers, semi-intensive steers, semi-intensive heifers, etc.), and can be calculated as follows:—

$$\frac{\text{Weight of carcase} \times 100}{\text{Weight of live animal}} = \frac{\text{K.O. \%}}{\text{or Dressing \%}}$$

Example:

Animal weighs 485 kg (1064 lb) liveweight
Carcase weight 272 kg (596 lb)

$$\frac{\text{K.O.\%}}{\text{or Dressing}} = \frac{272 \times 100}{485} = \frac{27200}{485} = 56\%$$

Calculating the killing out or dressing percentage for pigs is the same as for cattle.

Example:

Pig weighs 68 kg (150 lb) liveweight
Carcase weight—49 kg (108 lb)

$$\frac{\text{K.O.\%}}{\text{or Dressing \%}} = \frac{49 \times 100}{68} = 72\%$$

$$\left(\frac{108 \times 100}{150} = 72\% \right)$$

For a quick calculation of the carcase price per kg (per lb), standard or average killing out percentages are sometimes used. Standard killing out percentages can be relied upon where the animals being slaughtered are of a uniform type, weight and age, that is animals all of the same breed, cross or hybrid, reared under the same system of management. This would mostly apply to broiler chickens, turkeys and hybrid pigs on a strictly controlled feeding regime although intensively reared cattle will generally fall into a fairly narrow range of killing out percentages.

Calculating the carcase cost price of cattle, "sinking the offals" and using a standard killing out percentage is as follows:—

Example:
Liveweight price of cattle 86p per kg
   Standard K.O. percentage 56%

$$\text{Carcase price} = \frac{86 \times 100}{56} = \frac{8600}{56} = 153\cdot6\text{p per kg}$$

Where the carcase price per lb is required, the formula is as follows:—

$$\frac{\text{Liveweight price per kg.}}{2\cdot2} \times \frac{100}{56} = \text{Carcase price per lb}$$

Example:

$$\frac{86}{2\cdot2} \times \frac{100}{56} = \frac{8600}{123\cdot2} = 69\cdot8\text{p per lb}$$

The same formula is used for pigs.

Example:
   Liveweight price of pigs 68p. per kg
   Standard K.O. percentage 72%

$$\frac{68 \times 100}{72} = \frac{6800}{72} = 94\cdot4\text{p per kg}$$

   For carcase price per lb.

$$\frac{68}{2\cdot2} \times \frac{100}{72} = \frac{6800}{158\cdot4} = 42\cdot9\text{p per lb})$$

The first step in pricing meat or meat products for retail or wholesale sale, is to determine the gross profit margin. This must take into account the overhead costs (dealt with at the beginning of the chapter), coupled with the potential sales capacity of the business and the sales objective or trading objective; also the desired net profit margin.

SELLING PRICE

Once the gross profit margin has been decided, various methods can be used to calculate the selling price or sets of prices of cuts, carcases, offals, etc., but calculations must not become the hard and fast dictate of prices. Calculations will produce some essential information which will assist management in making pricing decisions but management must also consider the market situation, sales opportunity, probable consumer resistance to certain prices, and advantageous prices such as 39p instead of 40p; 49p instead of 50p. As stated above, calculations should aid and not dictate pricing decisions. They are nevertheless imperative in making the right decisions.

The usual practice in the meat trade is to calculate the gross profit margin in relation to selling price rather than cost price. A very good reason for working on the basis of profit on sales is that most efficiency ratios and other measurements will be related to sales, e.g. wages to sales ratio or percentage, rent and rates to sales percentage, packaging materials to sales percentage. Another reason is that sales figures are more quickly and more readily available than would be the cost of meat actually sold.

Selling Price and Cost

To calculate the selling price from the cost price to include the required gross profit on sales, the following method can be used:—

$$\frac{\text{Cost Price (C.P.)} \times 100}{(100 - \text{G.P. margin})} = \text{Selling Price (S.P.)}$$

Example:

Required profit margin—28%
Cost price of oxtails 74·96p per kg (34p per lb)

$$\text{S.P.} = \frac{74 \cdot 96 \times 100}{100 - 28} = \frac{7496}{72} = 104 \cdot 11\text{p per kg}$$

$$\frac{(34 \times 100}{100 - 28} = \frac{3400}{72} = 47 \cdot 22\text{p per lb)}$$

This can be rounded off to a selling price of 104p per kg (47p per lb)

When it is necessary to calculate the selling price of the boneless trimmed meat from primal cuts or carcases, using standard yield percentages, the following method can be used:—

$$\frac{\text{C.P. of whole}}{\text{primal cut}} \times \frac{100}{\text{Standard}\ \%\ \text{Yield}} \times \frac{100}{(100 - \text{G.P. margin})} = \text{S.P.}$$

Example:

Steakmeat C.P. 127p per kg (58p per lb)
Standard yield percentage of steakmeat—70%
Required gross profit margin—28%

$$\text{S.P. of boneless trimmed meat} = 127 \times \frac{100}{70} \times \frac{100}{(100 - 28)} =$$

$$127 \times \frac{100}{70} \times \frac{100}{72} = 127 \times \frac{125}{63} = \frac{15875}{63} = 251 \cdot 98\text{p per kg}$$

This would be rounded off to a selling price of 252p per kg

$$= (58 \times \frac{100}{70} \times \frac{100}{(100 - 28)} =$$

$$58 \times \frac{100}{70} \times \frac{100}{72} = 58 \times \frac{125}{63} = \frac{7250}{63} = 115 \cdot 1\text{p per lb})$$

This would generally be rounded off to a selling price of 115p per lb.

Where standard yield percentages are used and it is always the same gross profit margin used in pricing calculations, a pricing constant can be calculated for each primal cut. These constants can be useful for quick calculations of the selling price of boneless trimmed meat.

Yield/profit constants can be calculated as follows:—

$$\frac{100}{\text{Standard}\ \%\ \text{yield}} \times \frac{100}{100 - \text{G.P. margin}} = \begin{array}{l}\text{Yield/Profit Constant} \\ \text{(this figure should be} \\ \text{taken to two decimal} \\ \text{places)}\end{array}$$

279

Example 1:

Steakmeat; Standard yield percentage—70%
Required gross profit margin—30%

$$\text{Yield/profit Constant} = \frac{100}{70} \times \frac{100}{100-30} = \frac{100}{70} \times \frac{100}{70} = \frac{100}{49} = 2\cdot04$$

If steakmeats were costing 132·2p per kg (60p per lb) wholesale, the selling price of the boneless trimmed meat to produce 30% profit on sales would be:—

132·2 × 2·04 = 269·68p per kg (rounded off to 270p per kg)
(60 × 2·04 = 122·4p per lb (which would generally be rounded off to a selling price of 122p per lb))

Example 2:

Briskets: Standard yield percentages—56%
Required gross profit margin—28%

$$\text{Yield/profit Constant} = \frac{100}{56} \times \frac{100}{100-28} = \frac{100}{56} \times \frac{100}{72} = \frac{10000}{4032} = 2\cdot48$$

If briskets were costing 63·8p per kg (29p per lb) wholesale, the selling price of boneless trimmed meat to produce 28% profit on sales would be:—

63·8 × 2·48 = 158·22p per kg (rounded off at 158p per kg)
(29 × 2·48 = 71·9p per lb (rounded off at 72p per lb)

Selling Price of Cuts

Calculation of the selling prices of cuts from carcases is much more complex. There is not only trimming and boning loss but also the different values *and* the different proportions of each cut obtained from a carcase to be taken into account. There are so many variables involved that it is a common practice, and indeed a necessary one, to obtain standard percentages which incorporate the saleable meat yield of each cut and the proportion of each cut to the carcase (see p. 273). These standard cut percentages for carcases should be obtained by careful controlled cutting tests on a large number of carcases if they are to be reliable. Use of standard cut percentages will reduce the variables and a further reduction in these variables can be achieved by making the selling price differentials between each cut constant. If this method is adopted, tables can be compiled covering a wide range of wholesale price changes so that a complete list of selling prices are shown against each wholesale price:— these prices to include the required profit margin. Although this method is the quickest for pricing carcases, it gives little scope for manipulation of the various prices for maximum effectiveness in competitive pricing: there is less assistance in making the best and most effective decisions in support of an aggressive trading policy.

Using standard percentages but manipulating the price of each cut as best suits the market situation, consumer sales potential etc. can be carried out as follows:—

The first step is to calculate the overall selling price of the carcase thus:—

$$\text{C.P. of carcase} \times \frac{100}{100 - \text{required profit margin}} = \text{Overall S.P. of carcase}$$

The second step is to estimate the selling prices of each cut and then calculate the proportional value of each cut relative to the whole carcase, thus:—

$$\text{Estimated selling price} \times \frac{\text{Cut }\%}{100} = \text{Proportional value}$$

Next, total these proportional values and check this total against the required overall selling price.

If the total of the proportional values is higher than the required overall selling price, the prices of the individual cuts can be adjusted lower. If the total of the proportional values is lower than the required overall selling price, the prices of the individual cuts can be adjusted higher.

Example 1:

English lamb. C.P. 74p per kg (33·6p per lb)
Required gross profit margin—30%
Use percentages obtained in Example on p. 273 as standard percentages

$$\text{Overall selling price} = \frac{74 \times 100}{100 - 30} = \frac{74 \times 100}{70} = \frac{740}{7} = 105\text{p per kg}$$

$$\left(= \frac{33\cdot6 \times 100}{100 - 30} = \frac{33\cdot6 \times 100}{70} = \frac{336}{7} = 48\text{p per lb}\right)$$

	Estimated selling price per lb			Proportional value	Adjusted selling price per lb			Proportional value
Legs	56p $\times$	$\dfrac{25}{100}$	=	14	59p $\times$	$\dfrac{25}{100}$	=	14·75
Shoulders	45p $\times$	$\dfrac{22·6}{100}$	=	10·17	47p $\times$	$\dfrac{22·6}{100}$	=	10·62
Chump Chops	73p $\times$	$\dfrac{3·3}{100}$	=	2·41	79p $\times$	$\dfrac{3·3}{100}$	=	2·61
Loin Chops	69p $\times$	$\dfrac{11·3}{100}$	=	7·8	76p $\times$	$\dfrac{11·3}{100}$	=	8·59
Chump Ends	28p $\times$	$\dfrac{2·7}{100}$	=	0·76	32p $\times$	$\dfrac{2·7}{100}$	=	0·86
Best End Chops	48p $\times$	$\dfrac{9·9}{100}$	=	4·75	58p $\times$	$\dfrac{9·9}{100}$	=	5·74
Middle Neck	29p $\times$	$\dfrac{4·1}{100}$	=	1·19	34p $\times$	$\dfrac{4·1}{100}$	=	1·39
Scrag	27p $\times$	$\dfrac{4·8}{100}$	=	1·3	32p $\times$	$\dfrac{4·8}{100}$	=	1·54
Breast	15p $\times$	$\dfrac{8·6}{100}$	=	1·29	18p $\times$	$\dfrac{8·6}{100}$	=	1·55
Kidneys	55p $\times$	$\dfrac{0·7}{100}$	=	0·39	55p $\times$	$\dfrac{0·7}{100}$	=	0·39
				44·06				48·04

When pricing calculations have been completed and all the other relevant factors (consumer price consciousness, sales opportunity, etc.) have been considered, decisions on prices can be taken. The final step in the implementation of a pricing policy will be to inform customers and potential customers of those prices which are considered attractive.

The whole success of a pricing policy will rely on getting the message over. This can be done by notices on the shop window, writing on the window, notices inside the shop or store, price tickets in the meat on display or a combination of all. Whatever method is used to advertise prices to customers, and most important, potential customers, it is quite clear that the most attractive and aggressive pricing policy will stand or fall by the effectiveness of advertising prices.

Example 2 (metric):

	Estimated Selling price per kg			Proportional Value	Adjusted Selling price per kg			Proportional Value
Legs	123 ×	$\frac{25}{100}$	=	30·75	130 ×	$\frac{25}{100}$	=	32·50
Shoulders	99 ×	$\frac{22·6}{100}$	=	22·37	104 ×	$\frac{22·6}{100}$	=	23·50
Chump Chops	161 ×	$\frac{3·3}{100}$	=	5·31	174 ×	$\frac{3·3}{100}$	=	5·74
Loin Chops	152 ×	$\frac{11·3}{100}$	=	17·18	167 ×	$\frac{11·3}{100}$	=	18·87
Chump Ends	62 ×	$\frac{2·7}{100}$	=	1·67	70 ×	$\frac{2·7}{100}$	=	1·89
Best End Chops	106 ×	$\frac{9·9}{100}$	=	10·49	128 ×	$\frac{9·9}{100}$	=	12·67
Middle Neck	64 ×	$\frac{4·1}{100}$	=	2·62	75 ×	$\frac{4·1}{100}$	=	3·08
Scrag	59 ×	$\frac{4·8}{100}$	=	2·83	70 ×	$\frac{4·8}{100}$	=	3·36
Breast	33 ×	$\frac{8·6}{100}$	=	2·84	39 ×	$\frac{8·6}{100}$	=	3·35
Kidneys	121 ×	$\frac{0·7}{100}$	=	0·85	121 ×	$\frac{0·7}{100}$	=	0·85
				96·91				105·81

Hidden Losses

Whilst detailed costings are an essential part of successful trading, such guide figures will seldom agree with those actually achieved. In fact, a difference of 5% would probably be accepted as fairly good and, in general, the higher the turnover of the business the lower the percentage of hidden losses.

EVAPORATION

The first loss is that due to evaporation when cooling the hot carcase in the slaughterhouse. This will be influenced by the chiller temperature, rate of air circulation, humidity, and the type of carcase. A

283

heavy well-finished side of beef will, under similar storage conditions, lose a lower percentage of weight than one which is lighter and poorly finished. With beef sides, the weight loss over the first 24 hours chilling should not exceed 2% of the initial weight, and under good conditions this may be reduced to around $1\frac{1}{2}$%. A saving of $\frac{1}{2}$% on a 272 kg (600 lb) carcase, costing, say, 70·5p per kg (32p per lb) represents 96p.

The subsequent rate of weight loss should be considerably lower but will be influenced by the method of transport and the storage on receipt. With sides of beef or quarters, held under good chiller conditions at around 0°C–1°C (32°F–34°F) for 72 hours, prior to cutting, the weight loss based on the weight as received, would probably be about $\frac{3}{4}$ to 1%. Where extended maturing periods are employed, particularly with cuts having a large surface area in relation to volume, these figures will be considerably increased. For example, a rump matured for a period of 14 days, can lose 4% to 5% weight, even under good storage conditions. In breaking up sides/ quarters into their primal cuts, the loss should not exceed 1% and with good cutting standards this might be reduced to about 0·6%. Thus it may be found that the side of beef hot, to the weight of the primal cuts, would decrease somewhat in the following order:

	%
Initial hot weight	100
Cold weight (2% loss)	98
Storage, 3 days (1% loss)	97·02
Primal cuts (1% loss)	96·0498

Thus the weight loss from cold weight to primal cuts is approximately 2%.

ERRORS IN CUTTING

It is essential to remember that in the preparation of meat for retail sale, excessive fat, trimmings and bones, will only fetch a very small fraction of the cost price of the carcase, thus every effort must be made to reduce the quantity which is collected as waste. A weekly check of the amount related to the weight of meat handled is a worthwhile exercise.

Where a less popular portion adjoins a more valuable cut, care must be exercised in making the line of demarcation, in order to obtain the best yield without adversely affecting the saleability of the respective portions. The removal of a breast of lamb from the more valuable loin, the plate of beef from the ribs, the belly from a loin of pork and a shoulder from the ribs and neck, are typical examples. The following will indicate the effect of the cutting line when removing the thin flank from the sirloin of beef:—

	Loin %		Flank %
Standard cutting might give (by weight)	75		25
Value ratio	3	:	1
Thus $(75 \times 3) + (25 \times 1) =$	250		
Inaccurate cutting (5% error) might give	70		30
i.e. $(70 \times 3) + (30 \times 1) =$	240		
Loss	10		

$$\frac{10}{250} \times \frac{100}{1} = 4\% \text{—a loss of } £4 \text{ on each } £100 \text{ of sales.}$$

The yield of boneless trimmed meat from a side of beef will be influenced by the amount of suet, channel and fat trim, the amount of bone, and whether the majority of the meat is to be sold in the form of steaks or as roasting joints. The yield of boneless saleable meat from commercial sides of beef will usually average about 67% of the side weight, as cut. Individual samples may range from about 64% to, in exceptional cases, 74%.

The effect of preparing a primal cut for roasting, as compared with steaking, is shown from the following test carried out on twenty top rumps (thick flanks).

Top Rumps	Averages	
	Rolled	Steaks
	%	%
Prepared	77·90	35·90
Stewing meat	11·79	41·81
Fat, bone and loss	10·31	22·29
	100	100

In the case of rumps from fair quality beef, with hindquarters of around 63 kg (140 lb) weight, the bone content will usually be about

285

15%–16% and the rump suet and fat trim, slightly above these figures. It will probably be found that the yield obtainable will fall between the following figures, the higher yield being obtained from a fairly lean Friesian type carcase.

	%
Rump steaks	49 to 52
Fillet steaks	5 to 8 (undercut)
Rump skirt	5 to 6

By way of extreme contrast, a high class hotel concerned exclusively with eating quality, matures Scotch rumps under carefully controlled temperature conditions, for a period of 14 to 21 days. This involves a certain amount of trimming, which includes the first cut of the rump. With this type of operation the following yields are considered as being satisfactory, based upon an average weight of 11 kg (24 lb).

	%
Weight loss in maturing	3·90
Rump and fillet steaks	46·87 (well trimmed)
Skirt etc.	10·94
Suet and fat trim	13·28
Bone and gristles	16·94
Waste and weight loss	8·07
	100%

Where frozen meat is used for mince, due allowance must be made for the losses arising from "drip" and evaporation. If relatively small quantities are minced, the proportion of residual meat remaining in the mincer must also be considered.

Using frozen shins of beef, and mincing fairly large batches (about 27·2 kg (60 lb)), the following yields were obtained.

	%
Drip on partial defrosting	2·15
Gristles and skin	3·70
Mincing and evaporation losses	0·61
Prepared mince	93·54
	100%

The contract trade, supplying schools and similar institutions, is highly competitive and in submitting tenders careful evaluation of the losses incurred in preparation is essential. Many such contracts call for boneless legs and shoulders of lamb and pork, and subject to skilled preparation the figures below will provide a good guide.

IMPORTED LEGS OF LAMB (average 3 kg (6 lb))

	%
Boneless leg	78·5
Bone and waste	20·9
Loss	0·5
	100%

IMPORTED SHOULDERS OF LAMB (average 2 kg (4½ lb))

	%
Boneless shoulder	67·0
Knuckle (whole)	9·8
Bone and waste	14·3
Fat trim	7·9
Loss	1·0
	100%

LEGS OF PORK (average 6 kg (13 lb))

	%
Boneless leg	86·3
Bones etc.	13·1
Loss	0·6
	100%

PORK PICNICS (average 3·5 kg (7¼ lb))

	%
Boneless picnic	83·5
Bones etc.	15·4
Loss	1·1
	100%

VACUUM-PACKED CUTS (BEEF)

The quantity of drip exuding from vacuum-packed primal cuts of beef will be influenced by the condition of the meat, the rate of cooling prior to cutting, the temperature at cutting and the subsequent temperature at which the meat is held. The ideal temperature is about $-1°C$ (30°F) and small cuts will lose a higher percentage of drip than large portions and even different cuts from the same carcase will vary in their proportion of drip.

For example, long cut fillets may lose about $1\frac{1}{2}\%$ when held at a steady temperature of $-1°C$ (30°F) for 3 weeks, but this figure can be almost doubled if the holding temperature is increased to 7°C (45°F). In addition, where cuts are packed in layers, owing to the pressure, those in the lower layer will give a higher percentage of drip loss than those of the upper layer. Of the larger joints, rumps appear to be more prone to drip than most other portions. In general, with the exception of fillets, the percentage of drip will usually range from 0·5% to 1·5% of the initial weight of the primal cut.

Whilst such cuts are usually well prepared, some further trimming may be necessary, but with joints "fatted up" for roasting, the losses arising from drip and trimming may be partly balanced by the amount of fat added in rolling the joint. However, in the case of rump, the rump steaks as prepared for retail sale may represent a little over 80% of the gross weight.

FAT TRIM

The more important factors influencing yield of prepared retail joints include the kidney knob, channel fat and cod/mammary fat. The amount of excess fat etc. can vary from about 12% in the case of good commercial beef from steers and heifers to twice this figure in a wasteful poorly fleshed carcase. In the case of carcases from good quality young bulls the figure for excess fat can be well below the 12% level. Thus any increase above that required from the aspect of eating quality and juiciness will have a direct effect on the yield of saleable meat.

BONE CONTENT

Whereas joints of pork and lamb are usually sold with the bone in, the majority of beef is sold boneless. Consequently variations in

288

the proportion of bone will have a significant effect on the yield of boneless beef. With efficient de-boning the percentage of bone and gristle in sides of beef can vary from about 12% to 20%. (With manufacturing cow beef, the figure will usually exceed 20%.) These figures must however be considered in relation to the degree of finish. An increase in the amount of fat has the effect of reducing the percentage of bone, but could result in a greater loss of fat trimming prior to sale.

Beef and Beef and Veal Cutting

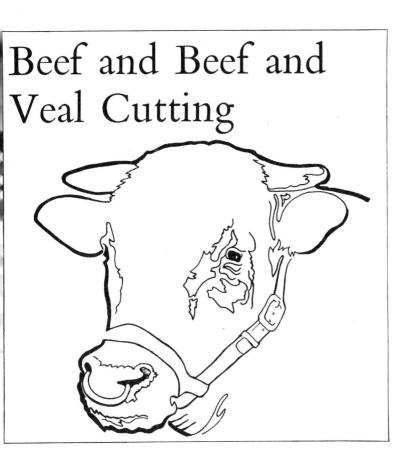

It has been stated that the history of cattle is the history of man, and there is more than a modicum of truth in this somewhat sweeping assertion. It is of interest that the basis of coinage rests upon the value of animals.

Lord Avebury in his work on "Coins and Coinage" wrote that the first silver pieces were made equivalent in value to a sheep, whilst the first gold coin was regarded as a token of exchange for a beast.

The word "pecuniary" is derived from the Latin "pecus"—cattle, and the term "chattels" which might be defined as personal movables, still persists in our legal jargon.

The line of demarcation between the live animal and its flesh, may reflect the original social distinction. For example, Steor, Sceap and Swin are Old English terms, but on reaching the festive board they become Old French, i.e. Boef, Moton and Porc.

Oxen, with sheep and goats form a sub-family of the Bovidae, the chief features being cloven hoofs, lack of upper incisors, a four-chambered stomach and hollow horns.

The bovine sub-family is represented by a number of species, including cattle, bison and buffaloes, and the Wild Ox (Aurochs) the ancestor of our domestic cattle, was widely spread throughout Europe. It survived in its wild state in remote forests, and on preserved estates, until the end of the fifteenth century.

In Great Britain a number of herds of Wild White cattle were enclosed and remained pure over many centuries, and at the present time five such herds remain.

	VERTEBRAE
Cervical vertebrae	7 bones
Dorsal ,,	13 ,, (5 segments, fused)
Lumbar ,, (sacrum)	1 ,, (5 segments, fused)
Coccygeal ,,	Variable, up to 21 bones, usually 2 or 3 are left on carcase

	PELVIC BONES
Ischium	Aitchbone
Ilium	Hip or part of rump

	RIBS
13 pairs (rarely 12 or 14)	8 pairs of true ribs
	5 ,, false ribs

	STERNUM
Breast bone	1 bone (7 segments)
	Xyphoid cartilage at hind end

HIND LIMB	FORE LIMB
Tarsals	Carpals
Tibia–Fibula	Radius–Ulna
Patella (kneecap)	Humerus
Femur	Scapula

Meat Structure

Meat has been broadly defined as the flesh from any beast used for food, but the butcher must also consider the tendons, ligaments and bones, as well as the muscle and fat, in the application of his craft.

The percentage of bone in a beef carcase may be as low as 10% of the total weight in a really fat beast, or over 20% in a poor animal. The former figure would mean that the carcase would require drastic trimming in order to make the meat acceptable to consumer demand.

In the absence of the head, the degree of ossification of the bones will provide an *approximate* indication of age. In the case of calves up to about one month, cartilage, not bone (ossa cordis) is present in the heart, and the shank bones are separated by boiling.

In rapidly maturing animals, the bones become ossified before those of animals with a slower growth rate. The cartilage over the aitch-bone (ischio-pubic symphysis) will provide a good guide, and up to about two years this can be split with a knife, but on ossification will have to be sawn through. Up to this age, the ends of the first four or five dorsal vertebrae are entirely cartilaginous. After this, the bone centres (buttons) appear and are practically completely bone at about the sixth year.

Whilst the fusion of the sacral vertebrae and the sternabrae are also considered as useful guides, the method of splitting the carcase renders them less reliable.

The "cherry red" colour frequently quoted by the trade as an

BONE

293

indication of youthfulness, is in general correct for young animals of about a year, subsequent assessment of the depth of colour becomes uncertain. It is broadly agreed that even with experience, it is difficult to give an accurate appraisal of age from the skeleton, to within about three months. The extremity of the blade bone consists of permanent cartilage, which remains as such throughout the life of the animal.

CONNECTIVE TISSUE

The butcher is mainly concerned with two types of connective tissues. The yellow tissues, those extending along the neck (paxwax) being typical, are extremely tough and contain the protein elastin. In addition to those in the lungs and blood vessels, a relatively small amount is distributed in the muscles, and it is not made tender by cooking.

The white tissues, such as those forming the Achilles tendon of the hind leg, are rich in protein collagen. Whilst in the raw state, the collagen fibres are tough, they can on boiling be turned into gelatin, which is tender. Thus, those joints which contain much white connective tissue are used for boiling or stewing, whilst those poor in collagen fibres are generally roasted or grilled. Collagen is also the chief protein of cartilage, found on the ends of bones.

FAT

Of recent years, consumer resistance to any excess of fat, has led to the production of animals with a relatively light finish. The edible portion of a good beef carcase might contain about 25% of total fat. However, the individual joints may vary from around 10% in the neck meat to almost 40% in fat brisket.

In the living animal, the fat cells consist of minute droplets of oil, and in the fattening process, the cells not only increase in number but also in size, which in combination with the decrease in the proportion of connective tissue will influence the colour. Conversely, if an animal draws on its own body fat, for example during lactation, the pigment will tend to become more concentrated, giving the characteristic yellow fat associated with old cow beef. In

294

some breeds, particularly the Jersey and Guernsey, the yellow colouration of the fat is a breed feature.

Marbling fat is considered as an indication of quality, the fat penetrating between the muscle bundles, frequently following the path of the blood vessels. On cooking, such fat exerts an internal basting effect which gives juiciness and introduces fat soluble flavours.

Histologically there are three types of muscle: (a) nonstriated, mainly associated with the digestive system; (b) cardiac muscle, which has cells of a distinctive shape; and (c) skeletal muscle (striated) or voluntary muscle, which is under the control of the will. It is of course the skeletal muscle with which the butcher is mainly concerned. Muscle is composed of bundles of fibres, held together by connective tissue. Each fibre consists of a minute elongated sac containing the meat proteins and certain salts and acids. The amount of connective tissue (largely collagen) in a muscle, is influenced by the activity of the muscle in the live animal, two extremes being the muscles of the shank and those of the fillet (psoas).

MUSCLE

Muscle colour can vary from almost white in veal to the very dark colour of mature bull beef. Even different muscles from a given animal will vary in colour, depth of colour being associated with age and activity, which are closely linked to the oxygen demand of the particular group of muscles. Muscle colour plays an important part in the initial selection of beef by the housewife, and is therefore a major factor in self-service establishments.

A freshly cut surface of beef is normally a dull purple red colour, but on exposure to the air, atmospheric oxygen is taken up by the myoglobin and the meat becomes brighter and more pink. Thus, when meat is wrapped (not vacuum-packed) it is usual to allow a few minutes for "blooming" in order to present an attractive colour. Depending upon temperature and humidity, further colour changes will take place due to the drying of the surface, the meat becoming progressively darker.

With drying, a surface layer is formed which differs from the underlying tissues, the layer having an increased transparency. This has the effect of increasing the depth of the penetration of light before reflection, giving a greater depth of colour. Dark cutting beef may occur from time to time and will show up unfavourably against the normal bright coloured beef. It would seem that "dark cutters" are more common at the time of the first frosts, and it is thought to be caused by cool conditions combined with fatigue, resulting in depleted glycogen reserves in the muscles. The condition is more likely to arise in nervous, excitable animals, with a lower resistance to stress. Meat from such carcases will have a high ultimate pH, and whilst the meat is usually tender, the higher pH will favour the growth of spoilage organisms.

On cooking, further colour changes occur, and the following will give some indication of the temperatures at which these take place in cooking beef by dry heat:—

Rare—little interior colour change takes place up to about 60°C (140°F).
Medium—pink interior 65°–70°C (150°–160°F).
Well done—grey/brown interior 71°–80°C (165°–175°F).

Types of Beef

Beef is classified into four main types—bull, ox or steer, cow, and heifer. However, of recent years there has been considerable interest in this country in beef from young bulls on account of their leanness, good growth rate and food conversion. Males castrated after the development of mature male characteristics, or those imperfectly castrated, are known as stags. The flesh from such animals is referred to as being staggy.

Free-martins are females born as a twin to a bull calf and which on maturity are usually barren, the carcase having the normal features of a heifer.

Beef from mature bulls is recognised by the heavy muscular development, particularly in the neck (crest muscle) and the shanks.

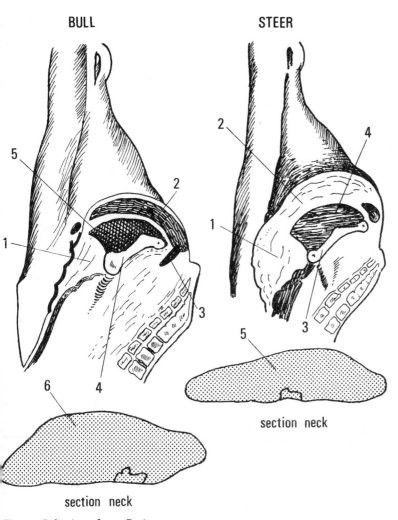

BULL   STEER

section neck

section neck

Fig. 1.   *Indications of sex—Bovine*

Bull: 1. Cod fat is scanty   2. Inguinal canal is not covered with fat   3. Retractor muscle developed   4. Aitchbone massive, particularly the round knob   5. Distinctive shape of gracilis muscle   6. Heavy neck muscles.

Steer or Ox: 1. Heavily lobulated cod fat   2. Inguinal canal covered with fat   3. Aitchbone slightly curved   4. Distinctive shape of gracilis muscle   5. Neck development moderate.

297

The flesh is dark in colour and it has been shown that the blood of bulls has a higher haemoglobin content than that of cows. There is usually a scanty fat covering, and the bones tend to be massive. The main anatomical features of the carcase are shown in Fig. 1, p. 297

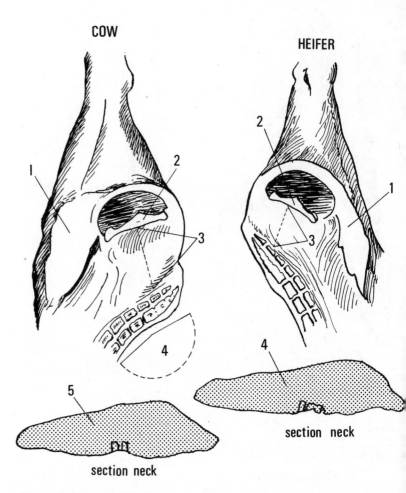

Fig. 2.   *Indications of sex—Bovine*

Cow: 1. *Pendulous udder, or if removed, leaving a triangular patch*   2. *Aitchbone flat and light in structure*   3. *Pelvic cavity wide and deep*   4. *Rump/loin angle wide*   5. *Neck light, poor development of crest muscle.*

Heifer: 1. *Mammary tissue firm and free of milk tissue*   2. *Aitchbone relatively flat and lighter than in the ox*   3. *Pelvic cavity tends to be proportionately wider than the ox, less wide than in cow*   4. *Neck development moderate*

Such beef is extremely popular on the Continent, where it attracts a premium. In this country, bulls slaughtered at around 12 months of age or a little older, produce a type of beef which appears to be popular in some areas. As a rule it is more likely to be less tender than beef from similar aged castrates, but on account of its leanness it is attractive to some purchasers.

BEEF FROM YOUNG BULLS

The quality covers a wide range, from the misnamed cow-heifer (which has produced one calf) to that obtained from an emaciated cow, past breeding and only fit for manufacturing purposes. It is usually assumed that deterioration of the carcase takes place rapidly from the third calf. Points of recognition are given in Fig. 2, p. 298.

COW BEEF

The vast majority of our beef supplies, both home produced and imported, is obtained from such beef. In the retail trade it is usually considered that ox beef will provide a higher proportion of saleable meat and possibly more bone, than in heifers. The major sex characteristics are shown in Fig. 1, p. 297.

STEER OR OX BEEF

This is usually thought by the connoisseur to provide the finest beef in terms of eating quality. It usually carries more fat than comparable ox beef and in beef breeds in particular, there may be a plentiful amount of marbling fat. It is possible that in many cases the eating quality is obtained at the expense of a somewhat wasteful carcase.

HEIFER BEEF

Comparison—Young Bulls and Steers

	Young bulls		Steers	
Age at slaughter approx	18 months		18 months	
	kg	lb	kg	lb
Live weight kg (lb)	513·9	1133	473	1043
Carcase weight kg (lb)	285·8	630	258	569
Dissection Results		%		%
Lean		68·7		62·9
Fat, internal and surface		14·2		19·7
Bone		15·7		15·7
Kidney and waste		1·4		1·7
		100		100

In general, on an equal weight basis, bull carcases will contain from 5% to 10% less fat and 5% to 10% more lean, and slightly more bone than steers. The eye muscles area is larger for bulls, but the percentage of hindquarters is better in steers by 1% to 2%.

Comparison—Steers and Heifers

	Steers	Heifers
Number of carcases	61	54
Weight range kg (lb)	156·5–317·51	160·57–317·51
	(345–700)	(354–700)
Edible Meat %	65·7	61·3
Excess fat %	16·2	22·1
Bone %	16·1	14·6
Edible to bone ratio %	4·08	4·20

Heifers fatten at a younger age than steers and at equal weights will have more fat in the carcase and less edible meat.

Cutting Methods

Whilst methods of cutting will vary in different parts of the world, or even within small localities, basically they are influenced by the methods of cooking employed.

However, the anatomical structure of the side results in some broad similarities in the methods of cutting the primal joints. In the case of the relatively hard long bones of the limbs, cutting between the joints with a knife is less strenuous than sawing through the body of the bone. There are a few exceptions, for example, for some banquets, marrow bones may be cut into rings, and when roasted the marrow is removed with a special spoon. In Scotland the Knap bones from the leg and shin may be sawn across for use in stock.

The formation of the muscle will influence the method of cutting, those joints or steaks intended for roasting, grilling or drying are invariably cut across the run of the muscle, rather than along them in order to reduce the length of the muscle fibres, to give the consumer an impression of tenderness.

300

With "moist" cooking, this is not as important, as the collagen connective tissue will tend to become gelatinous and tender.

When a side of beef is cut into a hindquarter and forequarter the two portions will take on different values per lb, the hind increasing and the fore correspondingly decreasing. Thus the line of demarcation is important in relating their respective values to that of the side.

A common method of division is to cut between the 10th and 11th ribs, leaving three ribs on the hind, but this may vary slightly in different parts of the country. Occasionally a 12 rib fore with a round cut forerib, leaving all the flank on the hind, may be preferred. In the case of intra-European trade a "Pistola" hind and fore is required (Fig. 4, p. 310). There is a growing tendency to market KKCF sides/hinds, from which the kidney knob and channel fat has been removed—details of the effect of this are given on Fig. 5, p. 310.

London and Home Counties Cutting

This has a higher percentage of roasting and grilling portions and a lower percentage of bone than the forequarter. The flank and the leg are the joints normally used for stewing, these primal cuts representing about 15% to 16% of the hind, whilst in the fore only about 30% will constitute good class roasting joints.

The hind may be cut on the block, which may obviate the loss of small pieces of suet, but it necessitates lifting the hind from a rail down on to the block. The following method is based upon the initial cutting being carried out on the rail.

THIN FLANK: This is removed by making an initial cut just below the cod fat, and then curving the knife downwards parallel with the chine to the first of the three ribs, which are sawn through. In deciding upon the line of demarcation, the difference in value between the flank and loin, and the necessity of maintaining a saleable loin, must be balanced. Flank may be used for pot roasting, in which case the inner skin of the belly should be stripped off, and

· 301

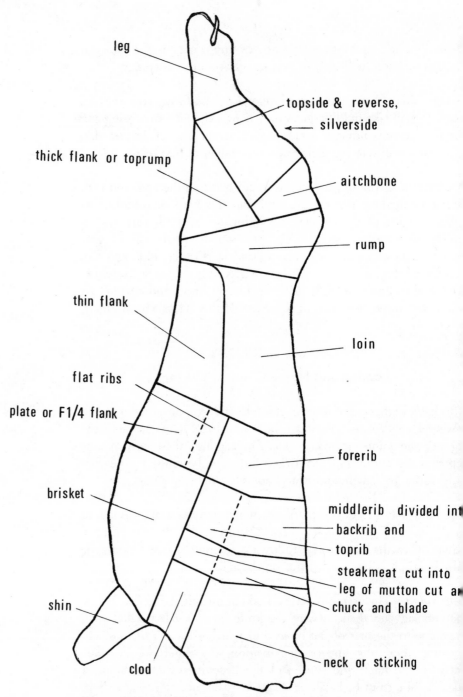

leg

topside & reverse,
← silverside

thick flank or toprump

aitchbone

rump

thin flank

loin

flat ribs

plate or F1/4 flank

forerib

brisket

middlerib divided int
backrib and
toprib
steakmeat cut into
leg of mutton cut a
chuck and blade

shin

clod

neck or sticking

Fig. 3. London and Home Counties Cutting (Primal Cuts).

302

the heavy gristle running between the muscles removed. In some cases the flank skirt is removed and the balance boned out for mince or sausage.

COD/MAMMARY FAT AND GOOSE SKIRT: The goose skirt is stripped away, and the cod/mammary fat cut off. The cod/mammary fat can subsequently be cut into strips and beaten flat for "fatting up" roasting joints. Milk or incipient milk tissue present in the udder must be trimmed away.

RUMP AND LOIN: This is divided from the top piece by sawing through the pelvic bone, just below the knob of the aitchbone, in a line slightly above the last bone of the rump. This is followed by a clean knife cut and this should divide the lymphatic glands (kernals) which are situated at either end of the rump. If the KKCF is present, it should be removed at this stage.

The method of separating the loin from the rump may vary slightly, but the following is suggested:—

Place the rump and loin on its chine, and locate the position of the cartilage on the end of the rump, with the point of a knife. Then using this as a guide, cut downwards through the cartilage in a straight line, followed by sawing or chopping through the bone. Some operatives prefer to open the rump from the loin and *joint* between the lumbar vertebrae and the sacrum. Whatever method is used, the thick end of the loin should be cut square.

The bones of the rump include part of the pelvic girdle, the sacrum and usually one or two of the tail bones, and the loin, six lumbar vertebrae, three dorsal vertebrae, part of three ribs and a small sliver of cartilage from the rump. In preparing the rump for steaks, the undercut (fillet) with its skirt is removed, followed by de-boning. When a "long" fillet is required, the complete muscle is removed extending from under the rump down into the loin, prior to splitting the rump and loin.

There are many methods of dealing with the loin, such as:—

Rolling with the bone in, which means that the size of the individual joints tends to be fixed by the position of the bones. On

303

account of the cost of such joints, this method is less popular than formerly, except perhaps for high class trades, the fillet being left in the loin.

A three bone wing rib may be removed and the loin rolled.

A more usual method today is to remove the fillet for steak, and then de-bone and roll the complete loin.

Striploin consists of the loin, ex the fillet, boneless and usually trimmed to give not more than 1 in. of "tops" as measured from the end of the eye muscle. In most cases the layer of gristle between the muscle and the external fat, removed. Such portions may be used for roasting or for cutting into entrecôte, or sirloin steaks.

T-bone steaks are cut from the thick end of the loin and include the fillet, the T-bone, and the tops are usually trimmed to under 2 in. in length. The American porterhouse steak is somewhat similar to the T-bone, but usually consists of four or five steaks *only* from the thick end, to ensure a maximum amount of fillet. Their club steak is cut from the rib end of the short loin and consequently possesses no fillet, but contains the bone.

TOPPIECE: The toppiece is usually cut on the block. Whilst some butchers prefer to remove the leg of beef prior to cutting the thick flank (toprump) the following is the more common procedure:—

THICK FLANK: The shank end of the stifle joint (patella) is located with the point of a knife and the knife inserted under the skin covering the bone, so that the joint is exposed. A straight cut is then made down on to the femur (silverside bone) picking up the skin covering the bone (periosteum) and stripping the bone clean. The thick flank extends under the bone and the cut should continue until the white membrane covering the silverside is reached. It is then completely detached by a straight cut through the external skin. After removal of the patella, the joint may be prepared for roasting, as braising steaks or pickled in preparation for spiced beef.

LEG: The leg is normally removed by jointing between the tibia femur articulation, care being taken to ensure that the cut is formed parallel with the base of the buttock. Occasionally a full cut leg may be removed by sawing and chopping through the end of the femur.

AITCHBONE: A fairly cut aitchbone joint will probably weigh about 10 to 12 lb and consequently is not popular for domestic use. To remove the aitchbone as a joint, a saw cut is made across the pubic bone, the buttock is then laid silverside bone downwards on the block, and a clean cut made with a knife, through the saw cut, topside and silverside, down on to the bone. This is then sawn/chopped across to separate the aitchbone.

Usually the aitchbone is removed to give a long cut silverside, in the following manner:—

With the buttock lying back downwards, the thin layer of muscle is removed from the front surface of the aitchbone, followed by cutting along the lower surface of the bone. The flesh showing through the hole in the aitchbone is loosened, care being taken not to penetrate deeply into the underlying flesh. Cutting from the rounded end of the aitchbone, the meat is released and the bone can be pulled away from the flesh and this will open up the ball and socket joint, permitting the tendon to be cut. The final removal of the bone necessitates cutting round the end of the ischium, which is deeply embedded in the meat. The whole operation can best be carried out with a thin boning knife. If removed cleanly, the weight of the bone can be used as a guide to the total weight of bone in the whole hindquarter, when multiplied by ten.

ROUND OR BUTTOCK: This has two distinctive groups of muscles and the femur bone. The topside, which is the inner muscle of the thigh and the silverside, to which is left attached the femur. The name silverside is probably associated with the sheen on the membrane which covers its inner surface.

The separation of these two portions may be made from the bone side of the buttock, by cutting round the bone until the seam is located, following through until the external fat is reached. This is cut through to remove the topside. As an alternative, some butchers prefer to split them from the outside surface. In this case the buttock

L

is placed bone downwards, with the leg end away from the cutter. The muscle at the top of the buttock, usually separated by a thin layer of fat, can be found at the shank end, the seam being carefully followed through, leaving the round "eye" muscle on the silverside. Following removal of the bone from the silverside, the heavy sheet of gristle adjoining the bone, should be trimmed out.

FOREQUARTER

In the normal retail trade this can best be dealt with by cutting on the block, and whilst there may be some slight variation in the order in which the primal cuts are removed, this is a matter of minor importance.

THIN SKIRT: Where the thin skirt (diaphragm) is left in the fore, this should be removed by cutting closely to the ribs.

SHIN: This can be removed more easily if the shin projects beyond the edge of the block. The elbow joint can be found with the point of a knife, and using this as a guide line, the ligaments around the elbow and the flesh can be cut through. A forward downward pressure on the shin will open up the joint, so that the underlying flesh can be cut to detach the shin. In fores from older animals, or in frozen beef, the joint can be more readily opened, if the tip end of the elbow joint is sawn through first. The bones of the shin consist of the ulna–radius and carpals.

CLOD AND STICKING: These portions are usually removed in one piece and they contain seven cervical vertebrae (neck bones) and the humerus (clod bone).

A protuberance will indicate the position of the scapula (shoulder blade) and humerus joint, or, in fat animals, this can be found with the point of the knife. A straight cut is then made on this line, the socket joint being opened with a boning knife, then with a steak knife, a cut is made down on to the neck bones. A small cut made into the clod fat as a "handle" will enable the weight to be taken, when the clod and sticking is detached by sawing through the neck bones. Care should be taken to ensure that the point end of the brisket is not cut and for this reason some butchers will remove the brisket and flank before the clod and sticking.

FLANK AND BRISKET: The two portions in one piece are sometimes referred to as a coast. The flank contains part of four rib bones and their cartilages, and the brisket has seven sternabrae, forming the breast bone and part of six ribs, attached. The guiding line in removing the coast is sometimes taken from about 1 in. from the end of the skirt at the rib end, to the first bone of the brisket. The cut is made, first with a knife over the ribs, followed by sawing through the ten ribs. When making this cut, in order to avoid cutting into the leg of mutton cut, this portion should be loosened, lifted backwards and secured with a hook.

The brisket is divided from the flank by cutting between the sixth and seventh ribs, a little extra pressure being required to cut through the costal cartilage.

FORERIB: This is the prime roasting joint of the fore and consequently should be cut as full as possible, the two ends being parallel to ensure that each rib of beef has its fair portion, or if boned and rolled, square cut edges. It is removed by a clean cut between the sixth and seventh rib, drawing the knife downwards on to the chine bone, the separation being completed by sawing through the chine. Frequently a narrow transverse strip is taken from the top of the ribs, the portion being termed flat tops.

MIDDLERIB: This is usually cut four ribs and in addition to containing part of the four ribs and chine, the major portion of the scapula extends into it.

The middlerib is cut from the remaining portion (a pony) by cutting between the second and third ribs, and over the scapula, with a knife. The scapula is sawn through and the division completed with a knife, down as far as the chine bone, which is sawn through. The middlerib is then cut into backribs and topribs, by cutting across on to the bladebone, sawing through and continuing the cut with a knife, down to the base of the ribs, which are sawn across, close to the chine.

It is customary to remove the blade bone from both portions in order to facilitate carving and frequently they may be completely de-boned and rolled.

307

STEAKMEAT: This is the remaining primal cut of the fore and it includes a part of two rib bones, the chine and the joint end of the blade bone. The leg of mutton cut can be used for pot roasting or for braising steaks. It is removed from the steakmeat by sawing across the two rib bones, close to the chine, and cutting downwards with a knife just missing the blade bone. It is usually sold boneless, but in some areas is sold with the rib bones attached, but in both cases the large blood vessels just below the layer of fat should first be removed. The portion remaining, the chuck and blade, provides very good stewing steak. The chuck contains very small portions of two ribs and the chine, whilst the blade bone, as its name implies, has a portion of the scapula. To separate the chuck from the blade bone, the natural seam between the two parts will indicate the division, the majority of the fat on the external surface is left on the chuck.

Kidney Knobs and Channel Fat

It is probable that the most important single factor affecting carcase yield of beef is the amount of kidney knob and channel fat. On average these internal fats constitute about one-third of the fat which has to be removed from a beef carcase, before it can be sold over the counter. Thus hinds/sides become much less variable in their composition and with suitable plant, these fats can be processed more efficiently, hot at the abattoir, rather than cold, following collection several days later, with other fat trimmings from the retail shop.

The reduction in carcase variability simplifies the negotiation of prices between wholesalers and carcase meat buyers.

In view of the foregoing the national organisations involved have agreed that beef carcases sold by deadweight may be weighed either including or excluding KKCF. In some circles it is contended that the suet fat provides protection to the valuable fillet, preventing excessive drying. For advisory purposes the various organisations concerned and the MLC have agreed an average yield of KKCF of 3·75%. Thus the *average* carcase without KKCF is taken to weigh 3·75% less than the average carcase with KKCF. To convert the price per lb offered for the average carcase with KKCF to an

308

equivalent price/lb for the same carcase ex KKCF, it is necessary to multiply it by 1·0389, irrespective of carcase weight. Conversely, to convert from a carcase ex KKCF to a carcase with KKCF, multiply by 0·9625. (Four decimal places are necessary to get the price/kg (lb) accurate to one-tenth of a penny.) Thus the producer of cattle with smaller than average KKCF will be better off than the producer of cattle with above average KKCF, as he will be paid for the heavier carcase.

With intra-community trade there is a demand for pistola hinds and fores, which has the effect of dividing the side into the more, and the less valuable cuts.

PISTOLA CUTS

In addition to the normal ten rib fore/three rib hind, pistola cutting is officially recognised for the purpose of "Intervention Buying". The pistola hind has eight ribs attached, leaving a five-rib pistola fore, with all of the thin flank (normally part of the hind) attached. For "Intervention Buying", in addition to the normal trim, the kidney, kidney and pelvic fat, and in females, the udder, must be removed.

Wholesale primal cuts bone-in (except topside)
Hindquarter (three rib)

In most carcase markets, wholesale cuts of beef are available in order to supplement those obtained from sides or quarters. Some butchers may require, say, roasting joints in excess of those available from the quarter beef, whilst others may want more stewing beef.

Trade terms for these cuts can vary in different areas, the following with their representative approximate percentages are generally applicable. Reference has already been made to sides ex KKCF on p. 310.

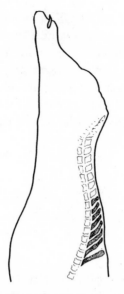
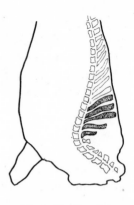

Fig. 5.   Pistola Cut Forequarter 5 Rib Cut.

Fig. 4.   Pistola Cut Hindquarter 8 Rib Cut.

HINDQUARTER (3 rib)		Approx %
H¼ X	Hindquarter less thin flank	93
H¼ XX	,, ,, ,, ,, and kidney knob	88
Tpce (top piece)	,, ,, ,, ,, and rump and loin	54
Rp & Ln	Rump and loin and kidney knob	40
Rp & Ln	,, ,, less kidney knob	35
Rp	Rump	15
Ln X	Loin less kidney knob	20
Leg		9
Flank		7
Aitchbone	(variable)	6
Topside		14
Silverside		13
Thick Flank		11

FOREQUARTER (10 rib)		
F¼ X	Forequarter less flank (plate)	90
F¼ XX	,, ,, ,, and brisket	78
Short F¼	,, ,, forerib and flank	79
Pony	Middle rib and steakmeat	41
Rib & Pony (crop)	,, ,, ,, and forerib	52
Coast	Flank and brisket	22
Middle Rib	(usually 4 bone)	20
Steakmeat	(usually 2 bone)	21

310

CS/S	Clod Sticking and Shin	26
Forerib	(usually 4 bone)	11
Brisket		12
Flank/plate		10
Shin		6

Vacuum-Packed Primal Cuts = Boneless

When the importation of bone-in beef from Argentina was prohibited, attention was focused on the possibility of shipping boneless
chilled cuts. Subsequently a considerable trade developed in vacuum-
packed primal cuts, transported at chill temperatures, a little above
the freezing point of meat. Following this, many firms in the United
Kingdom adopted this system for the distribution of beef cuts.

Two basic systems may be employed: (a) drawing a vacuum within
a film, followed by heat shrinking, in order to obtain a close surface
contact with the meat; and (b) drawing a vacuum, without heat
shrinking. There are some differences of opinion regarding the
merits of each system, but the latter is a "one step" operation and it
is considered to give a more rapid oxygenation when the packs are
opened. This process appears to be favoured for moderate holding
periods. For meat intended for prolonged storage, the additional
shrinking may be preferred.

Apart from pre-slaughter care, dressing and cooling, which are
dealt with on p. 201, temperature control throughout is a vital factor.
In addition to its direct influence on preservation, the effect of the
CO_2 formed in the pack increases as the temperature decreases.
Ideally, a temperature of $-1°C$ ($30°F$) should be maintained
throughout the chain of distribution and to the refrigerated show
case. In practice the internal temperature of the meat should not
exceed $4°C$ ($40°F$). Packing and evacuation should be completed
within 30 minutes of cutting, so that the CO_2 concentration builds
up in the pack and is not dissipated in the atmosphere. The re-
packing of faulty packs should be avoided, as the CO_2 will have
been lost and the subsequent control of bacterial growth will be
inadequate. When loading cartons with the packs, care must be

**TEMPERATURE
CONTROL**

311

taken to ensure that the fat surface is at the top, to avoid discolouration of the fat by the exudation of meat juices.

When opening a pack the film should be punctured at its lower level and the free fluid drained off, prior to complete removal of the film. Normally, when the meat is exposed to the air, it will take up atmospheric oxygen and acquire the characteristic bright beef colour in about 10 to 15 minutes.

Drip—Certain cuts will show a consistently high percentage of drip, and an investigation carried out at the Meat Research Institute on twenty sides of beef, packed 2 days after slaughter and held at 1°C (34°F) for 1 week, gave the following average results:—

		Weight		*Drip*		
		kg	lb	g	oz	%
Rump		5·5	12·1	73·7	2·6	1·35
Topside		8·4	18·7	62·3	2·2	0·74
Striploin		5·6	12·9	42·3	1·5	0·72
Silverside		6·6	14·9	48·1	1·7	0·70
Thick flank		5·2	11·8	28·4	1·0	0·54
	Total	31·3	70·4	254·8	9·0	0·80

CONSUMER PORTIONS

When beef is cut into steaks, each of a specific weight, the yield obtained will be markedly affected by the degree of trim, and the standard of presentation. Consequently the following figures can only be taken as a rough guide on account of the many variables involved.

		Side of beef—Good quality 136·1 kg (300 lb)					
	kg	lb	oz	kg	lb	oz	Approx % of side
Fillet steaks	2·04	4	8				
Rump steaks	3·4	7	8				
Sirloin steaks	4·9	10	13	10·4	22	13	7·6
Cubed steak for pies	15·7	34	8				
Braising steaks	15·9	35	2				
Rolled rib—boneless	5·4	12	0				

312

Rolled brisket—boneless	6·6	14 10			
Lean meat, mince, beefburger, etc.	26·1	57 9			
Ox kidney	0·425	15	70·2	154 12	51·6
Suet, fat and trim from guts	27·9	61 8			
Bones and gristles	25·3	55 12			
Loss and evaporation	2·4	5 3	55·5	122 7	40·8
			136·1	300 0	100%

Following, is one method of establishing a pricing pattern, assuming that average percentages of weight of the various portions are available and reliable.

Multiply the percentages for each cut by their usual price per kg (lb) and total them.

From this total deduct the required gross profit on sales.

This figure is then divided by 100 to find the maximum wholesale price per kg (lb) to include the required gross profit.

Divide the wholesale price into each of the prices of the various cuts in order to find the factor for each joint.

Thus by multiplying the wholesale price in pence per kg (lb) by the factors, the range of prices to include the required profit can be established, based upon any fluctuations in the wholesale price.

With the normal London and Home Counties method of cutting and preparation, with a good commercial side of beef, the following Desirability Factors are calculated to give a theoretical margin of approximately 30% gross on sales.

	Desirability Factor
Topside	2·32
Silverside	2·14
Thick flank (top rump)	2·07
Rump	3·25

313

	Desirability Factor
Fillet	4·12
Sirloin—boneless	2·76
Strip loin steaks	3·38
Forerib	1·44
Forerib—boneless	2·33
Brisket—bone in	0·82
Brisket—boneless	1·44
Steak meat (chuck, blade etc.)	1·61
Middle rib—boneless	1·43
Clod and neck (stewing steak)	1·37
Leg and shin meat	1·31
Flank—lean trim, mince	1·00
Suet and fat trim	0·08
Bones and gristles	0·03

Under practical shop conditions, losses will arise from cutting to weight, evaporation and drip, and the effect of these will reduce the actual cash taken by about 3% to 5% depending upon the efficiency of the management.

Culinary Uses

Topside	Roasting or braising steaks
Silverside	Braising, roasting, or if salted, boiled
Thick flank	Roasting, braising, or braising steaks
Rump	Grilling or frying steaks
Fillet	Grilling or frying steaks
Sirloin	Roasting, or grilling and frying steaks
Forerib	Roasting, or grilling or frying steaks (rib steaks)
H¼ Flank	Stewing, braising, or used for mince
Coast (rolled)	Braising, roasting, or if pickled, boiled
Middlerib	Roasting, braising
Steakmeat	Usually stewing, but "feather" steaks from blade bone may be braised
Clod & Sticking	Stewing, pie meat
Skirts	Stewing and pie meat
Kidney	Flavouring and pies

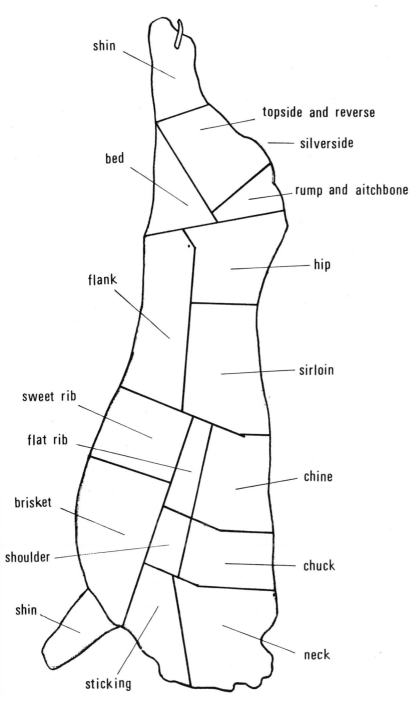

shin

topside and reverse

silverside

bed

rump and aitchbone

hip

flank

sirloin

sweet rib

flat rib

chine

brisket

shoulder

chuck

shin

neck

sticking

Fig. 6. Midlands Cutting (Primal Cuts).

315

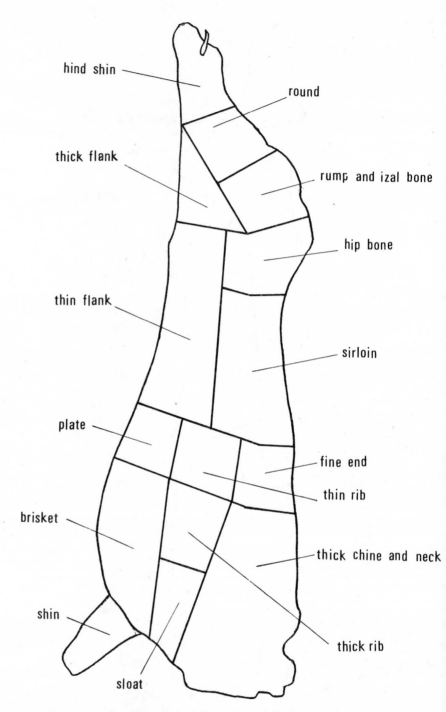

hind shin

round

thick flank

rump and izal bone

hip bone

thin flank

sirloin

plate

fine end

thin rib

brisket

thick chine and neck

shin

thick rib

sloat

Fig. 7.   *North and East England Cutting (Primal Cuts).*

316

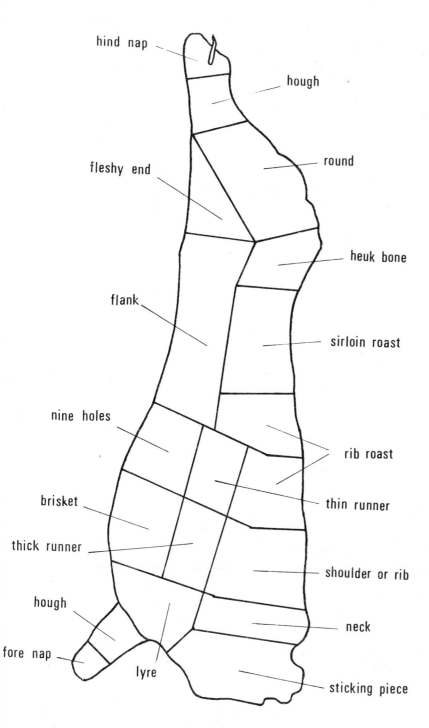

hind nap

hough

fleshy end

round

heuk bone

flank

sirloin roast

nine holes

rib roast

brisket

thin runner

thick runner

hough

shoulder or rib

fore nap

neck

lyre

sticking piece

Fig. 8.   Edinburgh Cutting (Primal Cuts).

317

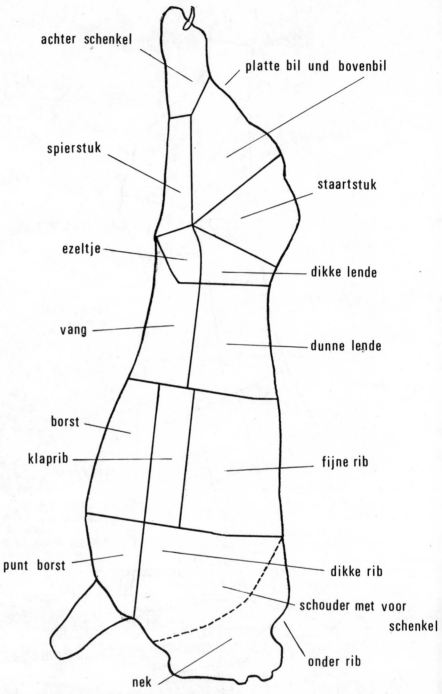

achter schenkel

platte bil und bovenbil

spierstuk

staartstuk

ezeltje

dikke lende

vang

dunne lende

borst

klaprib

fijne rib

punt borst

dikke rib

schouder met voor
schenkel

onder rib

nek

Fig. 9. Netherlands Cutting (Primal Cuts).

318

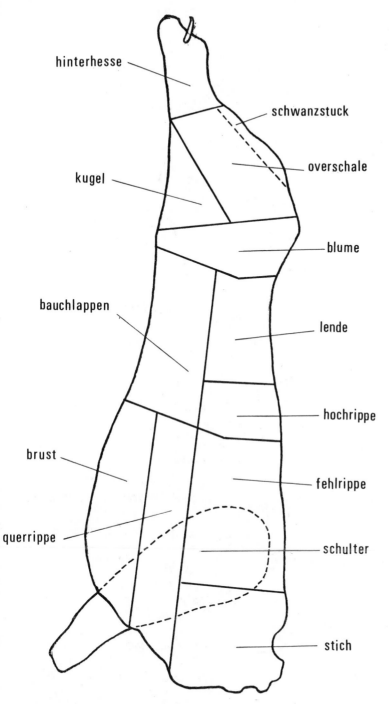

hinterhesse

schwanzstuck

overschale

kugel

blume

bauchlappen

lende

hochrippe

brust

fehlrippe

querrippe

schulter

stich

Fig. 10. *West German Cutting (Primal Cuts).*

319

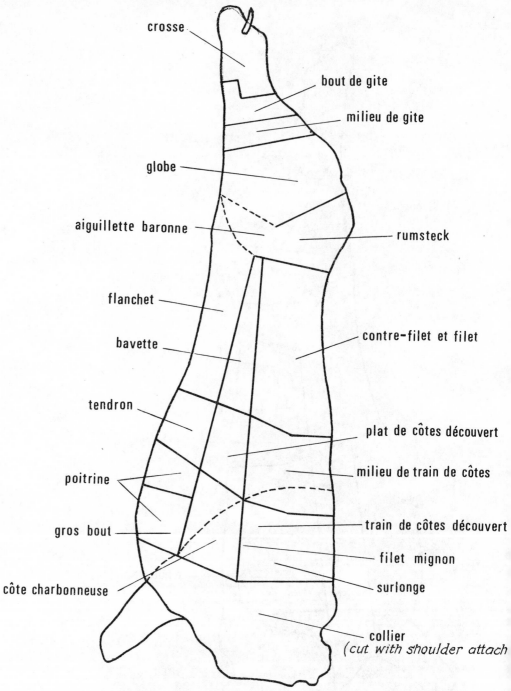

crosse

bout de gite

milieu de gite

globe

aiguillette baronne

rumsteck

flanchet

bavette

contre-filet et filet

tendron

plat de côtes découvert

poitrine

milieu de train de côtes

gros bout

train de côtes découvert

filet mignon

côte charbonneuse

surlonge

collier
(cut with shoulder attach

the shoulder including the shin is cut into 10 joints

Fig. 11. French (Paris) Cutting (Primal Cuts).

320

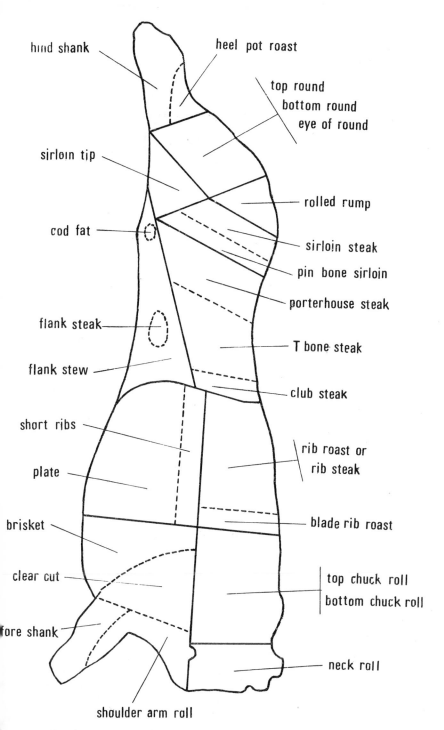

hind shank

heel pot roast

top round
bottom round
eye of round

sirloin tip

rolled rump

cod fat

sirloin steak

pin bone sirloin

porterhouse steak

flank steak

T bone steak

flank stew

club steak

short ribs

plate

rib roast or
rib steak

brisket

blade rib roast

clear cut

top chuck roll
bottom chuck roll

fore shank

neck roll

shoulder arm roll

Fig. 12. *United States—National Method (External surface, 12 rib Fore).*

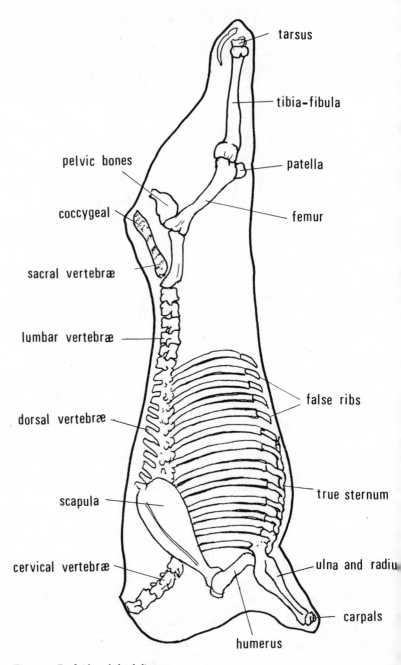

tarsus

tibia-fibula

pelvic bones

patella

coccygeal

femur

sacral vertebræ

lumbar vertebræ

false ribs

dorsal vertebræ

scapula

true sternum

cervical vertebræ

ulna and radiu

carpals

humerus

Fig. 13. Beef side—skeletal diagram.

322

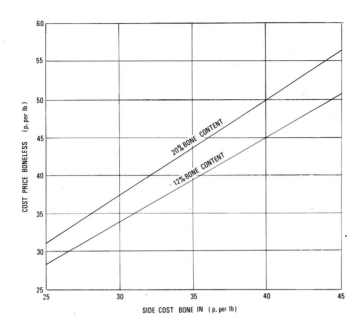

Fig. 14. *Effect of bone content on cost boneless.*

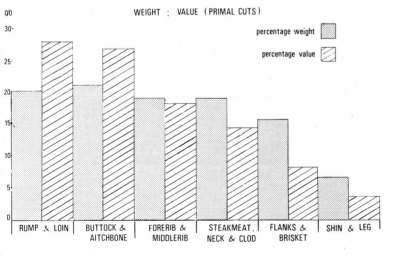

Fig. 15. *English side of Beef.*

Veal Cutting

Calves may range from "bobby" calves, slaughtered within a few days of birth and necessary by-products of the dairy industry, to specialist fed veal calves, yielding carcase weights of over 90·72 kg (200 lb).

Heavy fat calves, or in some cases hinds and ends, are imported from Holland and fetch high prices as a luxury article, mainly purchased for the hotel trade. Of recent years this trade has met a very active competition from a few producers in England, chiefly from south-west areas.

Market quotes for veal carcases are usually classified as English Fats, English Best, English Mediums and Bobbies. The difference in quality is reflected by the wide range of prices and English Fats may fetch three to four times as much per kg (lb) as Bobbies.

Good quality calves should be compact, with well developed legs, loins carrying a good amount of flesh, and the shanks and neck, short and thick. The kidneys should be well covered and the back and shoulders should have a reasonable deposition of fat, which should be white to creamy-white in colour. Firm flesh is important and whilst white flesh is usually quoted as being essential, a slightly pink tinge is common. The presence of some fat on the inside of the ribs is usually accepted as an indication of quality.

English Fats would be superior in terms of conformation and finish and the quality would be reflected in the price.

For the better class of trade, using heavy calves, the method of cutting fillets and oysters of veal is still practised (see Fig. 16), whilst with smaller calves most butchers will tend to follow that used for lamb carcases (see Fig. 17). With the smaller types of carcases, there is much to be said for completely de-boning the whole forequarter, followed by fatting up and rolling. When using this method some operators prefer to remove the neck, for use as pie veal, with the object of improving the eating quality of the roll.

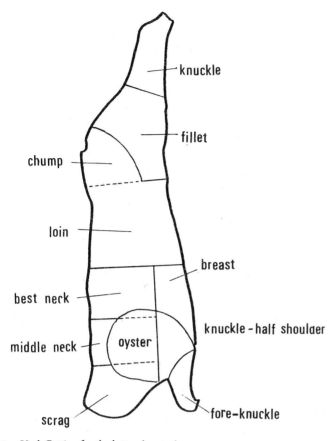

knuckle

fillet

chump

loin

breast

best neck

knuckle - half shoulder

middle neck

oyster

scrag

fore-knuckle

Fig. 16. *Veal Cutting for the better class trade.*

SPLITTING: The carcase can be chopped into sides by suspending it from a rail, or quartered, leaving one pair of ribs on the hinds followed by chopping down the hinds and fores.

THE HIND: As previously mentioned, with heavier calves, the following procedure may be adopted. A cut is made *above* the aitch-bone, leaving a loin and chump. The knuckle is removed and the leg portion ex the knuckle is known as the fillet. This may be sliced thinly to give veal cutlets, or in portions as a high grade roast. The more common method, particularly with smaller calves, is to cut the leg *below* the aitchbone.

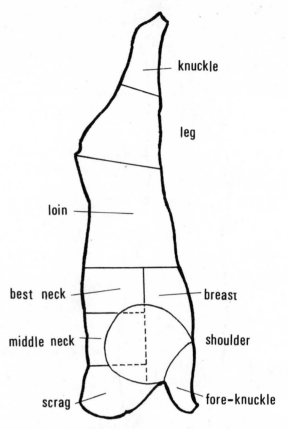

knuckle

leg

loin

best neck

breast

middle neck

shoulder

scrag

fore-knuckle

Fig. 17. *Veal Cutting as for lamb carcases.*

THE FORE: The shoulder is removed by finding the end of the scapula with the point of a knife, then marking round and completing the separation as in the case of lamb (see p. 347). The whole shoulder is usually too large for normal domestic requirements, but it may be sold ex the knuckle, divided into an oyster and knuckle end shoulder, or completely de-boned and rolled.

The breast is next removed by marking down with a knife and sawing through the ribs, and it can be sold bone-in, or de-boned and rolled for roasting.

The best end neck (usually six bones) is cut from the remaining whole neck and is chiefly used as veal cutlets.

326

The middle neck and scrag remaining is used for stewing.

ROLLED FOREQUARTERS: The following procedure is recommended:—

1. Remove the knuckle, leaving as small a hole as possible in the surface of the quarter.
2. Trim any bloody portions from the throat and remove the sticking bones (cervical vertebrae) as far as the first rib bone.
3. Cut back the outer flesh of the brisket, not extending beyond the attachment with the ribs.
4. Run the point of the knife between the brisket and rib bones. The thin skin on the flesh side of the rib bones must be left intact, otherwise the ribs will not "strip out".
5. Mark down the centre line of each rib, through the pleura, then beat back the flesh with the back of a light chopper. The ribs will then be forced away from the flesh, leaving them absolutely clean.
6. The rib bones can then be removed with the chine bones (dorsal vertebrae).
7. The clod bone (humerus) is tunnelled out and the blade bone removed, working from the inside of the fore.
8. Beat out the meat and trim to shape.★
9. Roll, placing a small portion of flare fat along the centre, and a thin layer of fat round the outer surface.

	Culinary Uses
Leg	Roasting, usually deboned
Fillet	Escalopes—grilling or frying
Loin	Roasted—or in chops, grilling or frying
Best End Neck	Cutlets, grilling or frying
Oyster	Roasting
Knuckle Half Shoulder	Roasting or stewing
Middle Neck & Scrag	Stewing
Breast	Stewing or boned & stuffed for roasting
Knuckles	Boiling

★ When trimming to shape, the neck may be removed for stewing

327

Veal Cutting Tests

Side A 20·4 kg (45 lb)	Approx %	Side B 14·5 kg (32 lb)	Approx %
Fillet	21	Leg ex-knuckle	23
Loin with chump	19	Loin	15
Best End Neck	6	Best End Neck	6
Oyster	16	Shoulder ex-knuckle	22
Knuckle-half shoulder	6	Middle Neck & Scrag	10
Middle Neck & Scrag	9	Shank Meat	6
Breast	9	Breast	9
Knuckles (whole)	13	Shank Bones	8
Loss	1	Loss	1
	100%		100%

Forequarter 7·7 kg (17 lb)	Approx %
Boneless Meat	60
Fore Knuckle (whole)	11
Bones and trim	28
Loss	1
	100%

Tenderising Techniques

Tenderness is probably the most important single factor influencing the consumer's assessment of meat quality. Apart from the tenderising changes which take place in the maturing of meat (see p. 153) many other methods are employed with the objective of increasing tenderness.

The three main meat constituents affecting tenderness are the white collagen fibres, a good example being the achilles tendon of the hindquarter, the yellow elastin as found in the ligamentum nuches (paxwax) of the neck and the muscle fibres. A number of enzymes are capable of influencing tenderising changes in these tissues.

Plant enzymes such as ficin from figs, rates very high with elastin nd high with collagen, but probably on account of its flavour, is not widely used except as an adjunct to Parma ham. Bromelin (pineapple) has a pronounced effect on collagen, a slight influence on elastin, but does not affect muscle fibre. Papain (pawpaw) is fairly widely used, giving good results with both muscle fibre and elastin and it is moderately active in the case of collagen.

Other substances including weak acids, such as lemon juice and vinegar, have been used, whilst sodium chloride in 2% concentration, will increase tenderness materially.

The direct application of enzymes may take the form of dipping in an enzyme solution, or dusting the surface with a powder enzyme preparation. The chief disadvantages of surface application include lack of uniformity in distribution, excessive softness of the surface layer and in most cases, discolouration of the meat.

INJECTION-SPRAY PROCESS

With the object of overcoming difficulties of distribution of the tenderising substance, a new method has been developed. The apparatus consists of a number of hollow needles which automatically penetrate the meat. The spacing of the needles can be arranged as required and they are retractable, which permits the pumping of bone-in cuts of meat. On the down stroke of the needles nitrogen is injected, which opens up the meat tissues and on the return stroke a gas/liquid phase is utilised, resulting in atomisation of the enzyme throughout the opened texture of the meat. The amount of solution injected can be controlled within wide limits, i.e. 3% to 22%.

In addition to the introduction of enzymes, other substances, such as onion and smoke flavour, can be incorporated in the solution.

American regulations stipulate a maximum of 3% "pick-up" in the meat and that enzyme injected producers shall be frozen.

THE PROTEN PROCESS

This was developed by Swift and Company Ltd in 1961 and involves the introduction of the enzyme papain into the circulatory system

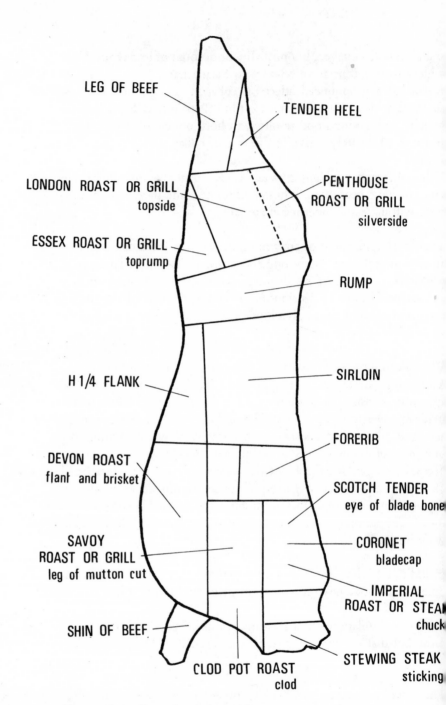

LEG OF BEEF

TENDER HEEL

LONDON ROAST OR GRILL
topside

PENTHOUSE
ROAST OR GRILL
silverside

ESSEX ROAST OR GRILL
toprump

RUMP

H 1/4 FLANK

SIRLOIN

DEVON ROAST
flank and brisket

FORERIB

SCOTCH TENDER
eye of blade bone

SAVOY
ROAST OR GRILL
leg of mutton cut

CORONET
bladecap

IMPERIAL
ROAST OR STEAK
chuck

SHIN OF BEEF

CLOD POT ROAST
clod

STEWING STEAK
sticking

Fig. 18. Proten Cuts and equivalent regular cuts.

330

f the live animal. The system was based upon the following onsiderations.

he vascular system provides an excellent method of distribution hroughout the body tissues, the heart being an efficient pump and he bloodstream acting as a diluent for the enzyme introduced. It vas found that the more active and therefore tougher muscles, with heir greater supply of blood, would become relatively more tender han the less active muscles. The injection of papain solution is arried out a few minutes prior to stunning, which allows the nzyme to circulate throughout all parts of the carcase. After pre-enderising, the raw meat contains about four parts per million, .e. 4 lb per million lb. Activation of the enzyme does not commence ntil the application of heat on cooking. Its activity commences vhen the meat attains a temperature of 50°C (122°F), reaching its naximum within the 54°C (130°F) to 71°C (160°F) range. It ecomes inactive at about 80°C (175°F) and is destroyed at 100°C 212°F). Any marginal residual is readily absorbed as a natural vegetable protein substance, by the digestive processes of the human ody.

With traditional beef, about 30% of the side meat is considered as eing acceptable when cooked by dry heat (grilling or roasting). Whereas with beef so treated, this figure can be increased to around ·5%.

To take advantage of this, a modified method of cutting and sug-zested names for some of the portions are shown in Fig. 18. The "Penthouse" consists of the round muscle of the silverside and the "Coronet" is obtained from the blade bone and is sometimes known s "Feather Steak".

t has been found that an increase in length with a corresponding lecrease in muscle fibre diameter, tends to result in improved tender-1ess.

When, following slaughter, a side of beef is hung in the traditional nanner from the hock, certain muscles are not under tension and

TENDERSTRETCH METHOD

331

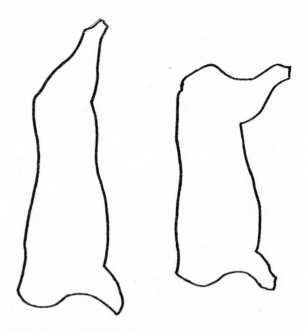

Fig. 19.   *Suspension of a side of beef—on the left from the hock joint, on the right from the pubic bone.*

consequently they are more liable to contraction on rigor morti
than those muscles under stress. The tenderstretch method aims a
maintaining the position of the muscles in a somewhat similar fashion
to that existing in the live animal. This entails suspension of the sid
of beef from the aitchbone (pubic bone) instead of from the hock
joint as shown in Fig. 19.

It is essential that this method of hanging be carried out within 1
hours of slaughter and the sides must be left in the chiller for at least
24 hours, suspended by the aitchbone. The improvement in tender
ness in the loin, rump, topside and cube roll (eye muscle of rib) i
equivalent to that obtained by 3 weeks' maturing at 2°C (34°F).

The silverside is not improved to the same extent as the cuts pre
viously referred to and the filler becomes slightly less tender, bu
not sufficiently to affect acceptability. Cutting methods must b
slightly modified to meet the alteration in the positions of the bone
and muscles. The topside undergoes considerable elongation, which
would not affect its use as steaks, but would adversely affect it
value if it is to be cut into roasting joints.

332

The reduction of muscle substance to particle size, as in the case of mince, steakettes, beefburgers and similar products, is an effective form of mechanical tenderising. Another method of breaking down the meat tissues, without destroying the general form, is penetration by mechanically operated rollers, fitted with sharp points.

COMMINUTED MEATS

A modification of the previous method is the preparation of fabricated steaks. For this purpose boneless meat from the less popular cuts is selected and then frozen into blocks, prior to flaking into thin slices. The slices are then pressed together and moulded to the desired shape and weight, prior to re-freezing. On compressing the flakes, the exudation of meat juices assists in holding them together on subsequent cooking, as the soluble protein coagulates on heating. The fat content can be adjusted as required and normally between 10% and 12% is preferred. Such steaks are cooked in their frozen condition, using an initial high temperature to sear the surface and thus seal the meat juices.

FABRICATED STEAKS

Hot De-boning of Beef

For some considerable time, the use of hot (pre-rigor) bull beef for manufacturing purposes, has been practised in Germany and America, mainly on account of its water-holding characteristics. However, in the case of butcher beef generally, the universal custom has been to hang the beef for 24–36 hours prior to cutting the sides.

The recent development of preparing primal cuts at slaughter establishments has created an interest in the practicability of de-boning the hot carcase and the direct preparation of the cuts.

Should this technique be widely adopted there could be a radical change in meat distribution, which could exert an influence on the design of abattoir/cutting units.

Whilst these techniques are still experimental and further investigation is desirable, they could lead to considerable savings in cost, probably combined with some improvement in quality.

The major advantages would include lower refrigeration costs, less drip, better muscle colour and higher boning-out yields.

The following is based upon the work carried out by the MLC and the Meat Research Institute, Langford.

The beef is removed from the bones of the hot hanging sides and immediately vacuum-packed on what is, in effect, an extension of the slaughter line, and before refrigeration.

Important parameters which have to be rigidly controlled are the time-temperature relationships to be followed, after the meat is removed from the bones, the level of hygiene throughout the hot de-boning process, and just as important, the carcase dressing and the cutting methods adopted.

The factors to be considered include:—

The time after sticking by which the process should be completed, and on an efficient line system, a beef carcase will normally be dressed and weighed in 30 minutes to 1 hour from sticking. The hot boning procedure must remove the meat from the bones and cool it quickly thereafter. However, the cooling must not be *too* rapid or cold shortening may result, causing tough meat. Present indications are that the temperature should not be allowed to fall below 10°C (50°F) for at least 10 hours.

The cutting method may be traditional, or as far as possible, it can be seamed in order to remove the muscles intact. The Meat Research Institute believe that there are good microbiological reasons which will dictate that the seaming method must be used.

On removal, the hot muscles or groups of muscles have to be vacuum-packed for ease of handling, possibly using different bag sizes and shapes for each. With this approach, muscles can be moulded into shapes most suitable for the preparation of retail joints. In theory, where appropriate, cold fat strips could be wrapped round hot muscles, before vacuum packing.

334

Some lower value muscle groups may be used for processing immediately. For example, they might be tumbled whilst hot, minced, or otherwise immediately prepared, before refrigeration.

The boning should be carried out in an air-conditioned cutting room held at 10°C (50°F). (Under EEC legislation this must be separate from the slaughter area and from cutting rooms for chilled meat.)

It is claimed that if the meat is held in this first refrigeration stage at 10°C (50°F) for 24 hours, the tenderness achieved is equivalent to conventionally chilled beef held for some 10 days of normal ageing at 0°–2°C (32°–35°F).

It is important that the vacuum-packed muscles should be cooled to 10°C (50°F), so that they should not be boxed or stacked after removal. Tray systems of cooling the packaged cuts need to be developed, but as the meat is vacuum-packed the relative humidity of the cooler is less important than with carcases conventionally cooled. After the pre-cooling phase the meat must be held at 0° ± 1°C (32° ± 34°F), until required and the shelf life should be better than that of cuts vacuum-packed in the normal manner.

Obviously, the initial bacteriological status of the meat must be very good. To derive the full benefit from the operation, a higher standard of bacterial cleanliness is required than is generally adopted in our abattoirs, to keep down the number of spoilage bacteria, and to prevent the growth of food poisoning bacteria on the surface or in the deep muscle.

Effective cooling occurs more rapidly than is the case with traditionally handled carcases or cuts. In the latter, towards the centre of the meat, the temperature does not fall sufficiently rapidly in relation to the fall in pH to prevent a pale appearance. In hot de-boned cuts, the cooling is more rapid and even the centre of the meat is cooled sufficiently quickly, giving a better colour uniformity.

One major advantage is the reduced drip in the pack, and with proper temperature control it can be greatly reduced. This in turn

335

results in a more attractive appearance through reduced staining o
the fat.

Should frozen meat become more popular, it may be possible fo
retail packs, prepared rapidly after hot de-boning, to be froze
immediately with little danger of cold-shortening. Bendall suggest
that provided these packs are held at the relatively high storag
temperature of $-3°$ to $-5°C$ ($27°$ to $23°F$) for at least 48 hour
before sale, there should be no contraction, and consequent toughen
ing on thawing by the consumer.

The meat industry will closely follow the progress of this technique
which could have a fundamental influence on the future pattern o
the trade.

Mechanised boning and cutting lay-out. Note monolithic polymer floor

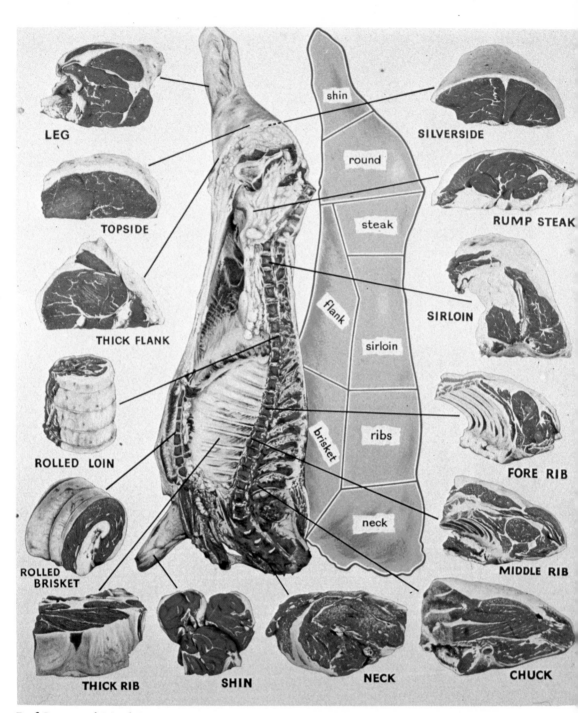

LEG

TOPSIDE

THICK FLANK

ROLLED LOIN

ROLLED BRISKET

THICK RIB

SHIN

NECK

CHUCK

MIDDLE RIB

FORE RIB

SIRLOIN

RUMP STEAK

SILVERSIDE

shin

round

steak

flank

sirloin

brisket

ribs

neck

Beef Carcase and Primal Cuts

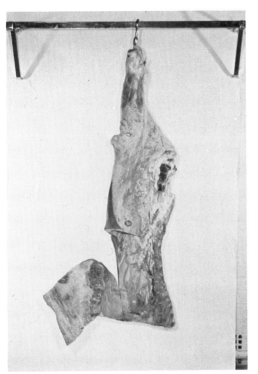

Square cutting of flank will ensure a good shaped loin

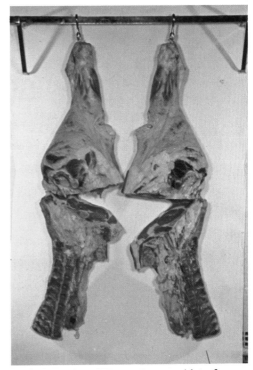

Correct line for separating rump and loin from top-piece

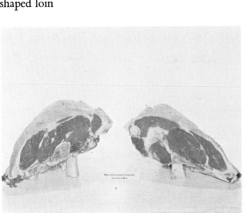

Note effect of position of cut on appearance of rumps

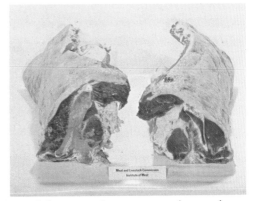

Loin, left, cut too close onto rump bone and "tops" uneven

Beef Cuts

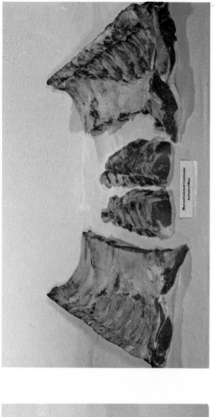

Fillets are valuable, loin bones should be left clean, correctly cut loin on left.

Rectangular appearance of well cut loin

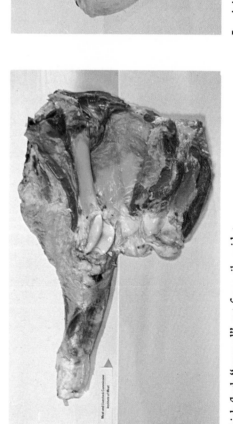

Leg jointed in straight line and parallel with lower cut surface of buttock

Thick flank "seamed" out from silverside

Silverside "seamed" leaving a clean surface, right

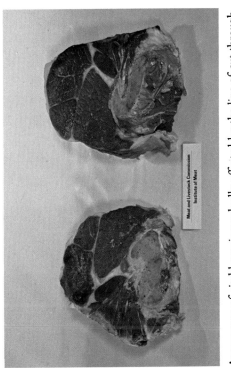

Appearance of aitchbone is markedly affected by the line of cut through the bone

The natural division between topside and silverside is clearly defined, left

Beef cuts

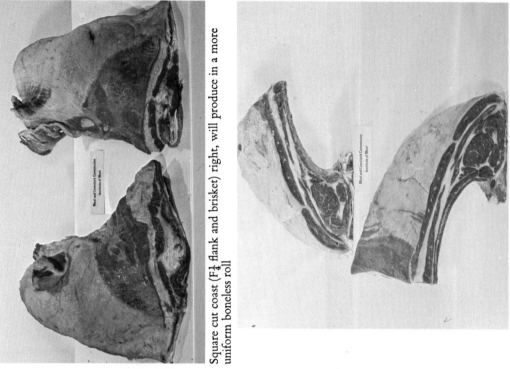

Square cut coast (F$\frac{1}{4}$ flank and brisket) right, will produce in a more uniform boneless roll

A four bone forerib is normally cut — best end and cristle rib

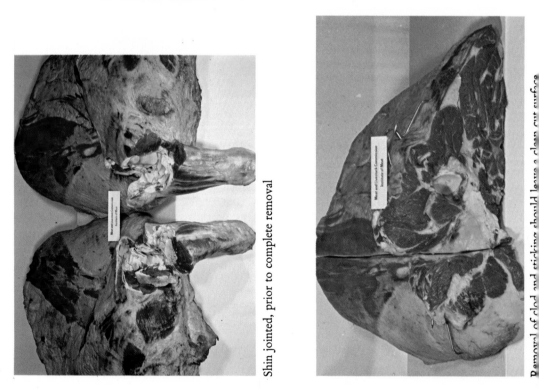

Shin jointed, prior to complete removal

Removal of clod and sticking should leave a clean cut surface

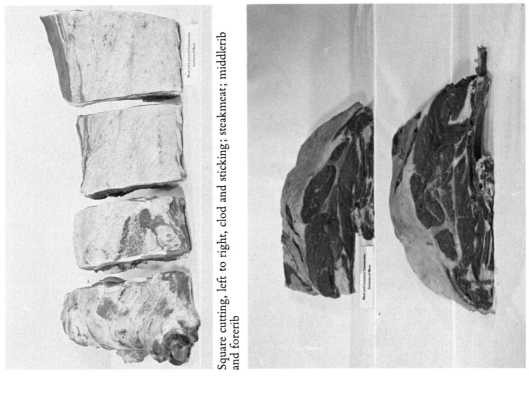

Square cutting, left to right, clod and sticking; steakmeat; middlerib and forerib

Steakmeat is cut into (left to right) blade bone; chuck and leg of mutton cut – upper photograph indicating the better cutting

Care in cutting, results in a more attractive middlerib, top

Beef cuts

Influence of a well cut middlerib, is reflected in the appearance of backribs and topribs

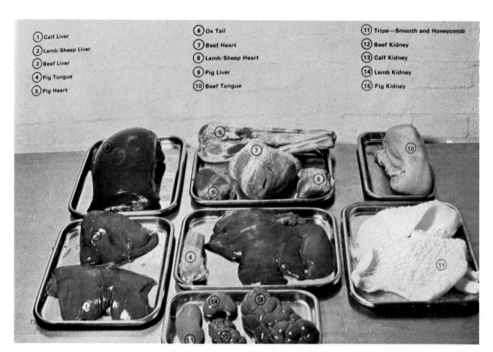

Edible Offals (Variety Meats)

Prepared dishes

Courtesy Joan Dando, MHCI, MIMA

Lamb and Lamb Cutting

Whilst various ethnic groups may have religious or ethical objections to the consumption of pork or beef, no such prejudices apply to lamb or mutton, except of course in the case of vegetarians. The United Kingdom is the only country in the northern hemisphere consuming any appreciable quantity of lamb/mutton, about six times more than the United States and four times more than France, per capita, per annum.

With their ability to provide meat, wool, skin and tallow, combined with their general docility, sheep were probably the earliest animals to be domesticated by man and their remains have been found in the Swiss lake dwellings of the Copper Age. They existed in the United

M

Kingdom from the earliest times and whilst the date of their introduction has not been established, it is known that the Roman
developed a flourishing wool industry.

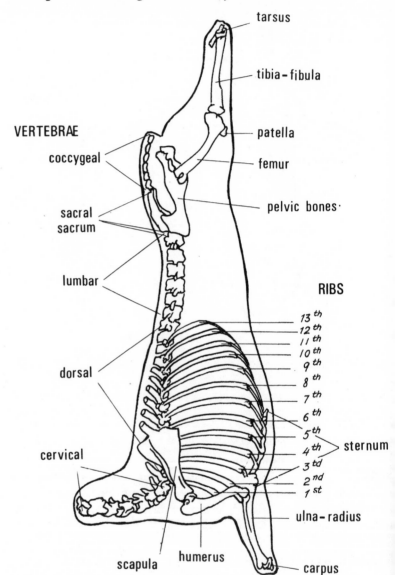

VERTEBRÆ

coccygeal

sacral
sacrum

lumbar

dorsal

cervical

scapula

humerus

tarsus

tibia-fibula

patella

femur

pelvic bones·

RIBS

13 th
12 th
11 th
10 th
9 th
8 th
7 th
6 th
5 th
4 th
3 td
2 nd
1 st

sternum

ulna-radius

carpus

Fig. 1.   *Skeletal diagram of lamb carcase.*

Sheep, as typical ruminants, have no upper incisor teeth but carry molars with strongly crescentic pillars and they have a four-chambered stomach. The wool which is a most valuable characteristic of sheep, can be considered as fine hair, the cells of which are finely notched, causing the strands to hold together, the hairy coat of wild sheep being somewhat like that of cattle. Horns are frequently present, often massive in the adult ram, and in contrast with the horns of cattle, they are triangular in cross section, not circular. The cleft snout and the smallness of the muzzle of sheep, allow them to graze much more closely than the ox. Four Horned Piebald (Jacob) sheep are not uncommon and apparently originated in Ireland and the Faroe Islands: a few flocks are maintained in this country, in order to ensure their survival.

The breeding season (oestrous cycles) is influenced by the changes in the rhythm of daylight and darkness hours, the breeding times being centred around the shortest day. Thus, as the seasons in New Zealand and Australia are opposite to those in this country, their supplies of lamb fit in conveniently with our periods of shortage. With the exception of some mid-European countries, lamb carcases are normally dressed without the head.

VERTEBRAE

Cervical	7 bones
Dorsal	13 bones
Lumbar	6 bones
Sacral (sacrum)	1 bone (5 segments fused)
Coccygeal	Very variable: up to 18 bones, as a rule about 6 bones are left on the carcase

With the exception of the cervical vertebrae, variations in their numbers commonly occur.

PELVIC BONES

| Ischium | Aitchbone |
| Ilium | Hip or chump bone |

RIBS

13 pairs, 14 pairs not uncommon, particularly in the Herdwick.

339

STERNUM   Breast bone   7 segments, sometimes 6

LIMBS

HIND LIMB	FORE LIMB
Tarsals	Carpals
Tibia–Fibula	Radius–Ulna
Patella (kneecap)	Humerus
Femur	Scapula

SEASONAL SUPPLIES OF LAMB

Climate will influence the breeding pattern and consequently the times at which the lambs are dropped. In areas such as Scotland and the uplands of Wales, if the lambs were born too early in the year, they would not withstand the cold weather and the oestrous cycles have evolved by natural selection to ensure the survival of the young.

Thus, in the United Kingdom the early lambs will come from the south whilst in New Zealand the killing season in the North Island will tend to be about a month earlier than in the South Island.

The following is a summary of the sources of home-killed lamb arriving at Smithfield, London:—

CHRISTMAS: A small luxury trade in milk-fed lambs from the west of England.

JANUARY UNTIL TOWARDS THE END OF MARCH: Dorset Horn, Dorset Down and other breeds from Dorset, Somerset, Hants, Sussex and the Isle of Wight.

EASTER: Down and Down cross breeds from the south and south-west counties and from Oxford, Suffolk and Cambridge.

APRIL TO MID SEPTEMBER: Regular trade of fat lambs from south and south-west counties and from Ireland.

MID JUNE TO MID DECEMBER: Scotch Hill or Border lambs, largely Blackface or their crosses.

340

The Australian killing season usually begins in the late spring (that is about September) and generally extends over six months. The vagaries of the Australian climate influence the date when the season opens, the length of the season, and the regularity of supplies. As a rule the first consignments of "new season" lamb are on the English market in November and are in good demand, as New Zealand lamb usually starts to arrive about a month after the Australian.

The New Zealand killing season usually starts in the North Island about November and in the South Island roughly a month later. The first "new season" lamb generally reaches this country in December or early January.

However, with their good long seasons, New Zealand killing may extend into August, so that lamb of that season may still be arriving late in that year.

RAMS: These are identified by the relatively large proportion of the forequarters, thickness of neck, and the development of the crest muscle. The pizzle is large in section, and if the legs are split the size of the retractor muscle is much greater than in the castrate. The flesh from the adult ram is dark and coarse in texture and it usually has a "full" flavour.

Owing to the small ratio of rams to ewes, the number of ram carcases reaching the market is relatively small.

EWES: These will vary considerably, from those which might be acceptable to certain classes of retail trade, to old emaciated carcases suitable only for manufacturing purposes.

With lactation, the mammary gland will be largely formed of spongy brown milk tissue, but usually in dressing the carcase the udder may be trimmed off. The pelvic cavity will be wide due to the production of lambs and frequently the inner belly wall may show a number of small veins. The most distinctive carcase characteristic is the large back passage, caused by the removal of both anus and

341

vagina. In general, the ribs will be well barrelled, the flanks thin and the neck slender and long.

MAIDEN EWES (Gimmers, Chilvers, etc.): In these the mammary organ will be smooth, oval and consist of firm fat. As compared with males, the back passage will be larger.

WETHER (Wedder, or in some parts Teg): These males, castrated at an early age, possess a lobulated cod fat, poorly developed pizzle and retractor muscle, and much lighter forequarters and neck, than in the ram.

At one time a number of high-class restaurants, specialising in grills and roasts, preferred carcases from old wethers and when hung for a period to mature they produced meat which had an excellent flavour, combined with tenderness.

LAMB

Carcase weight cannot be taken as an indication of age, as the carcase from an adult of some of the hill breeds might weigh only about 14·5 kg (32 lb), whilst that from some longwool breeds could weigh three times as much. Age specifications are frequently employed, and it is broadly accepted that lamb is derived from ovines of either sex, under 12 months of age. The vast majority of New Zealand lamb is slaughtered at from 4 to 6 months.

HOGGET

Hogget is somewhat more difficult to define with any degree of precision. Whilst it is agreed that it is derived from wethers or maiden ewes, one standard stipulates that "two permanent incisors have been cut, which teeth should not be fully grown" another requires "not more than two permanent incisors" (the central permanent incisors may be cut at about 10 months and fully up at 1 year).

In some cases for example (NZ) there is an upper weight limit for hoggets of 25·4 kg (56 lb), and carcases above this weight are classified as ewe or wether mutton, whichever is applicable.

Mutton is sometimes defined as an age basis as indicated by the dentition, i.e. "animals with at least two permanent incisors". Wether mutton from a castrated male ovine (showing no ram-like characteristics) and maiden ewes, may, for practical purposes, be grouped together.

Ewe mutton is derived from female sheep which have lambed.

Lamb carcases are usually described in terms of conformation, finish and quality.

The conformation is largely determined by the skeletal structure, the muscular development and the deposition of the subcutaneous fat.

In terms of retail yield, the bone content is probably not as important as in the case of beef, as the majority of the joints are sold with the bone in. The actual weight of bone in a "shanky" hill lamb may be somewhat similar to that of a compact Southdown at a similar weight. However, the shorter thicker bone of the Southdown will produce a plump stocky leg and on cooking, the shank meat will "draw up" less than in a shanky leg. Conformation is therefore influenced by the breed, age (as the shape changes with maturity), sex, and the plane of nutrition.

There is a strong consumer resistance to excessive fat, and this applies particularly to mutton fat, which even in rendered form has not the appeal of pork or beef dripping. In general, fat tends to be deposited on a lamb carcase in the following manner. It first develops over the tail-head and rump, extending along the back towards the neck, then around the barrel of the carcase and finally over the shanks. With long shanky legs, it is difficult to obtain a smooth even covering of fat. Some breeds tend to develop excessive fat in the tail-root rump area, whilst others may lay down a large amount of fat over the ribs. The fat should be a creamy white colour and fairly firm to the touch, although in young milk lambs the fat may be comparatively soft. On rare occasions a bright yellow fat may be

343

present, and excluding jaundice, is due to a recessive character in inheritance.

Quality is more difficult to define, but it is usually taken as being the effect of the cooked flesh (muscle, fat, connective tissues and meat juices) on the normal human palate. However, personal tastes may vary and the full meaty flavour of the flesh from a mature mountain wether might be preferred to the more delicate flavour of that of a milk lamb, whilst the strong flavour of "buck" mutton from rams has a limited appeal. To summarise, although some slight modifications might be required for hill lambs, the following is a standard for Grade A lambs, as stated by the Ministry of Agriculture, Fisheries and Food.

LAMB GRADE A

CONFORMATION: The carcase should be compact and uniformly fleshed throughout. Loins broad and well developed. Legs plump and full with the meat carried down towards the shank. Shoulders thick, full and smooth, neck short.

FINISH: The exterior covering of fat should be even, smooth and moderately thick over the back. Fat covering over the kidneys should be adequate, but a small proportion of the kidney may be exposed. Exterior and interior fat must be white or creamy in colour and firm.

QUALITY: The flesh should be firm and light pink in colour and there should be moderate fat streakings in the inside flank muscles and between the ribs. Rib bones should be narrow and red in colour, knuckle joints moist and red in colour.

As based upon a New Zealand standard for judging, using both visual and measurements for appraisal, a 13·6 kg (30 lb) "perfect" lamb carcase should have the following measurements.

   1. Bone length, measured from the tibia–tarsus
bone to the protrusion just above the stifle joint 162 mm (6·4 in.)
    2. Depth of rib eye at the 12th rib   35 mm (1·4 in.)
    3. Fat cover over rib eye   19 mm (0·12 in.)

344

The Revised Grade Symbols for New Zealand Meats
Revisions are operative from the start of the 1978/79 killing season

	Lamb grades			Mutton grades	
New symbol	Weight range	Colour code	New symbol	Weight range	Colour code
PL	8·0–12·5 kg 17·5–27·5 lb	Blue		Hoggets	
PM	13·0–16·0 kg 28·5–35·5 lb	Blue	HL	Up to 22·0 kg 48·5 lb	Blue
PX	16·5–19·5 kg 36·4–43·0 lb	Blue	HM	22·5–26·0 kg 49·5–57·5 lb	Blue
PH	16·5–19·5 kg 36·4–43·0 lb	Blue	HX	Up to 22·0 kg 48·5 lb	Red
PHH	20·0–25·5 kg 44·1–56·2 lb	Blue			
YL	8·0–12·5 kg 17·5–27·5 lb	Red		Wethers/Ewes	
YM	13·0–16·0 kg 28·5–35·5 lb	Red	ML1	Up to 22·0 kg 48·5 lb	Blue
OL	8·0–12·5 kg 17·5–27·5 lb	Blue stripe	ML2	22·5–26·0 kg 49·5–57·5 lb	Blue
OM	13·0–16·0 kg 28·5–35·5 lb	Blue stripe	MH1	26·5–30·0 kg 58·5–66·0 lb	Blue

Colour Coding

Symbol	Ticket colour	Symbol colour	Wrap colour			
				MH2	30·5–36·0 kg 67·0–79·5 lb	Blue
PL	Blue	Blue	Blue	MX	Up to 26·0 kg 57·5 lb	Red
PM	Yellow	Blue	Blue	MM	All weights	Black
PX	Red	Blue	Blue stripe	MF	All weights	Black
PH	Salmon	Blue	Blue stripe			
PHH	Grey	Blue	Blue stripe			
YL	Cerise	Red	Red		Rams	
YM	Gold	Red	Red	R	All weights	Black
OL	Blue	Blue stripe	Blue stripe			
OM	Yellow	Blue stripe	Blue stripe			

Over half of the supplies of lamb/mutton available in the United Kingdom are imported, and of these imports New Zealand makes the major contribution, usually between about 85% and 90%.

IMPORTED LAMB/MUTTON

Consequently the methods of grading New Zealand lamb for export is of considerable practical interest to the butcher.

NEW ZEALAND GRADING SYSTEM

All New Zealand Lamb exported is strictly graded to ensure uniformity within specified criteria. Grading allows the exact, and often varying requirements of particular markets to be met precisely and facilitates long-term contract arrangements around the world.

Carcases By Grade

P GRADE—P Grade lamb carcases should be well fleshed in the legs, loins and forequarters. They should carry adequate but not excessive fat cover.

Y GRADE—Y Grade lamb carcases are more lightly fleshed and carry less fat than P Grade carcases.

O GRADE—O Grade lamb carcases are similar to P Grade except that the legs are elongated with a long shank, giving an accentuated V shape at the crutch.

Standard Cuts Available For Export

Each *standard* cut is reference coded with a four figure number. Primal cuts are normally individually wrapped (IW) in shrink film or other protective wrapping, and packed in cartons up to about 25 kg (55 lb). The identification marks for lamb cuts in cartons are shown on the end of parcels and are colour coded.

Grade	End Panel Colour
P	Blue
Y	Red
O	Blue stripe

LAMB CUTS

The following cuts of lamb are available and in addition they can be obtained from hogget, wether and ewe mutton.

Lamb Sides—from carcases split through the spine. Side Boneless—product may be rolled, tied, netted or inserted into a polythene casing.

346

Leg—Chump on (Long Cut), boneless, chump on; chump off (Short Cut).

Hinds and Ends—the entire remaining portion of the bone in carcase after the removal of 5-rib forequarters.

Pistola—cut from bone-in carcase cut at right angles to the chine bone to leave 8 pairs of ribs on the pistola; flap removed.

Best End Neck (Rack)—seven ribs, the flap removed in a line parallel to the chine bone.

Loin—Chump-on (Long Loin); short loin, prepared from a bone-in side by the removal of the leg, chump-on and by the removal of the rack.

Loin, chump-off—taken from a bone-in side by the removal of a 5-rib forequarter. The flap is removed in a line parallel with the chine bone.

Saddle—take from a bone-in carcase, with the forequarters removed between the 5th and 6th ribs, and the legs, chump-on, removed at the ilium.

Forequarter, bone-in (Short F4)—prepared from a bone-in side, by a cut at right angles to the chine bone, between the 5th and 6th ribs. It includes the neck and shank.

Forequarter, Boneless (Shoulder Boneless)—Prepared from bone-in forequarters by the removal of all bones, may be rolled, tied, netted or inserted into a polythene casing.

Shoulder, Square Cut—taken from a bone-in forequarter by removing the shank and breast on a straight line parallel to the chine bone. The neck is removed by cutting between the 3rd and 4th cervical vertebrae.

Oyster Cut Shoulder—prepared from a bone-in forequarter and including the foreshank, arm bone and shoulder blade together with all the muscles overlying the shoulder blade and the first underlying muscle attached to the shoulder blade. These parts are separated from the balance of the forequarter along the first natural seam beneath the blade.

Other Portions—Breast and Flap—consists of the point-end breast and flap removed from a bone-in side or forequarter and
Flap—the bone-in flap is removed from the loin chump-on or loin chump-off, by straight cut parallel to the chine bone.
Shank—the bone-in shank is removed from the shoulder by a cut through the arm bone—the knuckle tip is not removed.

Neck Slices (neck, chops)—the full neck is derived from a bone-in carcase and consists of up to 4 cervical vertebrae and associated muscle tissue, cut into slices at right angles to the chine bone. Each slice is approximately 15 mm thick.

(Reference coding for the various cuts and illustrations are given in "An illustrated guide to New Zealand Lamb and Other Carcase Meat and Meat Products". *New Zealand Meat Producers Board*.)

New Zealand mutton is produced, processed and distributed to the same high standards as lamb. The cutting methods for lamb are equally applicable to mutton.

MUTTON GRADES

Mutton carcases are graded for the following characteristics:

Age—sheep are classed as mutton if they are over twelve months of age at the time of slaughter.

Sex—carcases are separated into:

(a) Hoggets (castrated males or maiden ewes with no more than two permanent incisor teeth in wear).

(b) Ewes (female ovines with more than two permanent incisors) and Wethers (castrated males with no ram characteristics).

(c) Rams (all uncastrated carcases showing strong male characteristics).

Weight—The weight range of the dressed carcase.

Conformation—The shape and muscling of the carcase is taken into account in assessing a particular grade.

Fat Cover—Carcases which are deficient in fat cover, with weak muscling are graded MM. Carcases carrying excess fat are graded MF.

Colour—Carcases having excessively dark lean meat or yellow fat will not be included in any of the carcase grades for export.

Boneless Mutton

Boneless mutton is produced from any of the grades listed. Fat content is expressed on a visual percentage basis. Most boneless mutton is bulk packed in a 27·2 kg (60 lb) carton as a single block, with a polythene liner, to 90% visual lean (VL). Further specifications to be stated if required: the particular grade(s) of mutton; the

348

percentage lean and method of measurement; and whether to be packed from part or full carcases.

Cutting Tests—New Zealand Lamb
London and Home Counties Cutting

Grades Gross Weight	PM 14·5 kg (32 lb)	YM 14 kg (31 lb)	YL 11·6 kg (25½ lb)
Percentages of gross weight			
2 legs	24·71	24·81	23·65
2 shoulders	21·48	23·60	22·18
2 best end necks	9·67	10·28	9·93
2 loins	21·97	19·16	21·82
2 breasts	10·65	10·28	8·70
2 scrags and middles	8·98	9·87	11·27
Totals	97·46	98·00	97·55
Loss, wrapper, drip etc.	2·54	2·00	2·45
	100%	100%	100%
Boneless Meat	76·96	75·56	74·02
Total Bone	20·50	22·44	23·53

Australian Grading

The weight classifications, with a few slight variations, tend to be similar to those used by New Zealand. Quality grades are identified by the tag colour and/or markings on the wrap or container in the appropriate colour.

1st Quality Blue; 2nd Quality Red and 3rd Quality Black.

The grading of carcase lamb is made according to the following characteristics:—

Age—no carcase can be classed as lamb if permanent incisor teeth have been cut.
Weight of dressed carcase.
Colour and appearance of bones, flesh and fat.

Conformation.
Fat covering.

Lamb—young animals within the dentition requirement, of either sex, 1st Quality; 2nd Quality and 3rd Quality.
Lamb cuts (bone-in and boneless)—cuts of lamb are graded in accordance with the quality of the carcase from which they are derived.
Boneless lamb (manufacturing type packs)—this type of lamb may be graded in accordance with the quality of the carcase from which it is derived.

Note Cuts of lamb of 1st and 2nd Quality may be mixed in one package, in which case they are classified as 2nd Quality and the trade descriptions are shown in red.

HOGGET

The grading of carcase hogget is made according to the following characteristics:—

Age—no carcase shall be classed as a hogget unless two permanent incisor teeth have been cut, which teeth shall not have fully grown.
Weight of dressed carcase.
Conformation.
Fat covering.

EXPORT GRADES ARE:

1st Quality; 2nd Quality.
There is no official classification of sex. Usually carcases may be of either sex.
Hogget cuts (bone-in and boneless) are graded in accordance with the quality of the carcase from which they are derived.
Boneless Hogget (manufacturing type packs). See Mutton section.

Note Cuts of hogget of 1st and 2nd Quality may be mixed in one package, in which case they are classified as 2nd Quality and the trade descriptions are shown in red.

350

The grading of carcase mutton is made according to the following characteristics:—

MUTTON

Sex.
Age—animals with at least two permanent incisor teeth.
Weight of dressed carcase.
Conformation.
Fat covering.

EXPORT GRADES ARE:

Wether mutton—adult castrated male animals.
1st Quality; 2nd Quality; 3rd Quality.
Other qualities.*
Maiden ewe mutton—adult female animals which show no signs of having been used for breeding.
1st Quality; 2nd Quality; 3rd Quality.

Note For practical purposes, it is usual to grade and ship wether mutton and maiden ewe mutton as one product, under the trade description "wether and/or maiden ewe".

Ewe mutton—adult female animals which have been used for breeding.
2nd Quality; 3rd Quality.
Other qualities.*

Ram mutton—adult uncastrated male animals (including stags).
One quality only, equivalent to 3rd or manufacturing qualities.

Mutton cuts (bone-in and boneless)—are graded according to the quality of the carcase from which they are derived.

Boneless mutton (manufacturing type packs)—this type of mutton is derived from all the above qualities and, when packed, may be either graded in accordance with the original quality of the carcase from which it is derived, or bulked together in one manufacturing grade of 3rd Quality.

* Other qualities refers to qualities lower than 3rd quality which can be produced for bone-in or boneless manufacturing purposes

351

Note Cuts of mutton of 1st and 2nd Quality may be mixed in one package, in which case they are classified as 2nd Quality and the trade descriptions are shown in red.

LAMB, MUTTON CUTTING

The sequence of carrying out the cutting of the carcase into primal cuts can vary considerably and the following indicate methods commonly used in many parts of the country.

The carcase may be suspended from a rail and chopped down into two sides, although where whole scrag "chops" are required, removal of the scrag by cutting round the neck, followed by sawing through the bone, may precede splitting.

As an alternative, the carcase may be divided into hinds and fores, usually between the 12th and 13th ribs, leaving one pair of ribs on the hinds.

The hinds are split by cutting between the centre of the legs, through the cartilage of the aitchbone. The tail is split, leaving as far as possible half the tail on each hind, followed by chopping down through the centre of the back bone.

The amount of breasts left attached to the fores will be influenced by whether or not the loin is to be sold whole or in the form of chops.

When splitting the fores, some butchers prefer to remove the breast before chopping down, as it permits greater freedom with the chopper. On the other hand, many will claim that a better shaped shoulder can be obtained if the breast is left on.

The following will apply mainly to fresh or defrosted rather than frozen lamb. Bandsaw cutting will be dealt with in a subsequent section.

Leg

With the hind lying back down on the block, cut a small section of the "flap" on to the leg and saw across the chump at a slight angle,

352

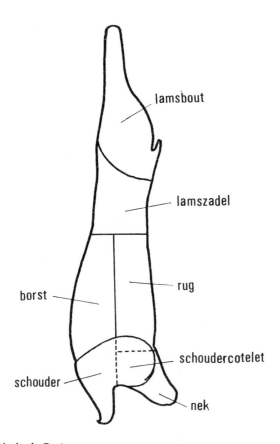

Fig. 2. *Netherlands Cutting.*

completing the separation with one clean cut of the knife. The line of demarcation may be influenced by the demand for chump chops, and to some extent by the weight of leg required.

Loin

If the loin is to be sold whole for roasting, the subsequent carving is facilitated if the chump bone is sawn through two or three times and the loin chined. Whilst chopping through the cartilages between the lumbar vertebrae may be satisfactory, if *properly* carried out, one sometimes finds that the "eye" muscle is chopped into and frequently the joints are missed.

It may be found that the purchaser will accept the kidney and suet with the loin, provided that the suet is not excessive.

353

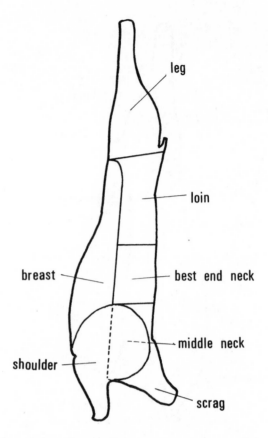

Fig. 3. *London and Home Counties Cutting.*

Shoulder

The shoulder may be round cut or square cut. The position of the extremity of the cartilage of the blade bone is situated below the 6th rib and where necessary can be found with the point of the knife.

A cut is then made from above the elbow round to the gristle end of the scapula continuing along the back of the middle neck to the scrag. Care should be taken to ensure that this cut does not penetrate into the underlying muscles of the neck and scrag. Slight pressure on the shank will lift the shoulder from the middle neck. Then stand the fore on its back and cut down as close to the ribs as possible, until the seam between the neck and shoulder muscles is reached. It is then possible to raise the shoulder and pass the knife between the shoulder and neck, cutting upwards at about 1 in. from the end of the neck. If carried out correctly, the butcher aims at showing the "five

354

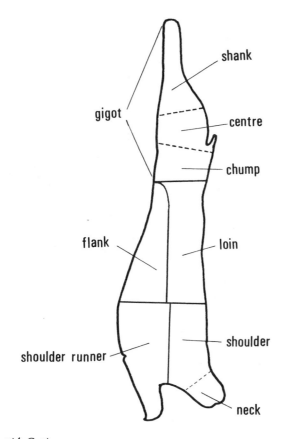

shank

gigot

centre

chump

flank

loin

shoulder

shoulder runner

neck

Fig. 4. Scottish Cutting.

fingers", five distinct marks caused by cutting through the thoracic muscles.

Target

The portion remaining from the fore after removal of the shoulder, is sometimes referred to as the target which consists of the best end neck, breast, middle neck and scrag. In removing the breast from the target, sloping the cut slightly will increase the exposed area, improving the appearance of the breast.

The best end neck is removed by cutting down to the back bone, followed by chopping the bone. Assuming that the carcase has been quartered between the 12th and 13th ribs, and that the shoulder is

355

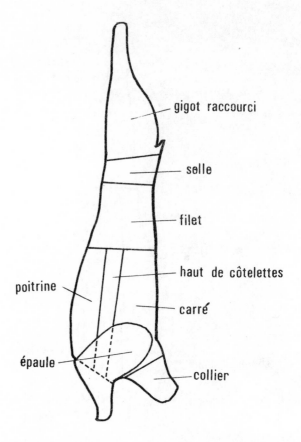

Fig. 5. French (Paris) Cutting.

cut fair, the best end neck will contain six rib bones. As in the case of the loin, it may be chined. The portion remaining consists of the scrag and middle neck and is frequently sold as one piece. As previously mentioned, the whole scrag (unsplit) may be cut from the carcase to provide neck cutlets.

Removing Bark

The bark or fell is a rough skin and for many types of trade it should be removed from the exterior surface of the loin and the best end neck, by stripping away.

BANDSAW CUTTING

When cutting imported lamb/mutton on a bandsaw it is desirable

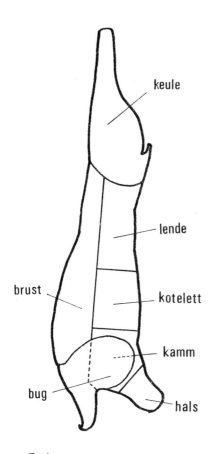

keule

lende

brust

kotelett

kamm

bug

hals

ig. 6.   *West German Cutting.*

at the carcases should be reasonably firm. One of the main diffi-
lties in using a bandsaw is that of obtaining a traditionally shaped
oulder. For this reason some firms prefer to mark round the
osition of the shoulder and chopping down to the ribs at the end
the scapula. This is followed by pushing the chopper under the
apula and pressing downwards to lever the shoulder away, the
oulder then being completely released by using a boning knife.
his method does not of course apply when blade and arm chops
e produced by sawing through the shoulder.

he carcase ex the shoulders, may then be dealt with in the following
anner:

357

a. The legs can be removed as a pair, aligning them so that the c
will be made just below the aitchbone.
b. The breasts can be removed from the jacket with the scrag en
nearest to the cutting edge and the breasts on the right. Th
cutting line should start at the 1st bone of the breast and the
angle slightly inwards towards the loin. Owing to the differen
in the value of the breasts and the chops, the line of demarcatio
must be accurate to ensure that an acceptable amount of "tops
is left on the loins and best end necks. As a rough guide the leng
of the tops should be about twice that of the eye muscle.
c. The chine and best end necks are separated from the middle nec
and scrag by sawing straight across between the 6th and 7th rik
d. The bark should be stripped off the loin and best end neck an
the suet (and kidney if present) removed, prior to splitting. Th
is carried out by sawing from the best ends towards the chump
following the centre line of the back bone. If intended for chop
the loin and best end neck can be left whole, rather than cutti
the six rib best end neck.
e. The pair of legs are separated by placing tail downwards on th
platform, care being taken to split the tail and the cartilage
the aitchbone. Where legs are intended for pre-packing, the shar
bone can be sawn through, but leaving it attached by the shar
muscle.
f. The scrag and middle necks can be split for stewing or sawn acr
to give cutlets. In either case it is desirable to remove the tip er
of the scrag.

Other cuts

CROWN ROAST: This decorative joint is prepared from a pair of six r
best end necks, as follows:

The necks are placed with the rib side upwards, and the ribs saw
across close to the vertebrae.

The vertebrae is removed with the boning knife, care being take
not to penetrate the outer fatty layer of the joint.

With the joint laid out flat, about 2 in. of the meat is removed fro
the end of the ribs, and the rib fingers trimmed away.

358

The joint is then turned inside out and shaped into a circle. With this method there is no opening at the bottom of the joint.

The crown roast can be secured by a single stitch at each of the two ends.

Decoration is largely a matter of personal taste, but tomato cut diamond fashion provides a good centre and paper frills or stuffed olives can be placed on the ends of the ribs.

SADDLE OF MUTTON: This is prepared from a good quality chine (pair of loins) with a well-developed eye muscle and preferably with the tail on. The following method is suggested:—

Remove the bark, pulling it from the fore-end to the tail, take out the kidneys and trim any excessive suet and the end of the flaps.

The tail is split and secured round the two kidneys by the use of two small skewers.

The final decoration can be carried out by plaiting thin strips of pork fat and forming a pork-fat rose.

FRENCHED LEG: There are alternative methods of dealing with this.

Traditionally the shank meat is removed at about 3 in. from its extremity and the end of the knuckle bone sawn off.

The leg can be treated as in the first method, but in addition the aitchbone and the tail bones are removed and the leg secured by stringing or by stretch netting.

In some cases a long cut leg is used which includes the chump meat.

ROLLED BREAST WITH SHANK: It is frequently found that rolled breast as such, does not provide an attractive joint. The lean content and the general appearance can be improved by cutting off a full cut breast with the shank of the shoulder attached. The shank meat can then be rolled into the breast.

359

SCRAG: Where the neck has been removed whole, double neck slice provide reasonably priced portions which can be braised or cooke en casserole. Such portions are not suitable for frying.

Pairs of slices are cut down from the fat side of the neck, but leavin the throat portion of the two slices attached, so that they can b opened up and laid flat as one portion.

WHOLESALE CUTS (LAMB)

Wholesale cuts will show some variation in different parts of th country, they will also be influenced by the amount of trim an method of cutting, particularly manual as against bandsaw.

The following figures are representative of those normally anti cipated from a good commercial quality lamb.

Cut	Definition	Approx % o carcase weigh
Hinds	pair of legs and loins	47
Hinds and Ends	pair of legs, loins and best end necks	58
Legs		25
Haunches	legs with chump ends of the loins attached	33
Chines and Ends	pair of loins and pair of best necks	34
Chines/Saddle	pair of loins	23
Ends	pair of best end necks	11
Fores	pair of $F\frac{1}{4}$s	53
Short Fores (with breasts)	$F\frac{1}{4}$s less best end necks	42
Shoulders		22
Sets	pairs of scrags, middle necks and breasts	19
Breasts		9
Scrags		10
Trunks	carcase, less legs	75
Whole Necks (targets)	pair of $F\frac{1}{4}$s, less shoulders	31

Desirability Factors (Lamb)

As based upon the London and Home Counties method of cutting the cost per kg (lb) when multiplied by the following factors for th

360

ppropriate cuts will yield a gross margin of approximately 30%
n sales (see p. 272, Beef).

English Lamb 18·14 kg (40 lb)

Portions	Desirability Factors
Legs	1·95
Shoulders	1·30
Chops (loin and chump)	2·14
Best End cutlets	1·89
Scrags and middles	0·88
Breast	0·59
Kidneys	1·03
Chumps	0·44

(Calculated on a cutting loss, fat trim, evaporation etc., of 6·7%)

Culinary Uses

Portions	Use
Legs	Roasting, braising, boiling; leg cutlets, frying or grilling
Loins	Roasting or in chops for frying or grilling
Shoulders	Roasting, braising
Breasts	Stewing or stuffed and roasted
Best End Neck	Roasting or as cutlets for grilling or frying
Middle Neck and Scrag	Stewing

Pork and Bacon Cutting

In contrast with cattle and sheep, the pig has incisor teeth in both upper and lower jaw, the central incisors extending forward to enable roots to be eaten.

As in the case of humans, the pig is omnivorous, the stomach a simple undivided organ and it is capable of dealing with animal as well as vegetable foods.

As it does not have a ruminant stomach, it cannot utilise large amounts of fibre. An important anatomical feature of the pig is the presence of a movable cartilaginous disc forming the snout, enabling the pig to "root" in the soil.

363

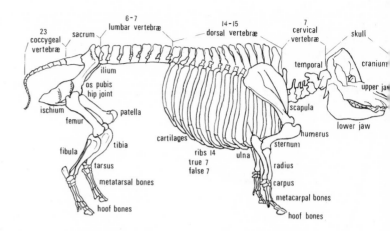

Fig. 1.   *Skeleton of Porcine Animal.*

Skill in cutting necessitates a good knowledge of the skeletal for
of the carcase, the positions of the bones and joints providing gui
points for the various lines of demarcation.

VERTEBRAE

Cervical	7 bones
Dorsal	14 or 15 bones
Lumbar	6 or 7 bones
Sacral	4 bones (fused together)
Coccygeal	20 to 26 bones (frequently 23)

PELVIC GIRDLE

Ilium
Os Pubis
Ischium

RIBS

14 or 15 pairs of ribs may be present, 7 pairs sternal and 7 or 8 pair
asternal. When the 15th pair of ribs is present, they may be com
pletely developed, the cartilage entering into the costal arch, o
they may only reach a length of as little as 1 in.

STERNUM

The breast bone consists of six sternabrae fused together.

364

HIND LIMB	FRONT LIMB	**LIMBS**
Femur	Scapula	
Patella	Humerus	
Tibia–Fibula	Ulna–Radius	
Tarsal (7 bones)	Carpus (8 bones)	
Metatarsal (4 bones)	Metacarpus (4 bones)	
Hoof bones	Hoof bones	

The main regions consist of: Cranium   **SKULL**
Upper jaw
Lower jaw

Dentition (adult): Canines, 4 teeth
Incisors, 12 teeth
Premolars, 16 teeth
Molars, 12 teeth

The hyoid bone is attached to the tongue, which as compared with that of the sheep has no central depression, is relatively long and thin and there are two or three vallate papillae towards the root of the tongue. It also has a large number of soft, long pointed papillae directed backwards towards the epiglottis.

MARKET TYPES

The vast majority of pigs coming on to the market consist of hogs (castrated males) and gilts. Of recent years there has been an increased interest in the production of young boars as they show good growth and food conversion rates and will usually produce a leaner carcase than will castrates.

At one time there was a traditional prejudice against such carcases on the alleged possibility of "boar taint".

Investigations into this problem have proved that such allegations have no scientific basis and certainly with pork and bacon weight carcases this difficulty should not arise.

The carcases from mature sows and boars are normally used for manufacturing purposes.

Various arbitrary classifications, largely based upon weight, are used for pork carcases which include porkets, pork pigs, cutters, bacon pigs and heavy hogs.

Broadly speaking, there is a fairly rapid decrease in the value per kg (lb) of the carcase as the carcase weight increases. The graph on p. 377 (Fig. 6) will give an indication of the general trend, as based upon Smithfield Market, London.

CARCASE YIELD

With the rapid extension of the Pig Carcase Classification Scheme (see p. 46) probe measurements of the backfat thickness on unsplit carcases provide a valuable indirect guide as to the amount of lean meat.

From this work it has been shown that the proportion of the joint and the distribution of the lean meat (by weight) in the carcase vary less than is popularly supposed, assuming that standard methods of cutting are used. In addition, the conformation of the carcase is not closely related to such characteristics.

However, what can vary is the *thickness* of the lean meat, that is, the same weight of lean meat when laid down over short thick bones, will be deposited more thickly than when distributed over longer thinner bones. Thus the plump carcase will generally have no higher lean meat content than the slimmer one of the same weight and fat thickness.

From this it may be concluded that the size of the eye muscle cannot be taken as an index of the lean meat content nor of the amount of lean meat in the higher priced cuts. However, from the aspect of display and consumer appeal, the thicker deposition of muscle may prove to be more attractive.

Excessive fat deposition is the most significant single factor influencing the yield of pork cuts as prepared for retail sale. It has been estimated that an increase of approximately 25·4 mm (1 in.) more backfat measured at a single point of the carcase will result in 3% less retail value than another of the same conformation and weight.

366

For the Wiltshire Bacon type of pig, a carcase weight of about 68–73 kg (150–160 lb) is preferred and from good pigs the weight of the trimmed sides as ready for curing will be in the region of 78% to 79% of the original carcase weight.

The "Heavy Hog" of about 90–113 kg (200–250 lb) carcase weight is frequently preferred by some of the larger organisations dealing with pre-packed bacon and the preparation of pies and sausages.

The intention of this type of pig production is to spread the initial costs over a greater carcase weight. The heavier carcase will yield a greater weight of lean meat but will also tend to have more fat, but the latter can usually be economically utilised by the food processor. With an increase in the weight there is usually an increase in the dressing out percentage, whilst the proportions of head, bones and kidneys will decrease.

The methods of cutting pig carcases in the wholesale markets may vary slightly in different parts of the country, but in general the following terms are widely recognised. The percentages given can

WHOLESALE PRIMAL CUTS

	Approx % of carcase (weight)
Pig ex (less head)	91
Loin (long, including neck end)	39
Short loin (loin less neck end)	25
Hog Meat (long loin less rind & some fat)	★
Short Hog Meat (short loin less rind & some fat)	★
Jacket (hands & bellies & long loins)	69
Legs (cut will vary depending on demand)	22
Legs & Long Loins	61
Hands & Bellies	30
Neck Ends	14
Fore-ends (neck ends & hands & springs)	27
Hands ⎫ (depending on cut)	13
Bellies ⎭	17
Head	8 to 9

★ The yield of hog meats will vary widely depending upon the quantity of fat on the loin and the degree of trim. One large catering group specifies that the backfat over the eye muscle shall not exceed 13 mm (approximately ½ an inch)

be taken as a guide of fairly representative yields which can reason
ably be anticipated from good commercial pigs of about 55 kg
(120 lb) carcase weight.

In some cases bellies of pork may be sold with the flare fat removed
this will of course influence the belly yield.

PORK CUTTING

Some variations may occur in the sequence of operations involved
in the cutting of pork. For example, in some wholesale markets, the
belly with the hand and spring attached may be removed from the
hanging carcase. The current trend towards marketing sides of pork
will eliminate the need for splitting at retail level and incidentally
it provides a better indication of backfat cover than can be judged
visually in a whole carcase. The following will indicate a method
commonly employed.

> The carcase is usually hung from the tendons of the right leg
> and the aitches are split by drawing the knife in an arc from the
> vent downwards, splitting the cartilage on the pubic bones.

> In carcases of young pigs it is possible to split the first few verte-
> brae from the tail, by downward pressure with a knife.

> The centre line of the carcase can be found by light finger
> pressure and a shallow cut made over the spinal process, through
> the skin and into the underlying backfat.

> With a right-handed cutter, the pig's left ear is grasped in the
> left hand and a cut made through the back of the neck to the
> first vertebrae of the neck, and continued following the jaw
> bone, so that the head is left attached to the right side.

> When splitting lightweight carcases, it is possible to take the
> weight of the left side on the forearm, but with heavier carcases
> it is desirable to support the left leg by its tendons prior to
> splitting.

Legs from heavier pigs are frequently used in the form of pork

368

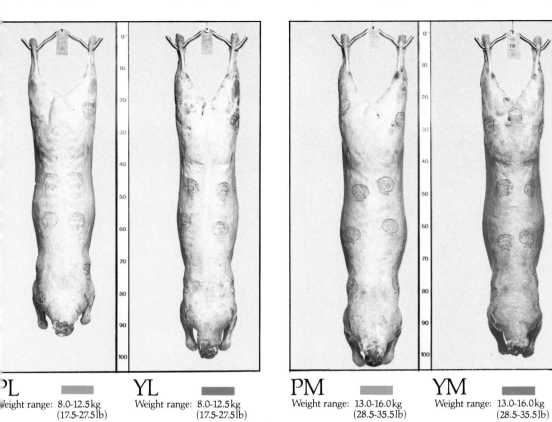

PL
Weight range: 8.0-12.5 kg (17.5-27.5 lb)

Ticket colour:
Symbol colour:
Wrap colour:

YL
Weight range: 8.0-12.5 kg (17.5-27.5 lb)

Ticket colour:
Symbol colour:
Wrap colour:

PM
Weight range: 13.0-16.0 kg (28.5-35.5 lb)

Ticket colour:
Symbol colour:
Wrap colour:

YM
Weight range: 13.0-16.0 kg (28.5-35.5 lb)

Ticket colour:
Symbol colour:
Wrap colour:

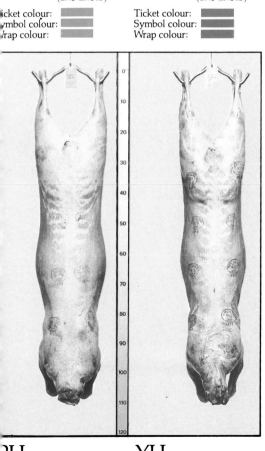
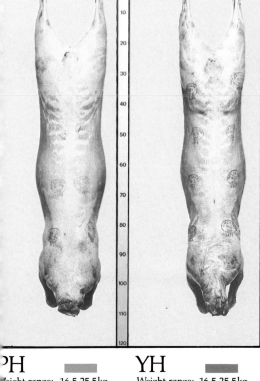
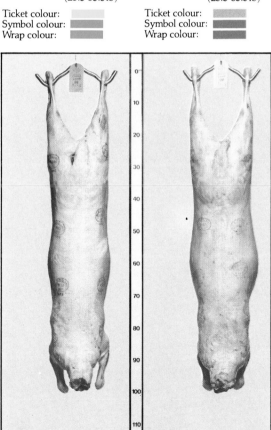

PH
Weight range: 16.5-25.5 kg (36.5-56.0 lb)

Ticket colour:
Symbol colour:
Wrap colour:

YH
Weight range: 16.5-25.5 kg (36.5-56.0 lb)

Ticket colour:
Symbol colour:
Wrap colour:

*OL
Weight range: 8.0-12.5 kg (17.5-27.5 lb)

Ticket colour:
Symbol colour:
Wrap colour:

*OM
Weight range: 13.0-16.0 kg (28.5-35.5 lb)

Ticket colour:
Symbol colour:
Wrap colour:

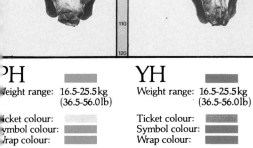
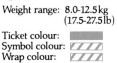

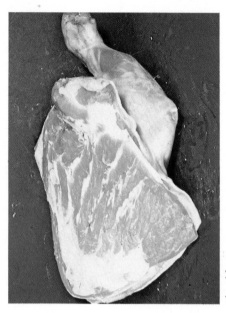

Shoulder

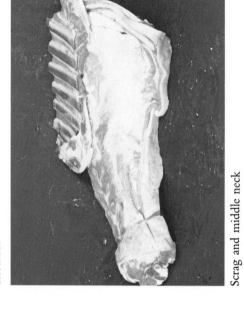

Scrag and middle neck

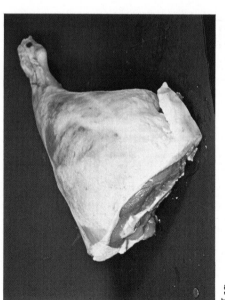

Leg

Breast

Lamb cuts

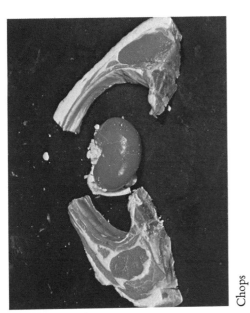

Chops

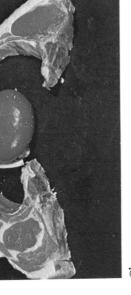

Loin chops

Lamb cuts

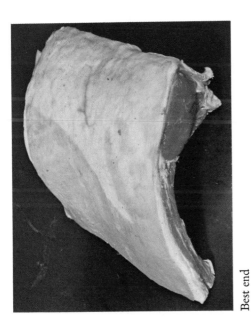

Best end

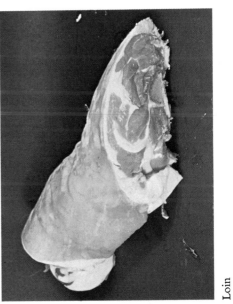

Loin

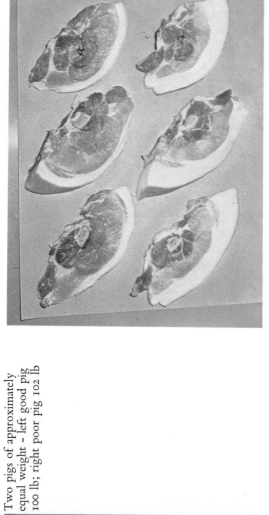

Attractive chump chops from good pig – saleable without trimming

Pork cutting

Loin is the most valuable cut – note greater length on good pig

Two pigs of approximately equal weight – left good pig 100 lb; right poor pig 102 lb

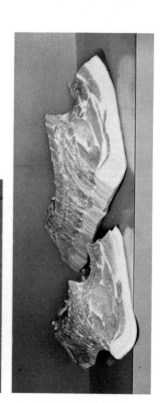

Loin of the good pig is acceptable without trimming

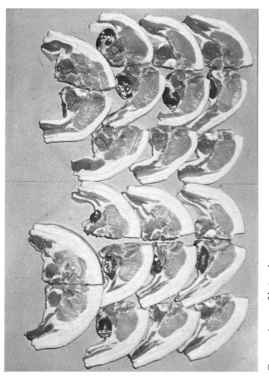

Comparison of loin chops

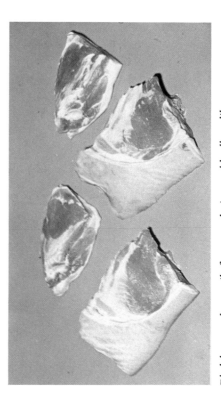

Bladebone and sparerib from good pig would sell readily

Pork cutting

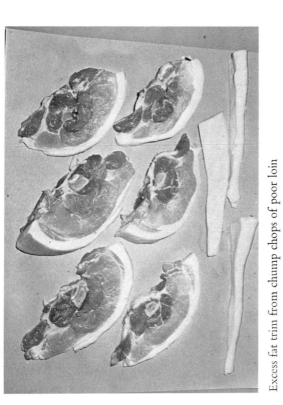

Excess fat trim from chump chops of poor loin

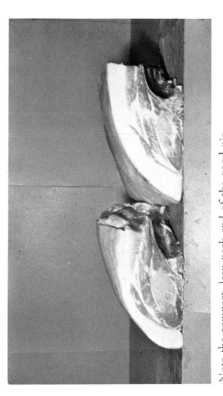

Note the compact, lean neck-end of the good pig

Excessive back fat on poor side and the "eye" muscle
lacking in both length and depth

Two sides of bacon of approximately equal weight,
left 52 lb right 51 lb

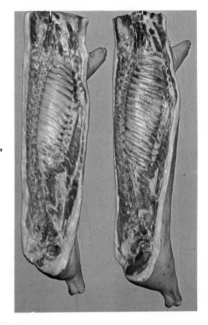

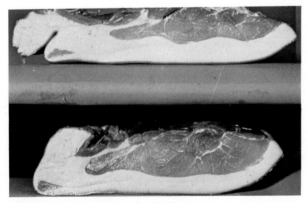

A cut through the long back emphasises the
excessive fat and shallow muscle development of the
poor side

Bacon cutting

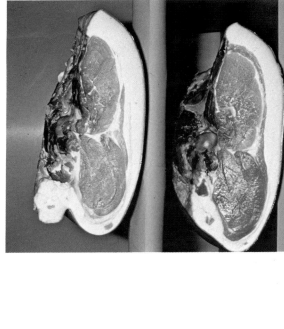

The cut surface of the gammon illustrates the contrast in the lean: fat ratio

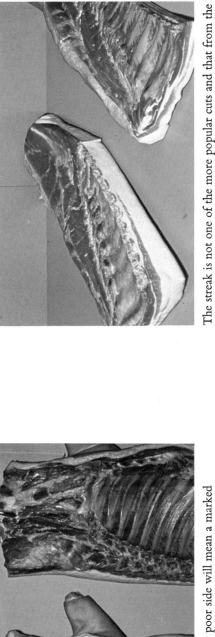

The streak is not one of the more popular cuts and that from the poor side would require drastic trimming to render it saleable

Bacon cutting

Note the shallow fleshing and the excessive fat cover of the gammon, from the poor side

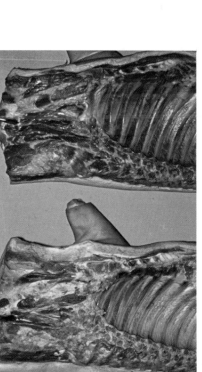

The heavy fore-end of the poor side will mean a marked reduction in the value of the side, whilst the amount of fat would make for difficult selling

Art of the Charcutier

steaks, or small rolled joints, in which case the knuckle is cut off and the bones removed from the remaining portion.

The hand and belly is removed from the long loin in the following manner.

Hand and Belly

> The joint between the blade bone and humerus can be found by touch, or confirmed with the point of a knife. It is then opened up by cutting from the neck end.
>
> Mark down over the ribs (usually fourteen) and from the last rib cut through the belly down to the chump end. The ribs are then sawn through, using a pork saw. If the left hand is placed under the back to saw the ribs, these can be cut through without sawing into the underlying meat. The hand and belly can then be divided from the loin with a clean cut, sloping the knife very slightly away from the belly.
>
> The flare fat "flick" or "flead", except in the case of very lean bellies, is usually removed prior to retail sale, by cutting lightly along its attachment with the skirt and stripping it towards the thin end of the belly.

In dividing the hand and spring from the belly, the knuckle should be pulled forward to avoid exposing the end of the shank bone. The hand is usually cut with the portion of one rib bone attached, but as there is now much more demand for the thick end of the belly, rather than hands, many butchers include all the ribs on the belly.

Hand and Spring

> A fairly light sharp chopper can be more easily controlled and it should be held at a slight upward slope from the handle so that the backbone followed by the spinous processes are split.
>
> With practice, a steady chopping rhythm can be maintained splitting the spinous processes, so that equal portions of the bone are left on each side.
>
> The left side is then detached. It may be necessary to cut some of the neck muscle in order to remove the side from the head.

N

369

The side is then laid on a cutting block and the hind trotter "broken over" by cutting through the tendons into the joint and pressing the trotter forward. In many cases where the leg is intended for retail sale, either whole or in halves, the trotter is usually completely removed. If the leg is to be used for salting by the arterial method, the trotter should not be removed until after pumping.

The Leg

The leg is removed by marking across with a knife about 13 mm ($\frac{1}{2}$ in.) below the knob of the aitchbone, at a slight angle from the tail to the flank side and the bones sawn through, using a pork saw. Care must be taken to ensure that the saw does not cut into the underlying muscle. The leg is then removed from the side with one clean sweep of the knife. This will give a "square cut" leg, with the tail attached to it (left leg).

In some areas, the traditional "round cut" leg is still preferred, in which the cut surface is markedly curved, leaving the tail attached to the loin.

Loin

The long loin is usually divided into a loin and neck end by cutting between the 4th and 5th ribs, which will leave a small portion of the gristle of the blade bone in the loin.

Neck

Methods of dealing with the neck end (blade bone and sparerib) vary, depending upon the relative demands for roasting joints or chops/cutlets.

In the case of the former, the blade bone joint is lifted from the sparerib by making a semi-circular cut through the rind, penetrating into the fat. Slight hand pressure on the top of the blade bone will enable it to be tilted away from the sparerib, which will ensure a clean cut without exposing the surface of the blade bone.

The complete neck end may be used for so-called sparerib chops by cutting through the blade bone meat, sawing the bone and including with the slices the underlying sparerib meat.

370

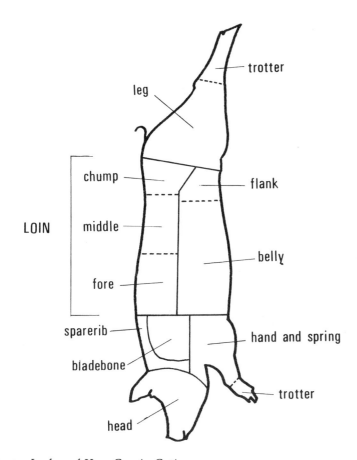

Fig. 2. London and Home Counties Cutting.

In dealing with the right side, the head is of course the first joint to be removed. The head is prepared by first chopping through the point of the lower jaw, care being taken to ensure that the tongue is not damaged. Secondly, the bone at the back of the head, where it adjoins the spine, is chopped through. The frontal bone can then be split, with a chopper, penetrating the bone only, so that the brain is not damaged. Thus on opening up the head the tongue and brains can be removed intact. The eyeballs can be removed by inserting the point of a fine knife behind the eyes, severing the ligaments; the eyeball can then be levered out from the socket, prior to cutting its attachments. Where head meat is to be used for manufacturing meat, the heads are left whole and the meat etc. removed from the skull.

THE RIGHT SIDE

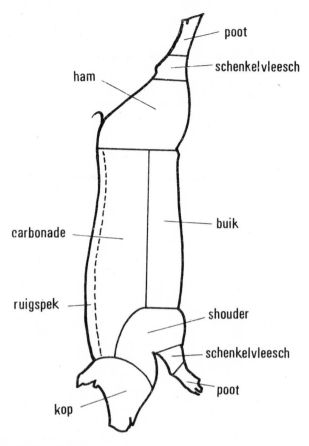

Fig. 3.   Netherlands Cutting.

Desirability Factors

As based on the London and Home Counties method of cutting, it will be found that a reasonably good quality pork pig of 45–50 kg (100–110 lb) will yield a gross margin of 30% on sales when each of the joints are multiplied by the following factors (see p. 273, Beef).

Portions	30% gross margin on sales
Leg (whole)	1·65
Loins (short, whole)	2·00
Hands and springs	0·85
Neck ends (blade bones & spareribs)	1·63
Bellies (whole)	1·45
Head, with tongue	0·15

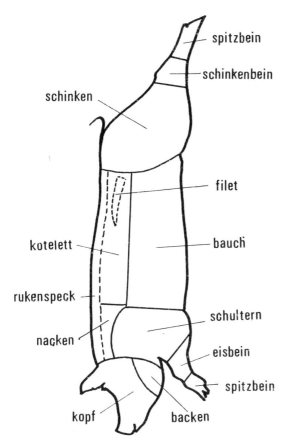

spitzbein

schinkenbein

schinken

filet

kotelett

bauch

rukenspeck

schultern

nacken

eisbein

spitzbein

kopf

backen

Fig. 4.   West German Cutting.

It must be stressed that these figures should be taken as a guide only. In the case of cutting the loin, sparerib, etc. into chops, removing the trotters and trimming fat, an appropriate increase in price must be made to allow for such losses.

Portions	Uses	
Legs	Roasting or with large legs, pork steaks, when salted, boiled	
Loins	Loin and chump chops for frying or grilling or in joints for roasting With some large pigs, particularly bacon pigs, the fillets may be removed and are a delicacy grilled or fried	**Culinary Uses**

373

Portions	Uses
Blade bones	Roasting, carving is greatly facilitated if the bone is removed and the pocket stuffed, prior to roasting
Sparerib	As a joint for roasting, or cut into chops
Blade bones and Spareribs	The blade bone, including the sparerib, is cut through in portions to provide large "sparerib" chops

PORK CUTTING TESTS

In establishing a costing system, it is essential that it should be based upon a good number of representative samples. In the case of pigs particularly, figures obtained from a single carcase can lead to serious errors if applied as a general standard. Also as far as practicable, the cutting must be uniform. To achieve this it is preferable to use anatomical guide points.

It is also emphasised that the theoretical profit arising from a "block test" will not agree with the cash actually going into the till, as indicated in "Hidden Losses" (p. 286).

Example 1 which follows shows the variation obtained in the different primal cuts, from pigs, which in general, appeared fairly uniform in type. It should be stressed that the example given are intended as a guide only. Each butcher should attempt to produce his own set of figures based upon his particular commodity and specific trade conditions.

Example 1:   6 Pure Bred Welsh Pigs
Average cold carcase weight 42 kg (92·9 lb)
Weight range 40·37–45·36 kg (89–100 lb)

	Average % Primal Cuts	Range %
Loins (short)	26·4	24·5–29·5
Neck ends	14·35	13·8–15·2
Legs (with feet)	23·10	21·5–24·0
Hands & Bellies	28·15	26·7–29·5
Heads	8·00	7·7– 8·3
	100%	

374

Example 2:   Good Cutter Pig
                Weight 54·5 kg (120 lb)

	Average % Primal Cuts
2 Legs (with feet)	23·7 (cut full)
2 Short Loins	25·0
2 Hands & Bellies	27·5
2 Neck Ends	14·6
1 Head	9·2 (cut fair)
	100%

Example 3:   Large White Bacon
                Carcase 68 kg (150 lb) Cold Weight

	%
2 Sides prepared for pumping	80·67
Head cut fair	9·17
Tongue with root	0·5
Flare fat	1·66
Fillets	0·71
Kidneys	0·41
Chine bone	2·50
Feet	2·00
Trim, tail, waste, fat & loss	2·38
	100%

Example 4:   Heavy Hog (ad lib feeding)
                Carcase weight 101·6 kg (224 lb)

	%
Fat	41·00
Lean Meat	34·51
Total for factory use	75·51
Head, bone (including feet) rind etc.	21·83
Loss, trim etc.	2·66
	100%

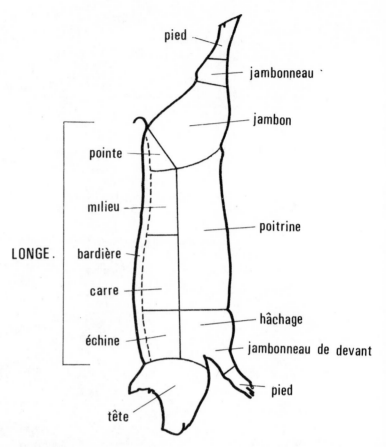

Fig. 5. French (Paris) Cutting.

Good presentation is the biggest single factor in successful bacon sales and this is largely dependent upon the initial preparation of the side.

BACON

All bacon should be examined on arrival and if necessary wiped with a clean paper towel to remove excess moisture. This applies to un-smoked rather than to smoked bacon. Moisture can give rise to sliminess and lead to rancidity in the fat. Special attention should be given to the blade bone pocket and the oyster bone as these are areas which may cause trouble from fly infestation during warmer weather. Whilst the removal of the surface bones should be carried out as soon as possible, the actual cutting of the side should be left

376

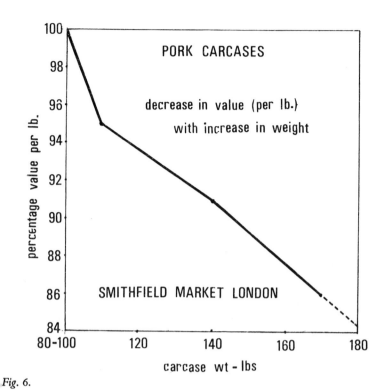

Fig. 6.

as late as possible in order to minimise weight loss from the cut surfaces. Bacon should be held in the chiller at between 2·5°C and 4°C (37°F and 40°F) in order to facilitate cutting and rashering.

The purchase of whole sides will usually be found to be more profitable than buying cuts, provided one can develop a good joint trade. In order to maintain a bright display, the volume must be related to the rate of sale and during quiet periods the proportion of vacuum packs can be increased. Correct lighting is important in the show-case to enhance the appearance, exposure to daylight resulting in fading.

An eight-day stock turn should be the objective, but where the amount sold warrants it, a twice-weekly delivery is recommended.

This particularly applies to unsmoked bacon, which is more attractive in appearance and palatability when fresh.

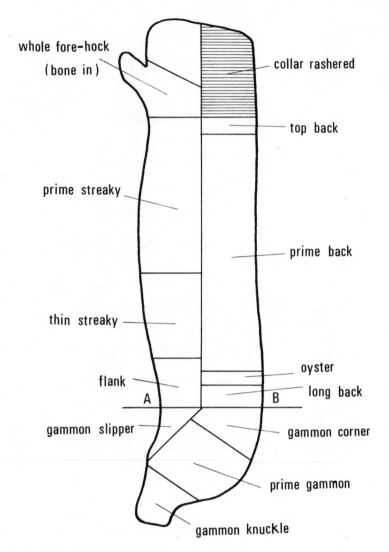

whole fore-hock
(bone in)

collar rashered

top back

prime streaky

prime back

thin streaky

oyster

flank

long back

A

B

gammon slipper

gammon corner

prime gammon

gammon knuckle

Fig. 7.   *Southern Trade bacon cutting.*

Whilst it is desirable to sell whole sides where at all possible, it may be necessary at certain times to buy cuts in order to balance sales.

Preparation of a "Wiltshire" Bacon Side

The surface bones consist of the chine, the ribs and oyster bone, and care in their removal from the side is reflected in the appearance of the rashers, after the middle has been cut on a bacon slicer.

378

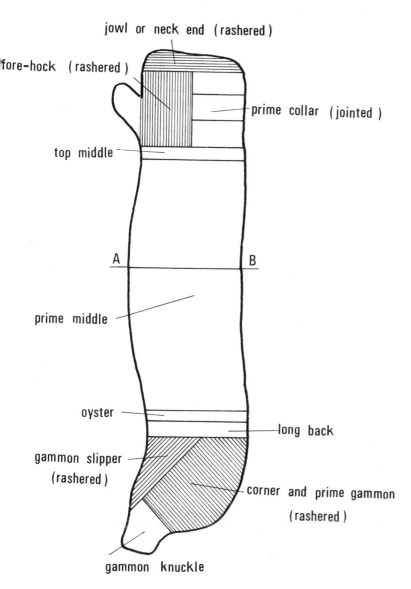

jowl or neck end (rashered)

fore-hock (rashered)

prime collar (jointed)

top middle

A   B

prime middle

oyster

long back

gammon slipper
(rashered)

corner and prime gammon
(rashered)

gammon knuckle

Fig. 8. *Northern Trade bacon cutting.*

REMOVING CHINE BONES: Those at the top of the ribs are sometimes deeply embedded and must be carefully removed, using
the top of a boning knife. Those in the loin area of the middle
are somewhat shallower and can be skimmed off much more
easily.

379

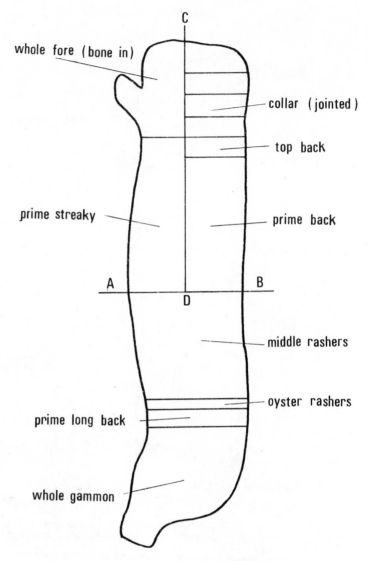

C

whole fore (bone in)

collar (jointed)

top back

prime streaky

prime back

A

B

D

middle rashers

oyster rashers

prime long back

whole gammon

Fig. 9. *A compromise between Northern and Southern form of presenting a Wiltshire side. By cutting from A to B and C to D the whole middle can be processed without using the last slice device, thus minimising waste.*

REMOVING SKIN (PLEURA) FROM RIB CAGE: As the meat on either side of the rib bones has to be loosened by a very slight incision, the removal of this skin from the inside of the rib cage makes this much easier. The skin should be stripped from the chine, downwards, towards the belly. This is not always possible with smoked bacon but is quite simple on unsmoked. It is also more important in the case of unsmoked bacon, as the texture of the meat is softer and more liable to knife penetration. In addition, in unsmoked bacon this is the first area to lose colour and its removal may improve appearance. This skin only weighs about 289 g (1 oz) and this is more than balanced by a higher standard of rasher presentation.

REMOVING RIB BONES: The meat is loosened on either side of the rib bones, particular care being taken at the top of the ribs as the underlying muscle (eye muscle) is the most valuable part of the side. The ribs can then be removed with a loop of string or a cheese-wire, passed underneath and pulling from the top towards the belly, whilst at the same time applying thumb pressure to the lifted end of the rib.

REMOVING OYSTER BONE: This is situated in the lower loin area of the middle, and in most cases it can be removed without the use of a knife, but care should be taken to protect the hands. The best method of detaching it is to lift the gammon end of the oyster bone with one hand and prise away the meat with the thumb of the other hand, until the bone is in a vertical position. As the bone has a sharp edge, a protective pad of clean cloth or paper towel should be used. The pad can then be placed on the exposed underside of the bone and the bone pressed down, snapping the tissue that surrounds it. Then from the pointed end, the operation of lifting and prising away the meat is repeated. As a rule the oyster bone can be completely detached, but in some cases there is a small tissue which must be cut with a knife.

Methods of cutting will vary in different parts of the country and even within local areas. Figures 7 and 8 indicate those frequently used in the south and north respectively and Fig. 9 is a compromise which is favoured by many traders.

Primal Cuts

One popular method is to start by cutting the side into three main portions in the following manner:

a. First cut through the side in line with the base of the second lowest rib. On the average side this is the approximate position where the thin streaky ends and the prime begins. This cut is made at a right angle to the back, with the knife upright.

b. To separate the collar and back from the fore-hock and prime streaky, insert a finger in the blade bone pocket to find the end of the fore-hock bone. Its position will provide a natural guide for the dividing line.

c. Continue this cut dividing the streaky from the back to ensure that a fair streaky is obtained, which will sell readily. A cut about halfway between the back and streak is a reasonable division.

Following removal of the knuckle, the hock can be boned and rolled whilst it is still attached to the streaky. This enables the streaky to be rashered towards the hock, which eliminates the use of the last slice device and the consequent pieces. Similarly, by stringing-up the collar, the back can be rashered in a like manner towards the collar. With the same objective in mind, the middle can be rashered whilst still attached to the gammon.

To prepare the gammon for joints and/or steaks, remove the knuckle by cutting at a right angle to the main bone, then remove the slipper (similar to thick flank in beef) by cutting along the main bone up to the knee cap (patella). This small bone can be quickly removed, leaving a compact slipper joint weighing 0·68–0·90 kg (1½–2 lb). Following removal of the main bone (femur) this portion should be tightly strung, ready to cut into steaks, rashers or joints.

PRE-PACKED BACON

Consumer acceptance of pre-packed bacon has increased rapidly during the last few years, mainly due to good presentation, value for money and full rashers with no bits or pieces. Many housewives prefer pre-packed bacon, as being hygienic and convenient. Whilst the gross profit may not be as much as with traditional bacon, the net profit is often more. Consequently it may be a good plan to stock both.

There are two main types of pre-packed bacon—vacuum-packed and film-wrapped. Most of the former is packed centrally and then distributed to retail outlets. Vacuum-packed bacon has a shelf life of about 3 weeks from the time of leaving the packing plant, but by the time it has reached the shop this period has been reduced to about 14 days.

This should be adequate if proper stock control is implemented, but consumer resistance may be encountered if a packet has only "one or two days to go". It is generally available in various weight consumer packs, and 3 and 5 lb bulk packs; Meta-pressed pre-packed joints are also available.

It has been shown that film wrapping can help the sale of bacon, provided that certain principles are rigidly adhered to. It must be appreciated that film wrapping is a form of *presentation* not *preservation* and it should receive the same care as that given to bacon displayed in the traditional fashion.

Packs can be heat sealed, using a hot plate or iron, or a shrink bag can be used and in both cases the shelf life is 3 to 4 days. It is essential that absolutely fresh bacon only be used for pre-packaging, and a system of code marking will assist proper stock control. As the objective is to get the film tightly on the surface of the meat, the bag should be as small as possible and the meat should be placed in the stuffer so that the lean end finishes up in the "nose" of the bag and the best face of the joint is not covered by the seam. Identity labels and price and weight tickets, should not obscure the more attractive areas of the joint.

When packing a number of joints, allow the bags some time to retract and thus expel any odd pockets of air, then smooth over by hand to remove excess air, prior to twisting and clipping. Boneless bacon should always be used and stringing is not necessary. As the joints can be cooked in the bag, the bag helps to mould the joint for economical carving.

Poultry and Game

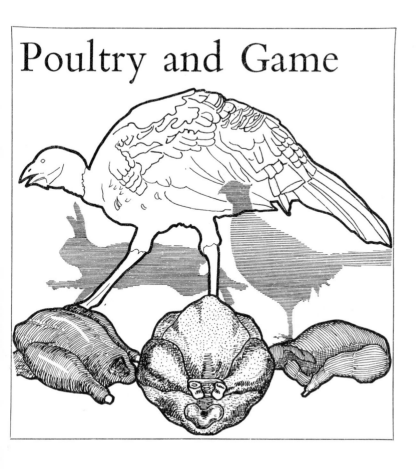

Since the last war and particularly over the past 10 or 12 years, the poultry industry has grown by leaps and bounds and now plays an important role in the country's food production. Gone are the days when poultry was considered a luxury. The chicken today is found in practically every butcher's shop, the supermarkets and the large multiples. It is bought as a stable item of diet by all sections of the community and, moreover, until fairly recently it has been much underpriced. It is hardly necessary to stress how important the sale of poultry has become to the family butcher. Before the war we imported large quantities of chicken from all over Europe, including Russia and also from America. During the Christmas period, truckloads of turkeys came from the Balkans, France, Italy and Canada.

385

Very good ducks, well-packed and graded, in boxes of six, came from China. Now, thanks to the efforts of our own producers, imports are negligible and practically confined to geese from Denmark and Poland at Christmas time. Indeed great credit is due to our producers who, by supplying us regularly with millions of chickens, turkeys, etc., have revolutionised the poultry trade.

The distributing trade does not always see eye to eye with the producer from the marketing angle but nevertheless we have much for which to thank him. As far as marketing is concerned the work of both wholesaler and retailer has been made relatively easy. With the exception of a percentage of fresh killed turkeys, all poultry is cartoned and graded to weights and quality and mostly under the producers special brand name. There are two types of poultry which have to be considered—the fresh and the oven-ready, the latter nearly always frozen. Generally speaking they are two different trades, the oven-ready sold by the supermarkets and the multiples, and the fresh by the butcher, for his more discriminating clientele.

The great bulk of poultry produced is oven-ready; it would not be possible to market the many millions in any other way, but there is a big market for the fresh article also. Fresh broiler chicken, strangely enough, are often thought by the housewife to be free range when she sees the fresh bird on show in the shops and who would want to disillusion her. One often hears rather disparaging remarks made of the broiler as to its lack of flavour. If this is true, and a chicken can easily be spoilt by over cooking or reheating, then it is equally true of a great deal of the food we eat today. Certainly, thanks to modern methods of production, nobody should ever get a skinny or tough bird as was often the case pre-war. You may be surprised and amused to know that a second quality bird in the olden days in Smithfield market was called rather condescendingly a "Butchers".

Turkeys

In the year 1952 a small group of turkey farmers formed themselves into a body that was to become known as "The British Turkey Federation". At that time the breeding of turkeys for the Christmas trade was a most lucrative business and the Federation expanded rapidly. It must be said that the situation is very different today

386

and the turkey farmer, because of extra food costs, is possibly losing money. About 1958 the oven-ready bird was first introduced into this country by Mr. John Lintern of Hoppers Farm, Buckinghamshire. So began the large turkey industry of today. The two types of turkey, the oven-ready and the fresh, constitute an entirely different trade. The former are by a long way the more prolific and as a rule the cheaper bird of the two. The industry is controlled very largely by the BTF whose members are mostly interested in the oven-ready production. Their best customers are the supermarkets and the large multiples, all of whom are very big buyers. The family butcher has the opportunity of selling the fresh killed article which many of his customers prefer. Turkeys, of course, are not confined to the Christmas period, there is a demand for them all the year round and a special demand at Easter and other holiday weekends. The oven-ready is naturally always available, though at times certain weights may be difficult while there can be few weeks in the year when the fresh bird cannot be seen on the wholesale markets. The growth of the turkey industry has been remarkable but it has had its ups and downs, and losses as well as profits. It must be said that the quality of the turkey has improved enormously during the past 15 years and producers have every right to be proud of their broad-breasted bird of today. It bears little resemblance to the turkey we handled pre-war.

THE CHRISTMAS TRADE: The fact that turkeys are sold all the year round has not taken away their popularity at Christmas when they still take first place in gracing the dinner table. If the butcher is concerned in selling solely the oven-ready bird his problems are comparatively few, but he is up against the supermarkets and multiples who have probably bought at rock bottom prices and it won't be easy to compete. Moreover he stands in danger of losing some custom to his rival round the corner, who has a display in his shop of fine fresh killed birds. Let there be no mistake there is still a large public who want the fresh bird always providing the price is not prohibitive. The turkey market at Christmas has always been liable to fluctuations and the weather alone can influence values considerably. One year is seldom like the next. But it is however a change from the ordinary routine. There is one rule which seldom lets the buyer down. If, at the start of Christmas trading, prices are high, then be careful, as in all probability there will be a slump before the finish. On the other hand, if the market starts at a reasonable level, there is every chance of a satisfactory finish. The popular weight bird for an average family is a hen between 5 and 6 kg (11 and 14 lb), though if prices are on the low side, there could be a run on a larger bird. It does not pay the producer of fresh birds to market a bird of 4·5 kg (10 lb) or under. Nowadays turkeys can be heavyweights with hens up to 9 kg (20 lb) or more. Large cocks, a good buy for the catering trade, are often in the middle thirties with the occasional bird touching 18 kg (40 lb).

Geese

Compared to the turkey the goose is of minor importance and its sale usually confined to the Christmas period or at Michaelmas if available. At best it is an uneconomic bird, the meat content being relatively small and it has become very expensive in recent years. There is no bird that has caused so much trouble to the retailer as the goose. No matter how careful you may have been in selection, the odd goose will let you down and eat tough. It has been said that a goose that has been frightened is the culprit but there is no proof of this. Certainly many a goose has been spoilt by being over-cooked but nevertheless the shopkeeper is the one that has to take the blame. Here are a few tips in selecting geese. Firstly the texture of the skin is a good indication. As with us humans the skin coarsens with age. But the best test of all is with the windpipe. Press the windpipe between the thumb and forefinger and if it crackles the bird is young. If on the other hand it feels rubbery, then discard it. The bulk of our geese come from East Anglia but to augment our limited supply, we get oven-ready geese from Denmark and Poland. The quality of these imported geese is very good. One last word in favour of the goose. From the eating point of view, in the opinion of the writer, it is superior to the turkey, although turkey producers might not agree.

Ducks

Like all other poultry, duck production has increased very con-siderably during the past decade. They can be bought throughout the year and have a snob value over other poultry. The butcher will probably not bother to keep them in stock but they are used exten-

388

ively by restaurants and are also in demand for dinner parties and banquets. They are marketed in cartons of six under the sender's brand and the weights run from 3 kg (5 lb) upwards. A duck under 3 kg (5 lb) is likely to be all skin and bone. Both fresh and oven-ready are to be had, mostly from Norfolk, Suffolk and Lincolnshire.

Last but by no means least is the chicken, which is the bread and butter of the poultry trade and its importance cannot be over-estimated. Like the turkey, the bulk of chickens, which are oven-ready, are sold by the supermarkets and the multiples but many butchers sell them both fresh and frozen. Though they are often subject to adverse comment, the oven-ready chicken is sold in its millions, year by year, so it must give satisfaction. All the same only the biased could claim that it is the equal of the fresh bird. The much maligned fresh broiler, a name no longer confused with boiler, gives the buyer few difficulties. It is marketed in cartons, under its brand name, and well graded in half pounds. Nearly always it is tied down for display purposes and, of course, all chickens are now bled. In hot weather it is advisable to take them out of the carton before putting them in the chiller to prevent them from turning green. The distinction between a chicken and a capon no longer holds good and the capon today is usually known as the larger bird of about 3 kg (6 lb) upwards. Though they are available throughout the year, it is at Christmas and Easter that they are chiefly in demand. At Christmas they have become very popular with the smaller family and are certainly a better buy than an immature turkey. At the other extreme, as far as size is concerned, is the poussin. Though it is not often seen in the average butcher's shop, it is an exceedingly popular line in some quarters, for large dinner parties etc. The majority are now marketed oven-ready and there are some fine packs available. Cartoned in dozens they are available from about 311 g to 510 g (11 oz to 18 oz) and graded to the g/oz.

One item in our trade that has not improved in quality in recent years is the hen. All are now battery laying hens and like chickens are marketed usually in cartons but the quality can only be described as fair. A sideline which should interest butchers with a high class connection is chicken liver. There is quite a big demand

Chickens and Capons

for it in certain areas. It is generally packed in cartons and sold frozen. We also get shipments from Denmark.

RETAIL HANDLING OF CHICKEN

It is accepted that the eviscerated deep frozen chicken simplifies handling at the retail level, but it is estimated that uneviscerated birds still account for some 20% of the UK market.

Official figures for 1973 gives 65,000 tons of chicken and 28,000 tons of old hens as being sold uneviscerated.

Subject to satisfactory distribution and storage conditions, the eviscerated deep frozen product will present very few problems to the butcher. Birds with the guts in are a different proposition and whilst the retention of the viscera may contribute to a "full" flavour, labour and some care is involved in the preparation, holding, evisceration, and trussing, for retail sale.

Although bleeding results in some slight loss of weight, it improves the keeping quality of the carcase and prevents the seepage of blood into the breast meat, a condition which is likely to arise if the neck is dislocated.

The table on p. 391 represents average figures obtained from good quality chicken, at around 2 kg (4 lb) liveweight, based upon the empty liveweight. In fasted birds of this weight, the gut content should not exceed about 4%.

Chickens are usually marketed according to weight classification, for example 2 to 3 lb, 3 to 4 lb, 4 to 5 lb and 5 to 6 lb. In some cases closer grading may be employed. It will usually be found that as the weight increases, there will also be an increase in the percentage dressed carcase yield.

At Christmas time there is usually a heavy demand for capons from 3 kg (6 lb) upwards, and whilst the tenderness of the flesh is widely appreciated some connoisseurs consider that the flavour is inferior to that of untreated birds. The term capon was at one time confined

390

birds which had actually been caponised, but it is now commonly applied to those over about 3 kg (6 lb) in weight.

Percentages of Empty Live Weight

	Pullet	Cockerel
Feathers and blood	11·8	12·0
Head and shanks	6·4	8·6
Digestive tract	9·4	8·6
Lungs	0·45	0·53
Heart	0·48	0·42
Liver with gall bag	1·7	2·1
Kidneys	0·55	0·50
Spleen	0·18	0·17
Neck	2·9	3·7
Testicles	—	0·09
Carcase	63·2	61·3
Loss	2·94	1·99
	100	100
Total bone in dressed carcase	15%	19·1%

Petit Poussins are sometimes referred to as "milk chickens" and usually weigh about ¼ to ½ kg (8 oz to 1 lb). A larger variety weighing from ½ to 1 kg (1 lb to 2 lb) is chiefly obtained from young cockerels of the light breeds. These types of fowl are principally used in the hotel trade.

Hens over 3 kg (6 lb) will usually fetch the highest price per kg (lb) and are generally referred to as boilers. Large numbers of live fat hens are required for the Kosher trade.

The most reliable test of age is the flexibility of the breast bone. In young birds the breast bone is easily twisted to one side, the beak is soft and easily bent, the feet are smooth, and the claws are sharp and slender in appearance. The comb is light and smooth and there is a downy touch to the feathers beneath the wings, whilst the flight muscle is soft and poorly developed.

With age the legs become rough, taking on a "horn-like" surface, the claws become strongly developed, and the spurs long. The flesh usually has a darker appearance, especially the legs, and they may have a certain amount of hair. The beak is difficult or impossible to bend and the breast bone is hard. There would be very little down present on the underside of the wing, and the flight muscle would be heavily developed and firm.

Quality

The main points considered as an indication of quality can be summarised as follows:

Flesh should be firm but pliable with a fine texture; there should not be too much fat especially in the abdominal cavity, although the Kosher trade does not object to fat hens. As a general rule a white skin is preferred, but this will depend on breed and feed. There should be no scores, cuts, "rubbed" portions on the skin, or blood patches. The breast should be straight, broad and well fleshed. The bones should be fine, with short legs well fleshed and with compact wings.

Cooling Uneviscerated Poultry

The shelf life will be markedly influenced by the initial treatment of the bird. Plucking should be carried out immediately after killing, as the presence of the feathers will tend to retain the high body heat. In addition, the feathers are more easily removed whilst the muscles are soft (pre-rigor).

The temperature of the preparation room will be much lower than the body temperature of the hot bird and the initial cooling rate may be somewhat similar to that obtained in a chiller. However, once the initial temperature has dropped, it is essential that the birds are placed in the chiller with a minimum of delay. Under no circumstances should plucked birds, awaiting stubbing, be piled on top of each other, as the inner layers maintain a high temperature, with a drastic effect on the ultimate quality of the product.

Greening first occurs in the region of the vent and crop, followed by off-flavours developing in the area adjoining the kidneys. With prolonged storage, seepage of bile will take place from the gall bag, resulting in a staining of the liver. Even with a chiller temperature of 2°C (35°F) it may take as long as 10 hours to reduce the centre temperature to 4°C (40°F) in a well-finished bird of around 2 kg (4½ lb).

With reasonably good storage conditions at about 4°C (40°F), some slight greening at the vent may develop in a period of around 5 days, whereas with storage at 0°C (32°F) similar birds will probably be in good condition after 8 days. The cooling rate is affected by the weight of the bird and the quantity of fat present. Thus, the probability of trouble arising from a lean roaster of 1·81 kg (4 lb) would be much less than in the case of a 3·18 kg (7 lb) fat hen.

Under normal chiller storage conditions, the weight loss during short period storage (excluding the hot weight loss) is usually in the region of 1 to 1½%. Tenderising changes take place more rapidly in chicken than in "red meats" and there is very little increase in tenderness after the first 24 hours.

Assuming a good initial condition, the International Institute of Refrigeration recommend that uneviscerated chicken held at 0–1°C (32–34°F) with a relative humidity of 85 to 90% should have a maximum desirable storage period of 7 days.

Under similar conditions, uneviscerated birds will keep better than those from which the viscera has been removed. Therefore as far as practicable, evisceration should not take place too early, and once eviscerated, the carcase should be held in the cooler.

Ideally, it is desirable to have a clean working surface every time a carcase is drawn, by wiping down the surface of the table with a suitable sterilising solution, after handling each bird. Edible offal should be separated at once from all inedible parts and the latter placed in a suitable container.

The drawing and trussing of poultry involves considerable handling, with the consequent risk of contamination being spread from the interior to the exterior surfaces. Therefore, the hands should be

Evisceration

393

frequently washed. Any technique which reduces the amount of handling will tend to improve the keeping quality.

The following procedure is suggested specifically for chicken but in general, can apply to all poultry.

REMOVE LEG SINEWS: These can be extracted by making an incision about 3·8 cm (1½ in.) in length below the knee, towards the foot. This will expose the sinews. Insert the trussing needle under the sinews, twist twice, and then push the foot forward on to the thigh. A slight pull on the needle will then remove all the sinews from the thigh. Then chop off the feet.

REMOVE HEAD, NECK AND CROP: Place bird on its breast; make a cut along the back of the neck towards the body, leaving sufficient skin to cover exposed neck-end, and if desired, for stuffing. Cut off the head and remove the neck close to the body and the crop can then be stripped away.

LOOSEN THORACIC ORGANS: Insert a finger into the chest cavity and with a rotary movement, break any attachments to free all the organs.

REMOVE WISHBONE: This operation is optional but it greatly assists when carving, particularly if slices of stuffing are required.

EVISCERATION: Make a cut round the vent. Do not puncture intestines. Insert two fingers into the body cavity and, with firm pressure on the gizzard, steadily withdraw all the entrails, leaving the kidneys and in the male, the testes. The kidneys are situated on either side of the backbone and closely attached. It is desirable that they should be removed as they can give rise to off-flavours. The testes (in the male) are easily removed. Should portions of the lung tissue be left attached to the chest-wall, they should be removed. The interior of the bird should be wiped out with a clean cloth.

Trussing

Methods of trussing can vary considerably and the following is one procedure, using a trussing needle and string. Break the backbone

394

n the region of the wings by a sharp tap with the back of a poultry chopper, so that the bird will lie flat. Place the bird on its back and raise the legs; insert threaded needle immediately behind the legs, through the bird, then pass the needle through the wing tip and the middle portion of the wing, threading through the other wing, in a similar manner. Thus, the string has passed through the body behind the legs, through the wings, and under the back.

The needle can then be taken off the string and the two ends tied together, bringing the wings close to the back and snugly into the sides. Finally, secure the legs by passing the string through them and tying them down to the tail.

Presentation is improved if any hairs present are singed off and the bird lightly dusted with flour.

Boneless poultry may be stuffed or used for galantines or similar products. Whilst this *can* be carried out without cutting the skin of the body, it is more easily accomplished in the following manner:—

Deboning Poultry

Remove the head, neck, shanks and wing tips and make a cut along the back of the bird from the neck end to the tail. Remove the skin and flesh over the back, and on reaching the wings and thighs, find the joints and cut through the ligaments, thus detaching them from the frame. Continue towards the breast, carefully trimming away the fillets attached to the breastbone. The carcase skeleton with the viscera can now be lifted away. The thighs can then be "tunnelled" out and the bone removed, with the exception of a small portion of the hock. These should be left in position, to retain the shape and facilitate stuffing. De-bone the wings to the first joint, the outer portion being removed. The bird is then ready for stuffing and shaping and if desired the back can be sewn up. In the case of galantines prepared in D-shape containers, various types of meat, such as diced tongue, can be incorporated in the filling.

The edible viscera of the fowl consist of the neck, heart, liver, spleen, kidneys and gizzard. In some cases, the feet, following thorough cleaning, may be used for stock.

Care must be taken in removing the gall bag from the liver and i may be safer to trim away a small margin of liver with the bag rather than risk breaking it. The neck should be wiped clean an jointed and any free blood should be squeezed out from the heart When preparing the gizzard, a cut should be made round the oute muscular coat, so that this portion can be peeled off, leaving th inner lining with its contents, intact.

GAME

The door is wide open for the butcher to take over the bulk of th game trade from the fast declining number of fishmongers, as h already has done the poultry. In a normal season many hundre thousand heads of game are shot and sold in this country, so it is branch of the trade well worth attention. Game, however, is highly specialised business though a most interesting one. Wherea the bulk of poultry under modern methods is cartoned and grade before it reaches the retailer, game has to be individually sorted an graded. This usually is the work of the wholesale gamedealer. great deal has been written, chiefly in women's magazines, on th subject of game, in order to help the housewife when she buys. Mos of it has been absolute nonsense and must have confused rather tha helped. It must be made clear at this juncture that practica experience in sorting game is the only way to learn and the writte word can only help to a limited degree. There is a great demand i the trade for people knowledgeable about game and moreover it is fascinating subject.

Handling Game

A grouse or a cock pheasant is a thing of beauty and deserves to b treated with respect. Never ruffle the feathers or handle the bir roughly. There is even a right and a wrong way to pick up a bird The head must always be directly in front of you. Hold the breas upwards in the palm of your hand and you can then lift the feather at the side of the bird, with your two fingers, to examine it fo freshness, etc. If you pick up a bird with the feet pointing toward you, a gamedealer will know at once you are a novice. Game generally speaking, improves with hanging but there is no hard an fast rule concerning this. Everything depends upon the weather an

396

he hanging facilities available. The great enemy is the fly which
an do much damage if allowed to lay her eggs on game. It is quite
 fallacy to suppose a bird must be high before being eaten, it is
urely a matter of taste. It is interesting to note that as far as the
vriter is aware, no case of ptomaine poisoning has ever been re-
orted through eating game. Grouse and partridges have to be
orted as young and old, the former to roast and the latter for cas-
erole and there can be no half measures; a bird is either young or
ld. Sorting grouse is far from easy and will be dealt with later.
To sell game, a licence is required and there is a closed season for
elling, and charts, with the various dates, are available.

The game season proper opens on 12 August when grouse shooting   **Grouse**
egins. The red grouse is the finest of all birds and one of which we
re extremely proud, as it is only found in the British Isles. Though
erhaps for some, an acquired taste, it is magnificent to eat and in
he opinion of the gourmet only rivalled by the young grey par-
ridge. Early in the season it need not hang as it retains the flavour
f the ling. To eat grouse on the "glorious twelfth" is both a
ashionable and expensive pastime. From the opening day of shoot-
ng until the middle of September, thousands of grouse are sent to
ondon daily and the wholesaler has a very busy time sorting and
rading. The expert way to tell a young grouse from an old one is
y the young feathers nestling on the breast. These gradually dis-
ppear as the season progresses and then you look for the pointed
ving tip on the first flight feather. This all sounds fairly simple but
nany old hen birds have feathers on their breast very similar to that
f the young bird and early in the season some old birds also have
he pointed wing tip. Hence many mistakes are made except by the
xpert and even gamekeepers are often guilty of mistakes. There
re many other tests which can be applied for the teaser but the only
ound advice to the buyer is to trust your wholesale gamedealer.
Grouse on the whole are good keeping birds and stand up well to
ot weather, but beware of the fly. When preparing for the table the
iver should be left in. Other birds of the grouse family are black-
ame and capercailzie. They are in comparatively short supply,
articularly the caper. The male of blackgame is the "blackcock"
nd the female "greyhen". A peculiarity of blackgame is that they

have layers of dark and white meat. The caper is the largest of all game and the cock bird can be compared to an eagle.

Partridges

The season opens on 1 September but little shooting is done until the end of the month or early October. From the trade's point of view this is perhaps just as well as during the first few weeks many of the birds are small and immature and these are difficult to sell. During the past decade the grey partridge has been diminishing in numbers and this has caused concern among sportsmen and land owners.

There are two types of partridges to deal with—the grey and the redleg or French partridge. The former is the more popular bird of the two, by both the sportsman and the gourmet and, though slightly smaller, always commands a higher price. The sorting of young and old does not present any great problem. As with grouse, the young feathers nestling on the breast are a sure indication of youth and the legs of the young grey are brownish while those of the old bird are white. Later in the season the young bird has a pointed wing tip on the first flight feather and that of the old bird is rounded. But early in the season some old birds carry the pointed wing tip and this fact does cause mistakes. The test of a redleg is quite different. The young bird has a small golden bar at the extreme wing tip of the first flight feather. If that is not visible then the bird is old. The old bird generally also has rather nobbly legs; this refers to the redleg only. In recent years a new industry of great importance has sprung up to counteract the shortage of the wild bird. Many thousands of the redleg type are being reared on game farms for home consumption. Though these birds have not the fine flavour of the wild partridge, they have certain advantages. Firstly they can be obtained in large numbers for banquets and large dinner parties, which is not usually possible with the wild variety and continuity of supply is important. Secondly they can be killed directly they reach maturity and are comparatively uniform in size. Because they are not shot, they are a good keeping bird. The demand for these farm-bred redleg partridges has become established in the trade.

398

The lordly pheasant is the most popular of all game birds and has a **Pheasants**
much wider appeal. During the first few weeks, when the leaves are
on the trees, shooting is only on a minor scale. It is when the estates
start shooting, about the end of September or early October, that
birds become really plentiful. After a few frosts pheasants put on fat
very quickly and improve in size and condition. Pheasants, perhaps
more than any other game, improve with hanging and during the
colder weather this presents no problem. Sorting, also, is easy enough
and only the old cocks and small birds have to be down graded.
Some estate pheasants shoot their spurs rather early and the spur of
the old cock is sharper and longer than that of a young bird. If you
press your thumb against the base of the spur and it breaks fairly
easily then the bird can be passed as first grade. Many pheasants are
now marketed oven-ready and this is a tremendous help to the
trade. They can be bought both fresh killed and frozen and are well
graded and packed in cartons, and naturally find a ready market.
There is no need to emphasise how helpful this new and important
industry is to the retailer and this alone should make it worth while
for the butcher to take out a game licence.

This is an item in our trade that surely should interest the butcher. **Venison**
It has increased in popularity and demand to a considerable extent
during the past few years, especially for banquets and dinner parties.
There are three main types, the doe deer, the red deer and the fallow
deer which is found in England only. Scotland sends many tons of
venison to London, chiefly haunches and saddles, which are the most
popular. Shoulders are also available but are less in demand and
therefore cheaper. There is a closed season for shooting but venison
can be sold all the year round.

Until fairly recently, hares have always been the most underpriced **Hares and Leverets**
meat in the country. Even now, when their value is more realistic,
they are still cheap meat. They are somewhat of an acquired taste
and more appreciated on the Continent than in this country, but a
jugged hare makes an excellent and economical main dish. For
show purposes, a hare must be kept clean, for a row of hares hanging
in a shop with their white bellies makes a fine show but a dirty

bedraggled lot on view is an unpleasant sight. An old clothes brush is useful in smartening up a row of hares. Be careful in handling them, for a prick from a broken bone can be poisonous. To recognise a young hare, run your thumb up the outside of the foreleg. Just above the first joint, a little way up the leg, you will feel a small knob which is the sign of a young hare. This disappears as the hare ages. A leveret is a small and very young hare and can usually be bought at a cheap price. A freshly shot hare can be recognised by the brightness of the eye.

Guinea Fowls

Before the war we imported great numbers from Hungary. They were very popular, extremely good eating and cheap. That is a thing of the past and we now rely on a few producers in this country for our supply. They are marketed oven-ready, well graded and packed in cartons. The majority weigh from $1\frac{3}{4}$ to $2\frac{1}{4}$ lb oven-ready weight. They can be sold all the year round.

Quail

The bulk of these are marketed oven-ready and they too can be sold all the year round. In the old days they all came from Egypt but production in this country is going ahead and the producers are doing an excellent job. They are in demand chiefly for banquets and dinner parties and they are sold singly.

Wildfowl

These are worth mentioning, though probably the average butcher will only stock them if requested by a customer. There are four main types of wildfowl—Wild Duck or Mallard, Pintail, Wigeon and Teal. Mallard are possibly the most in demand followed by teal. Wildfowl, like woodpigeons, are not good keepers in warm weather and are apt to turn green. Other game birds well known but by no means plentiful are Woodcock, Golden Plover and Snipe. All these birds, including wildfowl, should be fat to be of good quality.

Turkey and Turkey Meat

There is every indication that in the long term "red" meat will continue to increase in price, and that the demand for alternative animal protein sources will increase. This trend is evident in the greatly increased demand for turkey, the market (1980) value exceeded £250 million per annum.

The extensive research over the last decade into the genetical problems involved, has resulted in the production of birds which bear little similarity to their wild ancestors. Probably the most important improvement from the commercial aspect has been the development of the breast muscles (i.e. double breasted) and the increased growth and food conversion rates.

The modern method of killing involves electrical stunning, followed by sticking, the latter obviating the formation of a blood pocket when dislocation of the neck is employed.

UNEVISCERATED TURKEYS

With these birds, the highest price per kg (lb) is obtained for young hens from about 4 to 5 kg (10 to 12 lb) but there is also a strong demand for turkey cocks of over 11 kg (25 lb). In selecting such birds, the main considerations are, soft white flesh with plenty of breast meat, pliable pelvic bones, and the wing muscles not too heavily developed. Any greenness in the region of the vent or crop, will indicate a need for rapid sale.

PREPARATION: When purchased from a wholesale market, there will usually be some feathers on the neck and wings. Remove all feathers; this can best be carried out with the bird hanging at a suitable height, rather than having it twisted about on a bench, as this will mark the bird and affect the "bloom".

DRAWING SINEWS: This operation is most important, as, should the sinews be left in, they will spoil the flesh of the legs. Cut around just above the feet and break the leg, twist the foot and place it in a sinew hook, and with a steady pull on the leg, remove the sinews. Where large numbers are involved, a sinew extractor greatly assists this operation. The edges of the wing and the wing tips should then be trimmed away.

O

REMOVING HEAD, NECK AND CROP: Place the bird on its breast and cut across the neck in line with the shoulders and using the blade of the knife, remove the neck and head. Stand the bird on its tail, back towards you, and remove the crop. This is carried out by placing the left thumb down the inside of the bird and removing the crop by working the right thumb under and around it.

LOOSENING THE HEART, LUNGS, TRACHEA AND BLOOD VESSELS: Insert the index finger into the thoracic cavity, working in a circular motion.

EVISCERATION: Turn the bird upon its breast, holding the tail in the left hand, make a cut behind the vent so as to permit the back passage to be gripped. Cut carefully round the vent; do not puncture the gut. With the index finger, loosen the internal organs, insert hand, and pass fingers over the gizzard and steadily withdraw the viscera. Finally wipe out with a clean cloth.

BREAKING BACK: Tap sharply behind the shoulders with the back of a poultry chopper; this will break the vertebrae so that the bird will lie firmly on the dish for carving.

TRUSSING: Place the bird on its back, breast towards you, pull up the legs as far as they will go, and using a needle and string, pass the needle through the body of the bird, immediately behind the legs. Turn the bird on its side and sew through the thick part of the wing and then the thin portion, and continue through the other wing, the thin portion and then the thick part. Draw the string tightly and tie off. Turn the bird on its back, pass the needle through the hock and then through the back of the bird, picking up the opposite hock and tie the legs to the side of the bird. Make a small hole in the skin of the lower belly wall, in which to insert the tail. The appearance will be improved if the bird is singed off to remove any hairs, and then dusted over with flour.

GIBLETS: The edible viscera consists of the neck, heart, liver, spleen and muscular part of the gizzard. Great care should be taken in removing the gall bag from the liver, and it is safer to trim away a small margin of liver with the gall bag rather than risk breaking it. The neck should be wiped free of any blood and jointed, and any blood clots should be squeezed out from the heart. When preparing the gizzard, a cut should be made around the outer muscular coat, so that this portion can be peeled off, leaving the inner lining with its contents intact.

OVEN-READY TURKEYS

The rapid rise in popularity of oven-ready blast frozen birds has doubtless been influenced by the following considerations:

1. The skilled labour involved in the preparation of the un-eviscerated bird is eliminated.
2. The risk of potential contamination from the feathers and guts is removed.
3. They can be held in frozen form for a considerable period, thus smoothing out any fluctuations in supply and demand.
4. The wrapping film provides a good protection from contamina-tion, retards weight loss from evaporation, and the bright appearance is maintained.

Oven-ready birds are available over a wide weight range and various arbitrary weight divisions are recognised, a typical classification being 3 to 5 lb; 7 to 12 lb; 12 to 19 lb; 19 to 25 lb; and over 25 lb; and prices may be quoted on this or a somewhat similar scale.

It should, however, be appreciated that the most important factor affecting the oven-ready yield and the total meat on the carcase, is the liveweight of the bird. This is well recognised by the res-taurateur who prefers to buy the heavier bird and such birds are also attractive to the butcher dealing with cut turkey and boneless turkey roll.

The following table will indicate the effect of the increase in the weight and the sex on the average yields obtained from a large number of turkeys, drawn from two distinct strains of birds (cour-tesy Bernard Matthews Ltd).

Age	12 weeks		16 weeks		20 weeks		24 weeks
	Hens	Stags	Hens	Stags	Hens	Stags	Stags
Liveweight in kg	3·73	4·91	5·52	7·81	6·89	9·87	12·63
Liveweight in lb	8·23	10·83	12·17	17·21	15·20	21·77	27·84
% *Live weight*							
Blood, feathers, offal and loss	20·30	20·10	16·10	17·00	17·60	17·20	16·70
Oven-ready weight in kg	2·98	3·92	4·64	6·48	5·68	8·18	10·51
Oven-ready weight in lb	6·56	8·65	10·22	14·29	12·53	18·03	23·18
Killing out %	79·7	79·9	83·9	83·0	82·4	82·8	83·3
% *Oven-ready weight*							
MEAT—breast, thigh, drumsticks,							
wings and carcase	57·8	56·9	59·8	58·8	62·6	62·0	63·7
Giblets	9·0	9·8	8·3	8·7	8·2	8·2	7·2
Skin, wingtips and tail	11·6	10·9	11·6	10·4	12·5	11·0	13·5
Total bones	20·2	20·9	19·1	21·1	16·1	18·0	15·0
Loss in cutting	1·4	1·5	1·2	1·0	0·6	0·8	0·6
Totals	100%	100%	100%	100%	100%	100%	100%

From the above table it will be seen that increasing the oven-ready weight was associated with a marked increase in the meat percentage, a decrease in the giblets and a general reduction in the bone, particularly with the heavier birds of both sexes.

As would be expected, there is a downward trend in losses, arising from cutting and evaporation, with increasing weight.

PREPARATION OF CUTS

The butcher with his inherent skill in cutting, is well placed to apply his knowledge to the preparation of turkey cuts. It would be unwise to stipulate specific methods of dissection, as demands in terms of weight and the various portions, may vary somewhat from shop to shop. Therefore, the following methods are given as a guide only, although they have been found to be successful in a number of different areas.

In all methods of preparation, the initial operation is the division of the breast bone. This starts with a knife cut through the breast meat, following the line of the breast bone, which is then sawn through, followed by sawing through the backbone, to split the bird into two

404

halves. The breast cutlets are removed and the shanks and drumsticks, the remaining portion of each side being cut into joints, on the
bone. The following yield was obtained under normal commercial
conditions.

Turkey 11·57 kg (25½ lb)
(Joints bone-in except breast cutlets)

Portion	kg	lb	oz	% of net marked weight
Cutlets	4·21	9	4½	36·40
Shanks	0·65	1	7	5·64
Drumsticks	1·30	2	13¾	11·21
6 joints	2·12	4	10¾	18·32
Trim*	1·27	2	13	11·03
Giblets and bones	1·92	4	3¾	16·61
Cutting loss	1·11		3¾	0·79
	11·58	25	8	100

* Trim from neck, tail, mid wings and carcase used for burger meat.

Methods of presenting Turkey Cuts are shown in the colour plates.

Two methods of preparing boneless turkey roll are suggested. In
the case of *A* the drumsticks were removed prior to stuffing with
backfat and wet sage-and-onion stuffing. With this method it is
possible that the inclusion of the leg meat might influence the overall
eating quality. With turkey *B* the drumsticks were included in the
roll, but some care (and time) was involved in removing the sinews
and extraneous connective tissue.

TURKEY ROLL

A
11·57 kg + 0·91 kg (25½ lb + 2 lb) Stuffing and Backfat

	kg	lb	oz	% of total
Meat 6·75 kg (14 lb 14 oz)				
Stuffing 0·91 kg (2 lb)	7·65	16	14	61·36
Drumsticks	1·47	3	6	12·27
Burger meat trim	0·45	1	0	3·64
Giblets	0·74	1	10	5·91
Bones and loss	2·10	4	10	16·82
	12·41	27	8	100

405

B

12·25 kg + 0·74 kg (27 lb + 1 lb 10 oz) Stuffing and Backfat

	kg	lb	oz	% of total
Meat 7·30 kg (16 lb 1½ oz)				
Stuffing 0·74 kg (1 lb 10 oz)	8·04	17	11½	61·90
Burger meat trim*	1·42	3	2	10·92
Giblets	0·75		12	2·62
Bones and loss	3·19	7	0½	24·56
	12·99	28	10	100

* Trimmings from neck, tail, mid wing and carcase

Edible Offals

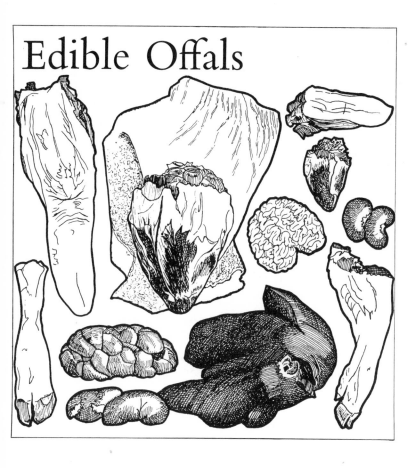

It is of etymological interest that the nouns beef, mutton and pork are derived from beouf, mouton and porc, whilst edible offals such as liver, heart and melt, originate from the Anglo-Saxon lifer, heorte and milte, respectively.

The line of demarcation between the festive board and the slaughter-house has now been erased and the edible offals provide a highly nutritious addition to our meat diet.

It is unfortunate that they should retain the name "of falls" which in its popular form offals is associated with waste, and in this respect the American nomenclature "variety meats" has much to recommend it.

407

In view of the current price of carcase meats, it is not surprising that the public are showing an increasing interest in edible offals and it is probably one of the few "lines" which are likely to provide the retailer with a satisfactory margin of gross profit.

In general, there is very little loss in the preparation of edible offals in terms of bones, fat and trim and it must be emphasised that in addition to the essential food constituents found in red meats, they are excellent sources of minerals and vitamins. Liver in particular provides many of the food factors necessary to health.

The following table indicates the protein value of some edible offals.

TABLE 1

Approximate Protein Content per 100 g (3·5 oz) (cooked)

Offal	Beef (g)	Veal (g)	Pork (g)	Lamb (g)
Brain	11·5	10·5	12·2	12·7
Heart	28·9	26·3	23·6	21·7
Kidney	24·7	26·3	25·4	23·1
Liver	22·9	21·5	21·6	23·7
Spleen	25·1	23·9	28·2	27·3
Sweetbreads	20·5	18·4	—	23·3
Tongue	22·2	26·2	24·1	21·5

(After Lilia Kizlaitis, Carol Diebel and A. J. Siedler)

Their price, with the exception of sweetbreads and possibly calves' liver, compares favourably with other meats and in general their low fat content will appeal to the calorie conscious consumer.

Some care is necessary in comparing the price per kg or per lb of plucks, as against buying livers and hearts and the value of the former can be enhanced where there is a demand for lites as pet food.

As edible offals tend to be less "durable" than other meats, more care is necessary in storage and handling. Whilst brains have a delicate flavour, they are highly perishable and it is suggested that in those areas where they can be obtained in an absolutely fresh condition, they should be rapidly frozen in waxed cartons.

Representative examples of the calorie and fat content are given in Table 2.

TABLE 2

Calorie and Fat Content per 100 g (3·5 oz) (fresh)

Offal	Calories		
	From protein	From fat	Total
Brain (calves)	44	61	105
Heart (lambs)	74	51	125
Kidney (ox)	73	27	100
Liver (ox)	88	43	131
Spleen (pig)	77	23	100
Sweetbreads (calves)	71	27	98
Tongue (ox)	69	159	228
Tripe	58	60	118

(After Kizlaitis et al.)

In costing offals, particularly frozen livers, the weight loss incurred due to drip can be considerable, therefore adequate allowance must be made for such losses, in arriving at a sale price.

Pig's fry was at one time a very popular dish in most areas, but is now relatively uncommon, at least in London and the South of England. It consists of pig's liver, heart, spleen with the caul fat and may include some mesenteric (gut) fat.

OFFALS OF THE OX—EDIBLE

Ox cheek and head trimmings	Stewing, brawn, processed meats
Liver	Braising
Tongue	Salting and boiling
Brains	Boiling, sauce and Liver sausage
Sweetbreads (a) Thymus	Frying or boiling
(b) Pancreas (gut bread)	
Heart	Baking, boiling, processed meats

Skirt (thick)	Stewing
Spleen (melt or milt)	Pies and flavouring soups
Tail	Soup and stewing
Kidneys	Stewing, pies and soup
Blood	Black puddings, blood and barley loaf
Blood plasma	As a binder in sausage
Stomach (Rumen)	Tripe
(Reticulum)	Honeycomb tripe
(Abomasum)	Red or black tripe
Feet	Cow heel, calves'-foot jelly
Fats (Suet)	Paste, pudding, mince meat etc.
(Other fat)	Dripping
Bones	Soup, gelatin
Udder	Boiled, salted, smoked or fried

OFFALS OF THE SHEEP—EDIBLE

Head	Boiling
Tongue	Salted and boiled or braised
Brain	Poached or fried
Pluck	
Liver	Frying, grilling, braising
Heart	Stuffed, roasted, braised
Lungs	Traditionally an ingredient of haggis
Sweetbreads	Fried, etc.
Paunch Honeycomb }	Tripe and container for haggis
Blood	May be used for black pudding
Lamb fries (testicles of young rams)	Fried or grilled

OFFALS OF THE PIG—EDIBLE

Head (is strictly part of the carcase)	Boars Head, salted and boiled or for brawn
Tongue	Salted and boiled
Brains	Fried or braised
Cheek	Bath Chap
Pluck	
Liver	Pâtés, stewing, frying
Heart	Braising
Weasand meat	Sausage etc.
Lungs	(Pet food)
Spleen	Used in Pig's fry or for flavouring

410

Stomach ⎱ Intestine ⎰	Chitterlings
Kidney (usually part of carcase)	Split through for grilling or stew
Blood	Black puddings
Feet ⎱ Tail ⎰	Salted and boiled
Fats	Lard
Skin	Rind emulsions

Ox—Elongated, thick at upper end, and tapering. A thinner left lobe indistinctly divided from the thicker portion. Small lobe (caudate) at thick end. This lobe is sometimes referred to as the "thumbpiece".

Sheep/lamb—Somewhat similar in miniature to that of the ox, but the thumbpiece is more pointed.

Pig—The connective tissue gives its surface a characteristic "nutmeg" appearance. Five lobes, tapering sharply at their edges.

Calf—Very soft texture, similar in form but much smaller than that of the ox, but the edges are more rounded and the thumbpiece has a blunted end. Evidence of the umbilical vein is present.

DISTINCTIVE FEATURES
Liver

Ox—Consists of fifteen to twenty-five lobes. With the change of diet from milk to a more bulky diet, the stomachs increase in size, displacing the kidney knob to the right and slightly backwards, usually causing a slight rotation of the left kidney. As a result the left side of beef is termed the "raison" or "open" side and as a rule it will contain less kidney suet than the right side.

Sheep/lamb—Has one lobe, is bean-shaped and dark brown in colour.

Pig—Larger than that of a sheep, single lobe, bean-shaped but relatively flat in cross section and reddish-brown in colour.

Kidney

Ox—Conical shape, three furrows, usually with a good amount of

Heart

fat in them and adjoining. Two bones are present in the aortic ring. Fat is usually white and firm, in old cows it may be yellow.

Sheep/lamb—Hard white fat, smaller, denser in texture and more pointed than in the calf. (Bones may be present in very old animals.)

Spleen   Calves—Relatively soft muscular tissue, blunt rounded point.

Pig—Fat soft and oily to the touch, and a rather round blunted point will help in differentiating it from the heart of a sheep.

Ox—Elongated oval shape, flat, bluish-grey in colour. White lymph follicles can be seen in the substance.

Sheep/lamb—Oyster shaped, reddish-brown in colour, usually left attached to the pluck.

Tongue   Pig—Extended tongue shape, red in colour, ridge extending along its length, where the omentum (caul) is attached. Triangular in cross-section.

Ox—Thick, tapering, end pointed, rough surface, black spots common. Six or more circumvallate papillae on each side. Usually marketed as long cut (complete tongue with root), short cut, with root trimmed, or as a blade consisting of the tongue muscle only.

Sheep/lamb—Thick, short, end tends to be hollowed out, smooth surface, black marking common. Nine to twelve papillae on each side. Depression running along the centre.

Pig—Long, relatively thin, pointed end. Smooth velvety surface. One papillae on each side.

412

Whitish-yellow in colour, lobulated and consists of two parts (pair). One portion from the cervical region is termed the neck bread, and that from the thorax, the heart bread. It is large during the period of growth and subsequently degenerates, being largely replaced by fibrous tissue. Thus the sweetbreads from calves, lambs and young cattle are in demand.

The pancreas is frequently referred to and sold as a gut bread. Its function however is to secrete fluids which assist digestion. It is brownish-yellow in colour and lobulated. In the ox it will weigh about 170 g (6 oz) and in sheep and pigs, about half this weight.

The average weights of the various edible offals from cattle, sheep and pigs will be found in Tables 5, 6 and 7 on pp. 244–246 in the section on slaughtering.

IDENTIFICATION OF OFFALS IN SAUSAGES

The Offals in Meat Products Order, S.I. No. 246, 1953, prohibits the use of certain offals in uncooked open meat products.

Modern histological methods of examining raw non-decomposed sausage meat can be successfully applied to the detection of some common offals, even after the materials have been emulsified in a colloid-mill.

Some tissues, after being finely comminuted, may prove more difficult to identify specifically than others. However, the presence of animal cells other than those of normal "meat" muscle, is readily recognised.

Meat and Nutrition

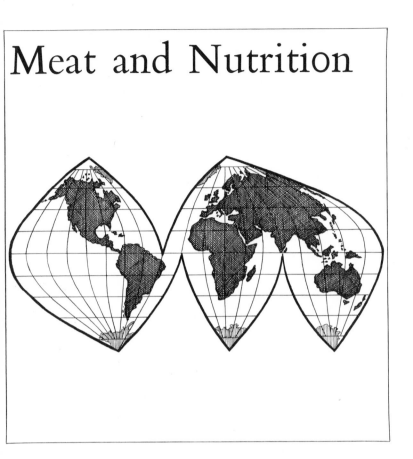

One of the really annoying things about dictionary definitions is that, all too often, the terms used to explain the meaning of words are confusingly interdependent. Unless one already has some prior knowledge of the "language" of the topic then the dictionary is unlikely to relieve one's ignorance.

For instance, take the three words "food", "nourishment" and "sustenance". They are all words whose meanings are familiar to most of us but suppose this was not the case, suppose one needed to know the exact meanings of these three words. Unless one was already aware of the real meaning of at least one of the three words then the dictionary would prove to be singularly unhelpful, viz:

415

"Food": (noun); "nourishment for animal or plant".
"Nourishment": (noun); "food, sustenance".
"Sustenance": (noun); "nourishment, food".

(Penguin English Dictionary)

Thus, anyone totally unfamiliar with any of these three terms and who seeks guidance from a dictionary is likely to remain in a state of frustrated and annoyed ignorance. Mere dictionary definitions simply refer back and forth between these three key words, words whose actual meanings were sought in the first place.

In order to gain any real understanding of the nature of "food", "nourishment" and "sustenance" one has to go beyond mere dictionary definitions and seek out the real nature, purpose and function of "food" (or "nourishment" or "sustenance") by entering into a wide-ranging study; into the field of Nutrition.

MEANING OF NUTRITION

Nutrition, as a proper, scientific study, is somewhat new and really only began in the 19th century. Most of the original work on the principles of nutrition came about as the result of a study of the food requirements of man himself, a study that was primarily concerned with health and the need to avoid faulty- or under-nutrition. In the course of human history it is probable that in most countries, at some time or other, the vast majority of the inhabitants have suffered malnutrition. Development of nutritional science has enabled us to recognise most of the dietary deficiencies that lead to malnutrition and to point the way towards improvement and recovery.

Animal Nutrition

More recently the emphasis of nutritional science has shifted and a new, off-shoot aspect, known as Animal Nutrition, has developed. This is because increasing world population (much of which has resulted from greater understanding of the principles of human nutrition) and the soaring costs of food production have caused nutritionists to turn their attentions to animal feeding in order that the food animals, mainly cattle, sheep, goats, pigs and poultry, may be produced as economically as possible. Proper application of the

416

principles of nutrition will eliminate wastage of resources due to over- or wrong-feeding and will ensure better birth and survival rates amongst such animals. In this way, further development of animal nutrition may help to solve many of the problems of human nutrition by making certain that food supplies will, eventually, equal food demands.

Furthermore, a wider knowledge of the principles of nutrition will enable us to concentrate production efforts on foods that most fulfil the requirements of an adequate diet and so avoid the wastage involved in producing poor quality or dietetically inadequate foodstuffs.

In this chapter we are concerned with the nutritional significance of meat, one food amongst many but one which holds a very special place in the nutrition of mankind.

In order to appreciate properly this special significance of meat one has first to consider the nature and the purpose of food in general; only then may the uniquely important nutritional properties of meat be understood.

The Nature, Purpose and Function of Food

Expressed in simple terms, food is that which, when eaten and absorbed into the body, promotes tissue growth and repair, and also provides a source of energy.

These are the *direct* functions of food. There are, in addition, certain *indirect* functions that food must perform; food must help the body to:

resist infection,
successfully adjust to unfavourable environments,
withstand the special strains associated with reproduction.

The chemical components of food which bring these functions, direct and indirect alike, are known as nutrients.

Altogether there are five such nutrients, six if one also includes water (below). Thus, before the term "food" can be properly applied to any substance or material it must contain one or more of the five following nutrients:

Carbohydrates, Fats, Proteins, Vitamins, Minerals.

Although not really a nutrient, water is vital if the body is to utilise the food properly. If water is not already present in the foodstuff then it must be taken separately in the diet. Given this vital role of water it is sometimes included as the sixth nutrient.

Some foodstuffs contain enough of each of the five nutrients to be able, by themselves, to fulfil *all* the five functions of food, i.e., supply energy, growth and repair materials, aid bacterial resistance, respond to environmental changes and permit reproduction.

Such foods are specially valuable and constitute a whole diet for an animal. Other foods, such as sugar, contain only one single nutrient (in this case, carbohydrate) and thus perform only a single nutritional function, i.e. provide energy. A single nutrient food such as this would constitute a wholly inadequate diet.

Most of our more familiar foodstuffs fall somewhere between these two extremes; they contain more than one nutrient but not all five nutrients. Conversely, they may contain all five nutrients but lack sufficient quantity of one of the nutrients to fully satisfy bodily needs. Most animals, including man, utilise these individually-incomplete foodstuffs. As a result, such animals require a mixed diet and, in this way, any nutritional deficiency in one individual food-stuff can be balanced by a richness of that same nutrient in another foodstuff, within the dietary range.

Where does meat stand in all this?

Meat as Food

Meat is considered to be one of the most nutritionally complete of all of man's foodstuffs, a point that will be more fully developed later. However, it is not a perfect foodstuff, it lacks vitamin C and

418

contains insufficient calcium mineral to be an entirely adequate, totally complete foodstuff. Nevertheless it comes very, very close.

On reflection this is hardly surprising. It will be remembered that one of the two essential reasons for eating food is to provide the basic raw materials for bodily growth and repair (the other reason being to provide energy). Now the body tissues of man and the food animals are almost identical. Human flesh has a chemical composition hardly distinguishable from bovine, ovine or porcine flesh (ox, sheep and pig respectively). It therefore follows that meat, being the flesh of animals and compositionally identical with the flesh of man, must be a well-nigh perfect sourse of the materials necessary for tissue growth and repair. Thus we utilise the flesh of animals in order to manufacture and sustain the flesh of man. It is simply a matter of converting one form of flesh into another.

It is only when we move away from meat to other, less flesh-like forms of food (such as vegetables, fruits and cereals) that we begin to encounter dietary difficulties. In chemical composition these non-flesh foods are so unlike human flesh that we must exert far greater efforts in terms of gathering, varying, digesting, absorbing and reassembling if we are to provide sufficient nutrients in sufficient variety to ensure adequate growth and sustenance of our body tissues. Perhaps an example will make this point clearer:

Non-Flesh Foods

Let us compare the digestive efforts of a purely carnivorous animal; the cat, with a similar sized, purely herbivorous animal such as a rabbit. Whilst they are roughly equal in body weight and the chemical composition of their tissues would be identical, their food processing organs are constructed on markedly different lines. Their dentition, stomach sizes and functions, digestive enzyme complexes, gut lengths, presence of appendix, faeces and urine composition are all highly dissimilar. Flesh, which forms the natural food of the cat, requires little in the way of gathering, variation, chewing, digesting, assimilating or final disposal. The rabbit, in contrast, must exert far greater efforts in all these functions, both physical and physiological, in order to thrive. Thus, in many aspects of adequate nutrition the cat is more efficient than the rabbit and is more conserving of effort.

419

It is also longer-lived. The inescapable conclusion is that flesh foods seem to be more dietetically valuable and less demanding of physical and physiological effort than non-flesh foods. Why is this?

At birth, humans, male and female alike, weight about 3 kg (7 lb) and have a total body cell-count of 2×10^{12} (2,000,000,000,000). At maturity the average human male, weighing about 76 kg (168 lb), consists of 60×10^{12} (60,000,000,000,000). Thus his body weight would have increased by 24-fold and his body cell-count by some 30-fold. Females, having smaller adult body weights than males, display cell-count and body weight increases of only about 20-fold and 18-fold respectively.

Dietary Needs

In order to attain such enormous cell- and body-weight increases the diet must contain sufficient quantities of all the requisite tissue building materials. In this way the dietary foods allow the tissues and organs to grow to sizes commensurate with the physiological tasks they have to perform within the body.

Even after growth is complete and the final adult-weight has been attained the need for growth materials in the diet continues. Indeed, it continues throughout life for there is an unceasing turnover in both cells and cellular materials, a turnover that commences at the moment of conception and ends only with death. Throughout this period each single body cell passes through its life cycle, dies and disintegrates and is replaced by yet another cell.

This process of self-destruction on the part of the body tissues, a process known as Catabolism, is complemented by a simultaneous process of tissue repair and replacement, a process known as Anabolism. As long as these two processes remain in balance whereby anabolic rate is equal to catabolic rate then the body will remain healthy and functional.

Protein Foods

However, the maintenance of this state of physiological balance is dependent upon an outside influence, namely the diet. It is from the foods forming the diet that the tissues draw the proteins and

420

protein-derivatives from which the new replacement cells are constructed. If the diet should fail to provide a sufficient amount of suitable protein material then the anabolic rate will falter, fall short of the catabolic rate and so cause the body to enter a decline and wasting phase. Sooner or later death will intervene; therefore we must eat in order to live.

The question then arises: "How much do we need to eat in order to live?" Leaving aside for the moment that portion of the food intake that is required for energy-supply purposes, the average man needs to eat a minimum of some 5–6 oz of "flesh" or "flesh-equivalent" food per day, say 130–150 grams since this is the daily weight loss resulting from the immutable catabolic processes.

Such figures are obtained by conducting an experiment designed to establish what is called the "nitrogen balance". Put simply, the experiment involves collecting and measuring the daily output of urea, a substance derived from catabolically-destroyed proteins and discharged from the body in the urine. At the same time a careful check is kept upon the amount of protein in the diet, protein that will be used in anabolic replacement of those tissues destroyed by catabolism and from which the urea was derived. By progressively reducing the intake protein levels a point is eventually reached whereby the intake protein level is exactly equal to the protein destruction level, this being the point of "nitrogen balance". The daily urea outputs of adults at the point of nitrogen balance show that some 25–30 grams of proteins are catabolically destroyed every 24 hours. This is "dry-weight" protein; expressed in terms of body tissues it represents some 125–130 grams of lost "flesh". This then is the amount that *must* be replaced each day by sufficient input of dietary protein.

Nitrogen balance

Moreover, mere quantity of protein is not enough, one must also have regard to protein quality. It is quite possible that a person may enjoy an adequate- even an abundant-level of dietary protein input yet still suffer anabolic insufficiency. For example, the protein gelatin may be eaten in quantities sufficient to achieve, even exceed,

Gelatin

421

nitrogen balance, yet, if gelatin be the sole source of anabolic protein, wasting will still occur. The reason for this apparent anomaly lies in the compositional nature of protein itself. Protein is not a pure chemical compound but rather a "collection" of chemical compounds, known as amino acids, of which there are about twenty-two different types. About half of these twenty-two amino acids are termed "essential" because they possess certain characteristics and qualities which render them vital to the anabolic processes. They cannot be replaced by the lesser, non-essential amino acids and, in this respect, they are not unlike vitamins, that is, they must be present in the diet if the body is to function correctly.

Returning now to gelatin: whilst this is undoubtedly protein in nature, gelatin lacks two essential amino acids, tyrosin and tryptophan, both of which are vital to the anabolic synthesis of body proteins. Thus, even when sufficient bulk of gelatin is taken in the diet, sufficient that is to achieve nitrogen balance, one is still unable to raise anabolic processes to the point of equality with catabolic processes, one is in a state of "protein debt" owing to the lack of tyrosin and tryptophan. Unless the situation is corrected by supplementing the gelatin with some other protein, one that is rich in the two essential amino acids, death must ultimately supervene.

Now let us turn to food as a source of energy.

Energy from food

Energy is required by all forms of life upon the planet Earth. The ultimate source of all energy is the sun, without which there could be no life. Even the most photophobic (light hating) subterranean animals of the deep soil and caves as well as the ocean beds, creatures which never glimpse the sun generation after generation, still rely indirectly, upon the energy which the sun pours out.

The sun radiates energy into space and some of this energy reaches earth to become trapped by the plants and converted into chemical energy, thus:

In sunlight—

422

$$6 \ CO_2 \quad + \quad 5 \ H_2O \quad = \quad 6 \ O_2 \quad + \quad C_6H_{10}O_5$$

Carbon dioxide	Water from	Oxygen passed	Starch stored
from the air	the soil	back into air	in the plant

The radiant energy from the sun, now converted and stored in the form of chemical energy—plant starch—is now utilised by animals grazing and feeding on the plants:—

$$6 \ O_2 \quad + \ C_6H_{10}O_5 \quad = \quad 6 \ CO_2 \quad \quad 5 \ H_2O \quad + \ Energy$$

Respired	Dietary	Exhaled in	Excreted as
air	starch	the breath	urine, sweat

The actual yield of energy is, of course, dependent upon the amount of starch oxidised (the name given to the latter type of chemical process). However, it is not only starch that provides the body with energy. Fats and proteins also make their contributions:

I gram carbohydrate yields	4 kilocalories of energy
I gram fat	yields	9 kilocalories of energy
I gram of protein	yields	4 kilocalories of energy

These are all average values. Whilst there is a minute variation between the energy yields of different carbohydrates such as wheat starch, maize starch, cornflour, sugar, etc. this is so very small that there is seldom any real need for differentiating between specific carbohydrates, nor indeed, between specific fats nor specific proteins, when calculating dietary calorie values. It is sufficient merely to identify the quantities of carbohydrates, fat and protein present in order to arrive at the total available energy supply.

Diets

Such energy yield figures are vital when calculating and designing specific diets. This, in fact, is the main task of dieticians, the drawing up of diets, so that the supply of energy available in the food intake is sufficient to meet the energy demands of the body but without any excesses which may, perhaps, be converted and stored as surplus body fat. Any excess of supply over demand results in obesity, an excessive and dangerous harbouring of fat. Conversely, any deficiency of supply results in weakness, lassitude and progressive loss of body tissue, the energy deficiency being met by increased catabolic destruction of one's own tissues.

The dual role of proteins in the diet should be noted. We have already discussed the vital part played by protein in the anabolic manufacture of tissues and this must always be the primary function of food proteins. However, once these anabolic demands have been met then any additional or surplus protein in the dietary intake can be utilised as an energy-source foodstuff, just like carbohydrate and fat. This duality only applies to protein; carbohydrate cannot be synthesised into cell-building material the way protein can. As a result neither carbohydrate nor fat can play any part in anabolic construction of body tissue, they are energy foodstuffs only.

We have now reviewed the nature and function of the main nutrients in food, the carbohydrates, fats and proteins, and we must now turn to the role of meat in the nutrition of man.

(Whilst the nutritional functions of vitamins and minerals has not been dealt with in general, the specific vitamins and minerals present in meat will be considered hereunder.)

The Nutritional Value of Meat

In the Food and Drugs Act "meat" is defined as:

> the flesh including fat, rind, skin, gristle and sinew of any animal or bird intended for human consumption and includes cured meat and offal but does not include fish, sausage meat, meat pie, pie filling, luncheon meat, meat roll or other meat products.

Whilst this clearly defines the legal concept of meat it is not altogether suitable for our purposes, namely the nutritional concept of meat. Firstly we have to sort out the various components that go to make up "meat", making allowances for the varying composition of different cuts and joints, examining the significance of cooking and cooking methods as well as the various forms of processing to which meat is commonly subjected. We must also include some account of the special dietary significance of offals.

424

Clearly, both from the Food and Drugs Act description as well as common usage the word "meat" is a generic term that embraces a large and variable number of foodstuffs.

Evaluation of the nutritional value of meat is therefore the sum of the nutritional values of all the various components that go to make up "meat", the lean and the fat and the connective tissues together with the relative amounts of these components in any particular cut, product or weight of meat.

Different retail cuts of meat vary enormously in the amount of inter- and intra-muscular fat deposits. They also vary enormously in the amount of outer covering fat present. For those reasons we have to distinguish between the nutritional value of lean muscle tissue and that of the associated fat.

If all the visible fat is removed from a piece of meat the composition of the lean muscle tissue that remains after the trimming is completed, is remarkably constant. This constancy exists not only between different species of meat animals (ox, sheep and pigs), but also between different muscles of the same animal.

However, in practice, this is somewhat irrelevant; we neither buy nor do we eat meat from which all visible fat is trimmed away. Inevitably fat is present in the meat both within the muscles as well as surrounding the muscles. Obviously, no matter what the retail cut or type of manufactured product the presence and quality of these fat desposits will affect significantly the overall composition and therefore the nutritional value.

TABLE 1

| | Components | | | | |
	Protein	Fat	Moisture	Ash	Calorific value
Lean muscle tissue	20%	9%	70%	1%	160/100 g
Meat, as bought	17%	20%	62%	1%	260/100 g

Hence, anyone faced with the task of designing a diet with specific amounts of any of the nutrients or the calorific contents must pay careful regard to the fattiness of the meat. Some cuts may contain so much fat that therapeutic diets designed around lean meat contents may be hopelessly, even dangerously upset.

"Bismark Steak" and similar presentations apart, we do not eat raw meats, only cooked preparations. As a result, nutritional values based on raw meats become significantly altered when the meat is finally cooked irrespective of the cooking method. We must therefore consider the nutritional values of cooked meats and the effects of cooking.

COOKED MEAT NUTRIENTS

When heat is applied to meat the proteins are denatured. This is a form of structural distortion that causes the protein to relinquish much of its water and also to become insoluble and altogether harder in texture. Furthermore, all these effects are directly proportional to the intensity and duration of the cooking.

TABLE 2

Meat cut and state	Protein	Fat	Moisture	Ash	Kcal/100 g
BEEF					
Chuck, raw	18·7	19·6	60·8	0·9	257
stewed	26·0	23·9	49·4	0·7	327
Rump, raw	17·4	25·3	56·5	0·8	303
grilled	23·6	27·3	48·1	1·0	347
PORK					
Loin, raw	15·5	24·5	59·3	0·7	287
roasted	22·5	28·5	48·1	0·9	353
Spareribs, raw	14·5	33·2	51·8	0·7	374
braised	20·8	39·7	39·7	0·6	440
LAMB					
Leg, raw	17·8	16·2	64·8	1·3	222
roasted	25·3	18·9	54·0	1·7	279
Loin, raw	16·3	24·8	57·7	1·3	293
grilled	22·0	29·4	47·0	1·7	359

After Price/Schweigert "The Science of Meat & Meat Products"

426

TABLE 3

The Effect of Cooking on the Nutrients in Offals

Offal type	Protein	Fat	Moisture	Ash	Kcal/100 g
Brains, raw	10·5	8·4	79·4	1·4	119
cooked	11·5	9·1	77·7	1·4	129
Heart, raw	17·1	3·6	77·5	1·1	108
cooked	31·3	5·7	61·3	1·1	188
Kidney, raw	15·4	6·7	75·9	1·1	130
cooked	33·0	12·0	53·0	1·2	252
Liver, raw*	19·9	3·8	69·7	1·3	140
cooked	26·4	10·6	56·0	1·7	229
Sweetbreads, raw	19·4	9·0	71·8	1·6	159
cooked	25·9	18·2	56·2	1·5	267
Tongue, raw	16·4	15·0	68·0	0·9	207
cooked	21·5	16·7	60·8	0·6	244
Tripe, raw	13·6	3·0	79·0	0·4	102
cooked	18·0	6·6	77·2	0·3	115

* Liver, being the mainstore of "animal starch" or glycogen also contains 5·3% carbo-hydrate

After Price/Schweigert "The Science of Meat & Meat Products"

Cooking also serves to reduce the moisture and fat content of meat, quite apart from the moisture that is lost due to protein denaturation there occur even greater losses due to the evaporative effect of the applied heat. Losses by evaporation are particularly acute when "dry" cooking methods, such as grilling, frying and roasting are employed.

The fat loss on heating follows a short period of softening and liquefaction of the fat and occurs not only from the outer covering fatty areas but also from the inner "marbling" fat deposits.

Thus, overall, the main effect of cooking is to cause the meat to shrink both in weight and in volume as the result of moisture and fat cook-out. In contrast, the protein content does not suffer reduction except for a slight loss of soluble proteins carried away when the meat juices drain out. However, this loss is far outweighed by the concentrating effect on the protein due to moisture and fat losses. As a result the protein content of cooked meat is invariably greater than that of raw meat. In the same way, whilst cooking causes loss of fat, i.e. the "dripping", the total weight loss due to cooking comprising both fat and moisture is such that the remaining fat in the cooked joint appears to have increased when subjected to compositional analysis. Since the volume of dripping obtained from roasted, grilled or fried meats shows clearly that loss of fat has occurred then the apparent increase of percentage fat indicated by the analytical figures is purely illusory.

PROCESSED MEATS   Here again is a term which requires some qualification and explanation. Processing can mean as little as simple comminution of raw meat whereby a "mince" is produced. In this case "processing" is limited to minimal mechanical treatment.

More often processing implies a treatment in which "foreign" or extrinsic materials are added to the basic raw meat, these additions including such things as:

Rusk, water, seasoning and spices, sometimes egg and milk or milk powder, as in sausage manufacture.

Curing salts (nitrate and nitrite), common salt, phosphates and sometimes sugar, as in ham, bacon, "corned" beef and salt beef preparation.

Vegetable additions such as onion, mushroom, etc. as in beef burger and pie-meats.

Stuffing or forcemeat as in "veal roasts", breasts of lamb, etc.

Other meats (kidney) or meat extracts such as gelatin, used in the manufacture of pie-meats and fillings.

Processing may involve no addition at all to the meats but simply be a form of cooking and sealing, as in canning.

These are the more familiar forms of meat processing; there are many other regional and national variations in the manufacture of "processed meats" particularly in the matter of sausages and sausage-like preparations.

Obviously, from a nutritional point of view, the addition of any extraneous food materials such as vegetable and cereal products will make considerable differences to the balance of nutrients as they occur in unprocessed meats. Indeed, in some cases the processed articles are subject to legal restraints affecting composition in order to avoid excessive "dilution" of the meat base. Thus, for example, there are minimum meat contents for various types of raw and cooked sausages, pies and canned meats sold in the UK.

In view of all this it would be impossible to discuss quantitatively the nutritional significance of processed meats in general since the nature of the "processing" is so variable as is the compositional nature of the finished product. Even if the survey is limited to a single type of processed meat the nutritional values are enormously variable, viz:

TABLE 4

Range of Figures found from the Analysis of 22 Samples of Beef and 47 Samples of Pork Sausage Purchased from Retail Establishments (1961–68)

	Beef sausage			Pork sausage		
	min.	max.	mean	min.	max.	mean
% Water	40·6	52·5	48·2	32·8	54·0	44·4
% Fat	16·6	33·2	24·9	22·7	37·2	31·2
% Protein	7·5	12·7	9·4	8·6	13·8	10·2
% Ash	2·2	2·9	2·5	1·1	3·2	2·1
% Carbohydrate	7·8	18·1	13·2	3·6	18·2	10·6
% Defatted meat	25·0	57·9	37·3	29·0	56·8	40·7
% Total meat	51·8	75·9	61·5	62·0	91·1	71·8

After Pearson "Chemical Analysis of Food"

Given such remarkable variation of product one can only assess the nutritional value of a single sample of a single product. Price & Schweigert have listed the nutritional values of a number of processed meat products, some of which are reproduced in Table 5 but, in reading these values, it must be appreciated that these are average

429

values only and the figures given are only valid for the products *as sampled.*

TABLE 5

Nutritional Values of a Selection of Cured Meat Products

Type of Product	Protein	Fat	Moisture	Ash	Kcal/100 g
Bacon, uncooked	20·0	14·4	61·7	3·6	216
Bacon, fried	30·4	52·0	8·1	3·2	611
Corned beef	25·3	12·0	59·3	3·4	216
Ham, as purchased	16·9	35·0	42·0	5·4	389
Pork brawn	15·5	22·0	58·8	2·7	268
Meat paste	17·5	19·2	60·7	2·8	248
Pork luncheon meat	15·0	24·9	54·9	3·9	294
Frankfurters	12·5	27·6	55·6	2·5	309

After Price/Schweigert "The Science of Meat & Meat Products"

PRESERVED MEATS

The preservation of meat and meat products is practised in order to conserve the meat against the depredations and decompositional effects of bacteria, yeasts and moulds as well as against auto–destruction by intrinsic enzymes (the latter being the catabolic enzymes which go on working long after the animal is dead).

There are, of course, a number of different methods of preservation available but perhaps only the commoner modes, canning, freezing and drying need concern us here. The effects of these three methods of preserving meats are briefly considered:

Canned Meats

As far as canned *"whole"* meats are concerned there is little or no change in nutritional value between pre- and post-heating of the contents of the can. Whilst the heating process itself may cause some cook-out of water, fat and soluble proteins (as described under "Cooked Meats") these cooked-out products remain trapped inside the can so that the nutritional value of the can contents, as a whole, remains the same. Hence, if both the solids and the broth or gravy are consumed the nutritional gain will remain undiminished. Obviously, if the canned meat is combined with other materials such as vegetables or cereal products then the nutritional value of the

430

meat will vary with the proportion of added, non-meat materials, a point already discussed under the section "Processed Meats".

Frozen Meats

When meat is frozen the moisture trapped in the interstitial spaces and fissures of the meat is changed to ice. As long as the freezing process is effected rapidly the ice crystals will be small and not unlike the crystals of salt. If, however, freezing is slow then the ice crystals will be long and needle-like and there is considerable risk of puncturing the delicate cell membranes. Thereafter, when the meat is thawed, much of the cellular fluids will be lost from the punctured cells, forming the so-called "drip". The drip loss from badly frozen meat can be considerable, causing not only a marked loss of moisture content but, far more important, the meat suffers a considerable loss of soluble protein, thereby reducing the nutritional value.

Unwrapped frozen meat, when stored for prolonged periods, suffers "freezer-burn". This involves loss of moisture by sublimation, the meat moisture being gradually transferred to the colder coils of the refrigerator to form "frost". However, during these sublimative processes only the meat moisture is lost, not the soluble proteins as in the case of "drip". Thus, freezer-burned meat displays no loss of nutrients, rather a concentration of gross nutrients as more and more moisture is withdrawn. Weight for weight, freezer-burned meat has a higher nutritional value than unburned meat. Unfortunately freezer-burned meat, after thawing, is dry, brittle, discoloured and most unattractive in appearance; it is useless for retail selling. If used in manufacturing processes such as sausage production, much of the lost moisture is re-absorbed into the meat tissues so largely restoring the original nutritional balance.

Dehydrated and Freeze-dried Meats

In much the same way, dehydrated and freeze-dried meats, when reconstituted with water, show little or no change from their original nutritional values, the latter being marginally better than the former in this respect. This slight difference is due to the fact that the moisture of ordinarily-dried meat is lost by evaporation from the surfaces of the meat. As the moisture evaporates any solutes present therein, such as soluble proteins, etc., are deposited as

431

a sort of case-hardening surface mass. Thereafter, when slow-dried meat is immersed in water to enable reconstitution to take place, much of the surface deposit of soluble material is floated off and dissolves in the "external" mass of water. It does not return to its original location within the mass of the meat. By this means some loss of nutrients is almost inevitable during the process of reconstitution.

In freeze-dried meats there is no loss of nutrients. The meat is first frozen and then subjected to rapid sublimative loss of moisture as ice. During sublimation the ice face shrinks inwards unlike the ordinary evaporative outward movement of water. Thus, in sublimative loss of moisture there is no surface deposition of solutes, all the solutes remaining in situ as the inward, sublimative shrinkage of ice occurs. During reconstitution water moves inwards and as such inward migration occurs, the solutes are carried along in dissolved state and therefore no loss can occur. Hence, reconstituted freeze-dried meats tend to retain more nutrients than ordinarily-dried meats and thus retain their original nutritional values.

We now turn to the vitamins and minerals in meat.

VITAMIN CONTENTS OF MEAT

In a relatively short chapter on the nutritional significance of meat it would be inappropriate to give accounts of all the various vitamins and their functions within the body. For such accounts a textbook of nutrition should be consulted. In this section we shall do no more than give brief consideration to those vitamins found in meat and meat offals.

Until about the turn of the century it was believed that all dietary requirements could be met by supplying adequate amounts of protein, fats, carbohydrates and minerals. Lack of knowledge of the presence and essential function of vitamins appears only in hindsight since, long before the establishment of nutritional science, it had been known that scurvy amongst sailors could be prevented by a supply of citrus fruits, principally limes (hence the appellation "Limeys", originally applied to British sailors). On the basis of protein, fat, carbohydrates and mineral contents, such citrus fruits are unremarkable and it now seems strange that no one ever questioned what specific property was possessed by the limes that they were able to prevent the development of scurvy, namely the presence of Vitamin C.

It is now known that, apart from the major nutrients, the body requires a supply of vitamins, substances which the body is unable to manufacture for itself but which are, nonetheless, vital in the manufacture and proper functioning of tissue enzymes.

The vitamins fall into two classes; those that are soluble in fats and oils, labelled A, D, E and K and others soluble only in watery solutions and labelled B and C. Meat contains vitamins from both groups. However, mere presence of vitamins in meat or in any other foodstuff is not enough, the quantity of such vitamins is equally important. Vitamins are substances whose chemical (i.e. molecular) structure is beyond the synthesising ability of the body. Therefore it is vital that these substances are present and available in the food in quantities sufficient to meet the demands of the tissues. If the vitamins are missing or present only in inadequately small quantities then the body tissues will be unable to carry out certain essential chemical reactions; some vital link in a chain of reactions will be lost and tissue function brought to a halt. Controlled laboratory investigations by nutritionists throughout the world have sought to establish the minimum daily quantities needed to keep a person in a healthy state. Whilst these daily doses are invariably small and vary from vitamin to vitamin a continuing supply of the vitamins is quite indispensable.

Lean Meat (Water Soluble) Vitamins

The so-called B-group vitamins were originally thought to be a single substance. It is now known that the original Vitamin B is really a complex of several different substances, all of which have a vitamin function but not all of which are found together in the same food source.

VITAMIN B includes such major vitamins as: Biotin, Thiamin (B_1) Riboflavin (B_2), Nicotinic acid, Folic acid, Pyridoxine, Pantothenic, acid, Cyanocobalamine (B_{12}) and maybe others, at present suspected

P

but, as yet, unidentified. Since most of these substances occu
together in the source-food it is almost impossible for anyone t
suffer a deficiency of just one of the substances listed above. A perso
suffering a Vitamin B deficiency is really suffering from a deficienc
of several vitamins thereby producing a confused medical pictur
known as a syndrome.

Meat is one of the more important sources for this "vitamin"
complex but the amount varies considerably with such factors a
species, age, fattiness, cut or joint as well as type of offal. Thus porl
contains five to ten times as much Thiamin as other species-meat
whilst liver contains about six times as much Riboflavin as any othe
offal or carcase meat. There is no constancy about the amounts o
the B-group components in meats and some idea of this highly
variable nature can be drawn from the figures given in Table 6
Note the variation of vitamin content with the age of the anima
when comparing the relative amounts present in veal and in beef.

TABLE 6

The Vitamin Content of Various Meats

Vitamin (units/100 g)	Beef	Veal	Pork	Mutton	Liver
A (Int. Nat. units)	trace	trace	trace	trace	20000
B_1 Thiamine (mg)	0·07	0·10	1·00	0·01	0·30
Riboflavin (mg)	0·20	0·25	0·20	0·25	3·0
Nicotinic acid (mg)	5	7	5	5	13
Pantothenic acid (mg)	0·4	0·6	0·6	0·5	8
Biotin (micrograms)	3	5	4	3	100
Folic acid (micrograms)	10	5	3	3	300
B_6 (mg)	0·3	0·3	0·5	0·4	0·7
B_{12} (micrograms)	2	0	2	2	50
C (ascorbic acid) (mg)	0	0	0	0	30
D (Int. Nat. units)	trace	trace	trace	trace	45

After McCance and Widdowson HMSO

VITAMIN C: As far as the other water soluble vitamin is concerned,
Vitamin C, meat comes off very poorly and anyone subsisting on an
exclusively carnivorous diet would soon suffer Vitamin C deficiency
effects or "scurvy". This is precisely what used to happen to sailors
subsisting mostly on salt pork during long sea voyages. Protein, fat

nd mineral requirements were more than adequately met by the pork diet yet the sailors frequently developed scurvy through lack of Vitamin C. In this respect meat is far from being a "perfect" food. Vitamin C (in other foods) is notoriously susceptible to destruction by heat. However, as far as the cooking of meat is concerned the B-group vitamins stand up very well to high temperatures and there is little or no reduction of vitamin levels in the cooked meats and offals compared with the raw meats. Thiamine is the most susceptible of the B-group complex but even after prolonged cooking the percentage reduction is seldom more than 25% although canned, cured meats may suffer severe Thiamine destruction (40%).

THE FAT-SOLUBLE VITAMINS, A, D, E AND K: Meat is relatively poor in respect of nearly all these vitamins. Thus, beef contains only about two International Units of Vitamin A per gram of fat, an amount that is totally insignificant in view of the fact that the average person needs some 5000 or so International Units per day. However, liver is much better than muscle meats and contains about 200 International Units per gram. Hence 25 grams of liver (about one ounce) will provide a person with all of his or her daily requirement of Vitamin A.

The situation with respect to the other fat-soluble vitamins, D, F and K, is hardly much better than Vitamin A in muscle meats. Unless one ate gross quantities of meat or subsisted mainly on liver one would be unlikely to obtain sufficient of any of the fat-soluble vitamins to satisfy the daily requirements. All these vitamins are so very much more abundant in other foods that to rely on meat or liver as the sole source of these vitamins would be to court medical disaster or financial collapse!

Finally, we consider the last of the nutrients in meat, the minerals. Analysis of human and animal tissues show that, mineral for mineral, there is little, if any, detectable difference. The mineral contents of animal and human tissues are almost entirely associated with the water and protein contents of muscle. This means that the leaner the cut or joint the greater will be the ash value on analysis, the ash being composed of all the many tissue minerals.

MINERAL CONTENT OF MEAT

435

The twenty or so minerals demanded by the body fall neatly into two categories. Firstly, there are eight minerals present in relatively large amounts in the tissues:

Sodium (N), Potassium (K), Calcium (Ca), Iron (Fe), Phosphorus (P), Magnesium (Mg), Sulphur (S) and Chlorides (Cl).

Apart from these, the major minerals, there is a second group of some ten to twelve present in only minute quantities (but nonetheless vital) including such elements as fluorine, cobalt, copper and zinc etc. In this second group are those known as the trace elements and the quantities present must indeed be small (hence "Trace") since many of them are considered to be extremely poisonous, including the four examples quoted.

For any detailed discussion of the significance of these minerals a reference book on the subject of nutrition should be consulted, the table below being no more than a brief summary of the nature and location of minerals in the body.

TABLE 7

The Principal Mineral Elements in the Body

Element	Atomic no.	Average amount in adults	Daily need	Location in the body
		g	mg	
Sodium	11	80	?	Body fluids
Magnesium	12	25	200–300	Bone
Phosphorus	15	600–900	800	Bone and tissues
Sulphur	16	170	?	Amino acids, skin
Chlorine	17	120	?	Body fluids
Potassium	19	135	800–1300	Cellular fluids
Calcium	20	1000–1500	500	Bone, teeth
Iron	26	4	10–12	Blood cells

The trace elements are mostly involved in activation of various cellular enzymes.

As far as meat is concerned we find all the major elements are present in available form. However, the quantities of such minerals present are not always adequate for our needs and we usually need to supplement the somewhat poor supply of certain minerals obtained from meat by eating vegetables, milk and other accompaniments. Thus, whilst meat, especially liver, will supply most of the iron we need, by contrast, we should fare badly if we relied only on meat sources for an adequate supply of calcium. In this respect a little "bone-dust" along with the meat would not come amiss!

What then is the overall status of meat as a food?

Clearly, as we pointed out at the beginning of this chapter, meat is an ideal tissue-building food; there is really nothing quite as good for flesh as flesh. Compositionally, human flesh is identical with animal flesh. All that is required is that the digestive system shall dissemble the animal flesh to a point whereby it can be assimilated into the blood and thence transferred to the cells. Thereafter the process is reversed and the assimilated components reassembled into human-type flesh.

Moreover, once the demand for tissue constructional material has been met the remaining portions of dietary meat can be utilised as a source of body energy. It matters not at all that there is no carbohydrate present in meat, there is always plenty of fat, even in the leanest of meats, and fat yields twice as many calories, gram for gram, as carbohydrate. Not only that, even the surplus proteins can be subjected to preparative chemical changes so that they too may provide the body with energy, again, marginally more than any carbohydrate on a weight for weight comparison.

It is only when we turn to vitamins and minerals that meat fails to measure up to the description; "perfect food". This it is not. It lacks too many essential support nutrients, substances such as calcium, vital for the building and repairing of bones and teeth. Liver apart, meat lacks nearly all vitamins save the B-group. Even if the other vitamins are not missing altogether then the amounts available are miserably small. Given an exclusive diet of meat, over a prolonged period we should gradually move into a phase of lowering health and increasing susceptibility to bacterial infection, reproduction rate would decline and infant mortality increase. This then is the real nutritional failing of meat, excepting the B-group, an inadequacy of most vitamins as well as one or two essential minerals. Curiously, the carnivorous animals do not suffer any apparent deficiency effects but then they are less finicky than man when it comes to selecting and eating the parts of the carcase. They go for the liver, the spleen and the guts, organs which are significantly less deficient than muscle meats. Moreover, eating the entrails means that the carnivore, unwittingly, reverts to an omnivore by consuming the semi-digest of plant material within the victim's gut and so repairs most of the dietary deficiencies of meat.

This, perhaps, is the secret of a good diet, plenty of meat but with a few vegetables too!

The Retail Shop

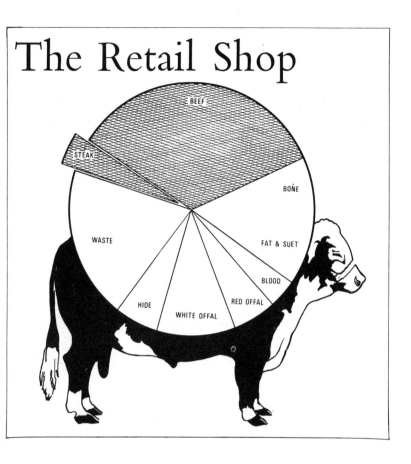

Choosing a site in which to trade is a gamble and as in all gambling there are varying odds which can be obtained. The factors which enter into the choice can never be reduced to a scientific formula. But it is possible, by carefully considering as many as possible of the factors involved, to reduce the odds down to those which any prudent businessman, say a bank manager, would consider acceptable as a commercial risk.

It is worth emphasising that there is no foolproof formula because this fact gives both hope to small men starting for the first time and brings essential vitality into trading. If there was a method of so assessing site potential as to be always right then we would not see

439

large companies make the mistakes which from time to time they do.

Types of site

There are two types of site to consider; that which exists as a trading venture, as a butcher's business, and a new unused site. Between these two lies the case of an existing shop trading in some other business which it is intended to fit and open as a butcher's shop. Many of the factors relevant to a new site or a conversion apply equally to the purchase of an existing business and for that reason we will consider the shop already trading.

Before contemplating the purchase of an existing butcher's shop there is one vital question which must be asked: "Why is the present owner selling?" Some of the answers to this question are easily provable, for example death of the proprietor. Some can be verified without too much detective work, for example ill-health, retirement. Sometimes the question will not be answered. But in any event credence should never be given to any story which you cannot verify. This is particularly true with regard to profits of the business. If they are not on the balance sheet then as far as the purchaser is concerned they do not exist. It is worth remembering that any reasons you are given for sale may be true. But if you cannot verify them it may well have cost a lot of money merely to prove that they were false.

Position of the business

Once having decided that you may proceed with a purchase, the position of the business should be examined as fully as possible. In this regard be cautious of the vendor who places too many restrictions upon you. Whilst both you and he may wish not to panic staff into leaving, it must be realised that you cannot examine a trading site on Sunday morning or on the half-closing day. And examine it you should, always bearing in mind the following points.

How many people are there who pass the shop day by day and are these a reasonable number bearing in mind the time of the year and weather conditions?

440

What opposition exists, what quality meat do they sell and at what price? Are the shops clean and busy looking or are they run down? What supporting traders exist such as bakers, greengrocers, grocers, post office, laundrettes, general hardware stores and public house? These are the minimum for a fair trading position although one or two non-food trades could be absent, provided there is not too much other meat opposition.

What is the position regarding bus stops and car parking facilities? Do the bus stops carry several routes or only one and do these routes run to and from areas of high housing density?

What is the make-up of the immediate area? Are there sufficient people within walking distance of the shop to support a reasonable trade and can they reach the shop easily? Do heavy traffic roads intervene? To cite an extreme example, it is no good having a shop near a housing estate if there is a motorway between you and the estate.

Are there any local authority developments in hand for the area? Normally this information is best obtained directly from local planning departments or from local estate agents. The type of search made by solicitors in conveyancing often does not reveal the sort of information required under this head. Local planning is not always on the grand scale. Much loss of trade can be caused by a decision to widen a road and hence narrow the pavement or worse still erect pedestrian barriers. There are several main shopping streets in London and presumably in other cities where the evidence of decay in trade caused by these barriers can be seen very clearly. Talk to as many neighbours as possible. This should be done without revealing your interest both to avoid embarrassing the vendor and to ensure as far as possible the unbiased nature of the replies. It is surprising how much information can be obtained merely by asking the question "How's trade" or by observing that you had been thinking of acquiring a shop locally. When engaged on this exercise it is good advice to disbelieve half of what is said and three-quarters of what is written. The gamble you are taking is in deciding what to believe.

Reputation of the business

In all the foregoing there should be a definite attempt to establish the standing and reputation of the vendor or if the business is being sold by a multiple, that of the recently current manager or managers. This without doubt is the most difficult information to obtain and assess. It should however be kept well in mind that a shop which has obtained an unpleasant reputation is often difficult to trade in and that the magic words "Under New Management" seldom have much effect.

Long distance opposition

Having assessed local opposition and support, a final point is to consider long distance opposition. This takes two forms either as an established market or High Street which attracts customers to consumer durables like clothes and furniture on fine days or provides one stop shopping in bad weather or a distant butcher who may serve your potential customers by supplying their deep freeze requirements. To these nine points doubtless others might be added but the experienced buyer of shops might consider any more to be merely gilding the lily. Having given careful consideration to these points the purchaser should be in a position to formulate his valuation of the potential trade the shop can do. He can decide whether the trade is or can be of the style and quality in which he is experienced. Against these assessments he will set the facts which the next section of investigation uncovers.

Capital required

Should the site be a shell or a different trade which is to be converted then no more remains to be done but to examine the capital required to fit or refit. If the business is trading as a butcher's shop or has been one then more must be done. Having assessed the exterior the interior must be examined.

Whether the shop is closed or open the equipment is of vital importance. Every secondhand item with a useful life will represent a great saving of capital. Again the best method is to list the following and make appropriate notes on each.

CONDITION OF THE BUILDING

Is it such as will bring requirements from the landlord or local authority for structural repair and decoration or from the public

442

control office or public health for specific equipment such as sinks and washhand basins? It is a well-known fact that a change of ownership has a galvanic effect upon local PHIs and they invariably seem to require more of a new owner than they did of the old one. Under this heading should be considered the possibility of un-authorised tenants such as rats and mice. Apart from the inevitable clash with the public health authority it may lead to great cost in renovation of floors and refrigeration insulation.

This should be assessed bearing in mind the need for dry storage and staff requirements. There are two schools of thought regarding shop size. One believes that there cannot be too much room and the other believes that this means that rubbish accumulates to fill the available space. To whichever school the intending purchaser belongs, he must assess the future needs and satisfy himself that he has enough room for his needs.

SPACE AVAILABLE FOR WORK

The importance of this will obviously vary with purchasers but even if such accommodation is not needed personally, the possibility of deriving an income from unused rooms should always be borne in mind and undoubtedly this will form part of the value which the vendor believes he is selling. Where such living accommodation exists it is obviously of greater attraction to the tenant and greater convenience and security to the shopkeeper if it has or can be given separate entrance facilities.

DOMESTIC ACCOMMODATION

Under this heading should be listed as a scheduled part of the contract all those items which are being purchased. Where there are numbers on body parts of cash registers and scales, these should be noted for it has been known for these things to be removed or exchanged after completion. These and a number of items of small equipment vary so enormously from shop to shop that no comprehensive guide could be given.

EQUIPMENT

Since meat is sold by net weight obviously scales are vital. They are also extremely expensive to replace and some guide to the present owner's attitude can be gauged from whether or not they are under contract for maintenance. Scales, cash registers, slicers, and refrigerated counters may be subject to hire agreements or hire purchase agreements. In either case careful examination will determine if the continuing hire or purchase charge is worthwhile or whether it would be better to pay a slightly increased charge for new items.

Blocks

Blocks are another source of high expenditure and even if worn they may have been well maintained and cleaned by a careful butcher.

Small Equipment

In aggregate the small equipment, knives, choppers, saws, hooks, tickets, trays, tins and block brushes, cost a lot of money and whilst it is not necessary to count every hook, a schedule which says "various small trade items such as hooks etc." is of little use when half have been removed. Although the legalities of purchase are covered elsewhere, it is worth bearing in mind that the essence of the final contract is the verbal agreement reached between purchaser and vendor as to what is being bought and sold. If you intend to pay for all that you have been shown then it is as well to take steps to make sure you receive all of them.

Ability to assess the working value of the small equipment and its suitability for the present and intended volume of trade must of course depend upon the trade experience of the purchaser. The variations are enormous and perhaps the most obvious example is the need to acquire or retain a small second scale if the sale of cooked meats is envisaged. Under this heading should also be considered the standard of lighting in the shop and window and the number of power points available. The general services such as electricity, gas, water and sewage should be the subject of a separate inspection by surveyors proficient in that work. The equipment mentioned above is by no means exhaustive. Bandsaws, trade bicycles, motor vans, running rails, window rails, window floats, shop blinds, window blinds, are a few additional items which come easily to mind as does the small electric cement mixer which was part of the equipment of one shop in the author's experience. An intending purchaser

444

Fig. 1. a. Dagger shape shop knife, with brass finger guard   b. Boning knife, brass mounted or plain beech handle   c. Newcastle pattern, brass mounted butcher knife.

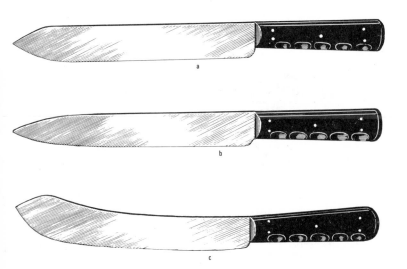

Fig. 2. a. Dagger type, with a somewhat sharply pointed end   b. London pattern, with tapering point   c. Scimitar, with slight curve from about 2/3 down the blade.

445

should however make a list prior to seeing the shop, of those items which his external inspection has convinced him are essential. These he should tick off when seen. Other extras may be worthwhile purchases or merely white elephants.

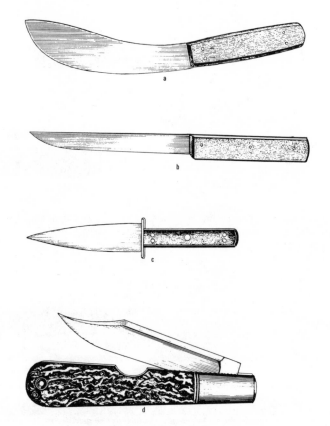

Fig. 3. *a. Green River pattern skinning knife   b. Green River pattern boning knife c. Pig sticking knife, double edged with guard   d. Inspector's knife.*

REFRIGERATION

Two broad considerations are involved here. Firstly the state of the plant and structures and secondly the suitability of the capacity for the trade to be done. The first is a subject for specialist examination. The second is part of the buyer's gamble and the odds will be in his favour if his experience in retail trade is wide.

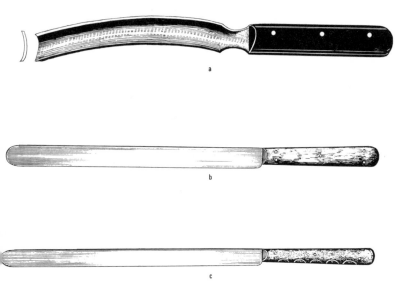

Fig. 4. *a. Ham/Gammon boner b. Beef knife c. Ham knife.*

Fig. 5. *a. Sliding knife with rosewood handle b. Breast knife with leather handle.*

Both internal and external delivery areas are of vital interest. The presence of parking restrictions outside, particularly those associated with a pedestrian crossing, can cause difficulties in delivery as also can the shortage of suitable hanging space or standing space for delivered meat. Here again the purchaser's trading experience is his only guide, provided that he has noted the items on his list and examined the shop front and rear for its virtues and vices in this respect.

DELIVERY AREAS

TRADING ASPECT

Having made the above assessments the purchaser should know how firm his intention is to acquire the business and if that intention is clear he should finally examine the trading aspect. This cannot be done without the active co-operation of the vendor and if this co-operation is absent then the purchaser should beware. Here again examination should be systematised.

Accounts

Trade, profit and assets should only be accepted on the basis or certified accounts. Grubby exercise books with pencilled entries may be of interest but that is all. They may well be inaccurate.

Type of Trade

The accounts should reveal ready money, credit trade, catering trade and contract trade. If takings are not subdivided under these headings and evidence exists of such trading, then the accounts are suspect. It is seldom possible to examine bank sheets but if the opportunity occurs it should be remembered that an established business without a book debt should not need to be overdrawn. If it is, then the profitability or the prudence of the vendor is suspect.

The ready money trade might be a mixture of counter, delivery and home freezer trade. Here again shop books should record the three aspects and excessive delivery is a suspicious factor in these days of high delivery cost. In fairness to a vendor it should be pointed out that some of the above "suspect points" do admit of straightforward explanations. For example an overdrawn account might well be due to a recent capital purchase or to an unusual book debt.

Similarly a high proportion of delivery might be due to the fact that there has been a deliberate policy to offer this service, which if sales prices have been adjusted is quite satisfactory.

Finally the purchaser's own accountant should always be asked to comment on the accounts and these should cover a minimum period of the past three years. Since accountants deal in history and sometimes in archaeology it should be expected that accounts for 3 years will only cover a period up to about 18 months prior to purchase. If they are much more ancient suspect them. Bear in mind also that

448

much could have happened to the business in the period since the last accounts were prepared. In this respect the daily shop books should be examined to see if there has been much alteration in the volume or structure of the trade. Suspect an unusual increase and any recent alterations.

It is difficult in most cases for an intending purchaser to make frequent visits to a shop when it is open and working. Neither side generally wishes the staff to be aware of an impending change and visits in the guise of insurance agents, family friends, or travelling salesmen usually are not very satisfactory.

CONDUCT OF THE TRADE

Despite the difficulties, however, it is vital to see how the shop is run, to assess the quality of the meat sold, the prices and the personal or impersonal way in which the business is conducted. The bigger the shop the more difficult is this task. Often a shop for sale is not a "happy shop". Staff know or sense that a change is impending. The seller himself may be sad at the decision he has made, perhaps forced on him by health or financial problems, or he may be apprehensive because in order to obtain his price he has had to lie and live the lie. Matters like this call for observational powers which only experience can bring. Yet it is probably true to say that most young butchers buying their first shop are tradesmen enough to "feel" that something is wrong even if they could not say exactly what it is. In trying to gauge this "feeling" it is better to visit a shop when it is busy than when it is slack.

In the final analysis it is this question of "feel" which should be paramount in your decision. The concrete factors which have been dealt with in the previous paragraphs can be listed and evaluated at leisure and it is fair to say that a shop worth thousands would not be rejected because it was deficient in a few "S" hooks. When, however, those concrete matters have all been considered and passed as satisfactory the final evaluation must depend upon the expertise and instinct of the purchaser just as its ultimate success will depend upon his enthusiasm, his knowledge, his hard work, and of course his luck.

Introduction to Planning and Construction

In contemplating the initial steps necessary in planning a retail butcher's shop, care and consideration of all the factors involved is of paramount importance, there is no set formula or constant principles that provide a rigid guide.

The area and shape of premises must have a bearing upon the early deliberations, together with accessibility to site, that is goods received at rear or from street, which would affect shop layout.

PLANNING CONSENT

Local Authorities must be consulted on such matters as sanitation and staff facilities, fire precautions, etc. and their approval of the proposals must be obtained. In addition to this, planning consent is required in all instances.

Additional factors apply to premises within a new shopping complex where the architects for the development may exercise some measure of control in the use of materials, colours and the design of the shopfront to maintain a degree of uniformity.

Under these circumstances any attempt to define a standard form of procedure, or cover the complex range of possible designs and layouts would only confuse the layman.

A reputable shopfitting firm, preferably one specialising in the food trade, should be consulted at the earliest possible stage, when the basic requirements and possibilities may be discussed preparatory to the preparation of scheme drawings. Normally the shopfitter will liaise with the relevant authorities, incorporate their requirements in drawings and obtain the necessary approvals.

Basic elements

Certain basic elements however may be followed and materials, with suitable alternatives, suggested to provide guide lines when contemplating a new venture.

For the purpose of providing a general line of approach to this subject, a structure of say 4·6 m (15′) wide by 18 m (60′) deep by 3 m (10′) high to be fitted out will be envisaged and a suggested shopfront and interior layout described. The shopfront will be of simple design with a high proportion of glass and other materials that require minimum maintenance, simplication of cleaning and eye appeal.

The sales area will be 8 m (25′) in depth where a cross screen would be positioned, ample space allowed for staff free movement behind counter and to rear, and for customers to enter and leave.

Beyond the cross screen the cold stores will be sited, preparation area provided, and staff room and toilets with rear access to service road.

The importance of an attractive if simple façade needs no emphasis. as the initial impact may well have a lasting effect upon customers.

The elements of construction comprising such a unit will now be dealt with and described in detail, broken down under main and sub-headings, comprising Shopfront, Sales Area, Store and Preparation, Display, Electrical, Ventilation, together with alternatives and advantages.

SHOP FRONT

WINDOW SASH FRAME: Constructed from aluminium, stainless steel or bronze, of light section, to receive virtually full height glass with small granite or marble base at pavement line, from 10 cm (4″) to 23 cm (9″) in height. This is applicable only if a portable refrigerated display unit is decided upon. At the top edge of glass a gap is formed for ventilation, or an adjustable louvre vent incorporated. Alternatively, the sash could be constructed from polished hardwood. This is attractive but requires greater attention and maintenance. In the case of a fixed refrigerated display top or conventional slab being decided upon, the stallriser (i.e. base to glass) would be from 59 cm (23″) to 69 cm (27″) high from the pavement with cill rail, plastic panel with granite or marble base member, or backing wall faced with frost-proof tiles.

ENTRANCE: A pair of 76 cm (30″) wide metal framed doors to match window sash, positioned at frontage line so that one door forms window end when open, with adjustable louvre vents, lay or vertical, incorporated in transome frame opening over doors. Alternatively, doors of armour plate type, i.e. all glass, or of polished hardwood, could be used. If the floor is above the pavement, a step at frontage line can be avoided by setting the doors back and sloping lobby floor to suit. However, this has the disadvantage of obstructing the customer shop space unless made to open outwards. The foregoing provides a 3 m (10′) window display with 1·5 m (5′) wide entrance.

SUNBLIND: This is best positioned immediately above the window sash so that the fascia is visible when the blind is in use. The operating arms may be of patent foldaway, trellis or plain arm type. The first type is unobtrusive and neat. The canvas should be plain but a wide choice is available. Alternatively there is the continental canopy type, or one of rigid aluminium construction. This type, however, although attractive, will not provide the same degree of protection as its projection is somewhat limited. Should the premises be within a new development, a blind may not be necessary.

FASCIA: A simple and clear design is preferable to one of elaborate decoration. To this end, a plain panel with clear lettering is recommended. The panel may be of plastic with applied perspex letters of clear bold type indicating the name in contrasting colours chosen from the wide range available. Alternatively, the fascia may be illuminated, in which case the panel would be of opaque plastic with tubular lighting concealed behind so that the entire panel glows, emphasising the lettering. Again, the letters could be of box section with concealed neon tubes, in perspex, to illuminate the individual letters. There are many variations of the above and it is suggested that these be discussed with the shopfitter in conjunction with the authorities, as planning consent must be obtained.

SALES AREA

FLOOR TREATMENT: Assuming a concrete foundation exists, a surface of terrazzo or tessellated tiles will produce a durable and easy to clean finish to the shop area. The terrazzo may be laid in situ, but preferably as 30 cm × 30 cm (12″ × 12″) pre-cast tiles with cove member

452

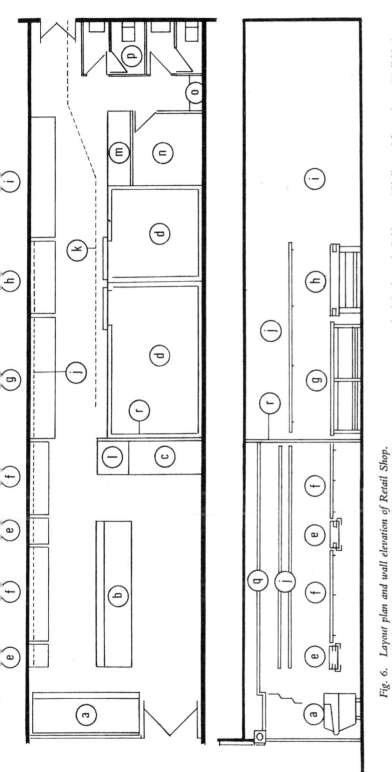

Fig. 6. Layout plan and wall elevation of Retail Shop.
a. Refrigerated portable window display b. Refrigerated serving counter c. Frozen food display d. Cold stores (Chiller and freezer) e. Wall blocks on cantilever irons f. Wall shelves on cantilever irons g. Work table h. Block and stand i. Machinery area (and/or blocks/work tables) j. Wall rails k. Ceiling rail (with travelling hooks) l. Slicer or display stand m. Block or work table n. Staff room o. Sink unit p. Toilets q. Line of suspended ceiling over shop area r. Cross screen.

formed to walls 10 cm to 15 cm (4″ × 6″) in height. Such tiles are from marble chippings, available in white and many colours. Tessellated tiles are available in a wide range of colours. They have flint chips incorporated. Normally 15 cm×15 cm (6″×6″) in size and coved upstand tiles 10 cm×15 cm (4″ × 6″) in height are provided for use at base of walls. Other floor surfaces are thermoplastic tiles and jointless composition compounds. They are less expensive and suitable should the floor to be treated be of timber construction for hygiene. Coves to walls should be incorporated wherever possible.

WALL TREATMENT: Glazed tiles continue to hold favour for ease of cleaning and attractive appearance. There is an extensive range of colours and combination of colours. Modern tiles have cushion edges giving a slight quilt like effect, obtainable in sizes 10 cm×10 cm (4″×4″) and 15 cm×15 cm (6″×6″). The smaller have a more pleasing finish but are more expensive to fix, there being eighty-one per square yard against thirty-six with the 15 cm (6″) ones. An alternative is the use of 1·25 m (4′) wide laminate panels such as "Formica" bonded to plywood backings and secured to wall battens, with neat aluminium vertical joint cover strips. There is a wide colour range. The bottom edges rest on floor cove member. They are obtainable up to 3 m (10′) high.

CEILING TREATMENT: Plaster finished with enamel paint is suitable, but a suspended ceiling about 2·5 m (9′) above floor is most attractive over sales area, masking any beams and electrical work, etc. These ceilings consist of bevel edged tiles normally 61 cm×61 cm (24″×24″), of fibrous material with asbestos incorporated to provide the required fire resistance. They are supported by a light alloy grid on ceiling hangers and decorated with emulsion paint, white or tinted. The tubular lighting may be secured to the underside or concealed in troughs with opaque perspex panels to underside, flush with ceiling tiles. There is another form of suspended ceiling, in general as described above, but with larger panels and all of translucent material like fibre glass, with tubular lights over, so that the entire ceiling becomes illuminated. It is most effective but rather more expensive.

CROSS SCREEN: As previously mentioned, a screen would be erected

454

about 7·6 m (25′) back from the frontage line to terminate sales area. For a tiled finish this would be on pre-cast block construction, with 1·25 m (4′) opening for access to the rear in line with the gangway behind the serving counter, with ventilation slots provided at suspended ceiling level, which could be fitted with fixed type louvre vent units. Should the wall treatment be of laminate panels then the screen could be of timber stud construction.

LAYOUT—SALES AREA: The components of this area will be separately described. They consist in brief of wall shelf approximately 6 m (20′) long, from back of window display to cross screen with two 61 cm (24″) long blocks incorporated, then a 1·25 m (4′) wide hang-way to the back of a serving counter which would be from 3 m to 3·5 m (10′ to 12′) in length by 1 m (3′) wide, leaving a space of about 2 m (6′ 6″) between the front of the counter and opposite wall. It is well to consider that inconvenience is often created by the tendency of customers to congregate at the window end of the counter which in turn causes congestion at the entrance. With this in mind it might be advisable to dispense with a cutting block in this position and form some sort of attraction towards the rear of the shop. This might be achieved by having an "L" shape counter, or a frozen food display cabinet with display shelves over and possibly an attractive canopy separately lit. Such a layout may not be possible if there is no rear access for the receipt of goods.

RAILWORK—SALES AREA: The present trend of pre-cutting reduces the need for an elaborate arrangement of display rails. Small joints can be effectively displayed on two wall rails over the shelf behind the counter, 1·83 m and 2·13 m (6′ and 7′) above the floor, 7·6 cm and 18 cm (3″ and 7″) out respectively. For window display three rails should suffice, positioned 1·767 m (5′ 6″), 2 m and 2·33 m (6′ 7″ and 7′ 8″) above the floor, carried by ceiling hangers with a 7·6 cm (3″) set between the rails to give a forward or backward slant as desired. All sales area railwork should be in polished stainless steel of light section rails of say 3·8 cm×0·9 cm ($1\frac{1}{2}$″×$\frac{3}{8}$″) material.

WALL SHELVES: White laminate bonded to blockboard is recommended, positioned on to the wall behind the counter, 0·86 m (2′ 10″) above the floor, supported by cantilever irons to leave the

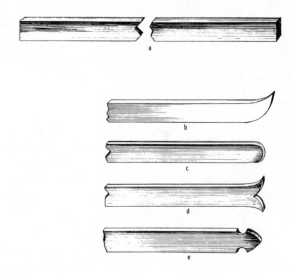

Fig. 7.   Rails
a. Plain end   b. Turned-up end   c. Rounded end   d. Fish tail end   e. Spear end.

space underneath clear for ease of cleaning. Shelves should have a matching backboard, to protect the wall face, 30·5 cm (12″) in height with the top edge 22·9 cm (9″) above the shelf, and secured with plated roundhead screws for easy replacement. The back edge of the shelf would be set about 3·2 cm (1¼″) clear of the backboard to form a cleaning gap. As an alternative the backboard could be of 3 mm (⅛″) thick "Darvic" or similar white plastic. Width of the shelves may be from 38 cm to 53 cm (15″ to 21″) to suit requirements. Should the sale of tinned goods be envisaged, sets of display shelves could be positioned where considered convenient and visually attractive, preferably towards rear of shop. Sets would comprise three or four 92 cm×23 cm (36″ by 9″) shelves carried by wall bars and adjustable brackets. A set may be positioned at window end with mirror backing.

CUTTING BLOCKS: With pre-cutting and a refrigerated display serving counter, a block and stand at end of the counter is not considered necessary. Small blocks behind in line with wall shelf should suffice. They should be about 61 cm (24″) or 46 cm×46 cm×18 cm (18″×18″×7″) thick with stainless steel corner plates. These would

456

Fig. 8.   "Albion" reversible hardwood block: all joints morticed, tenoned and dowelled.

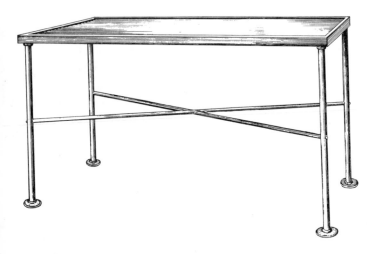

Fig. 9.   Stainless steel table: suitable as a filling out table for sausage—centre driphole carries off moisture. Adjustable base flanges compensate for uneven flooring.

457

be supported by cantilever angle irons grouted into brickwork with backboards and cleaning gaps as described for wall shelves. The width of blocks should be the same as the shelves to form a straight line.

CASH-DESK—OFFICE: It may be desired to have a desk in preference to taking the cash over the counter, and this could be incorporated in the cross screen with its front flush with the screen. This would house the telephone, records, cash, and centralise ordering, and would clear the counter and expedite service. Such an arrangement would however reduce the possibility of having an "L" shaped counter.

REAR AREA

Beyond the cross screen would be positioned the cold storage units, a chiller of say 93 m² (1000 sq ft) and a freezer of 46·45 m² (500 sq ft). These capacities are of course adjustable to suit individual requirements. Sliding doors are preferable to avoid obstruction. They should be positioned within reasonable reach of the shop and preparation area, which would be situated just the other side of the 1·25 m (4′) opening to sales area. This would accommodate a block and stand say 1·83 m×0·61 m (6′×2′) and a large work table about 3 m×0·61 m (10′×2′) with terrazzo or white laminate top. The floor treatment of exposed area would be granolithic, slightly roughened and coved to walls. This area would be provided with a galvanised rail for receipt of carcase meat, of 5 cm×0·9 cm (2″×⅜″) material, positioned 2·5 m (8′) above the floor and 23 cm (9″) clear of the wall (or cold-room) face and carried by braced wall brackets or ceiling hangers. An alternative to the fixed rail is one that carries swivel travelling hooks, providing an easy means of moving goods to the desired position, that is the cold store or chopping block. Also, it is often possible for the rail to commence at, or even extend beyond, the entry door to rear of premises, thus reducing the handling to an absolute minimum.

The toilets would be sited at the extreme rear and a staff room formed adjacent to it, having a sink with a hot water supply incorporated. Provision must be made for adequate ventilation to ensure a free flow of air from the front right through to the exterior of the

458

ear wall. Another factor is the siting of the refrigeration equipment, sometimes positioned on the top of the cold-room, but for preference, if possible it should be sited externally in a suitable louvred housing, to reduce undue heat build-up internally, particularly when the shop is closed. The units serving refrigerated displays are normally housed within them.

The window display will indicate the class of trade and standard of hygiene. The arrangement should be attractive and, as far as possible, maintained this way by constant attention during trading. A portable refrigerated display top on a base fitted with castors is recommended for hygiene and efficiency in use. The floor treatment would extend under and up to the window sash. This simplifies cleaning not only of the floor but also the inside of the shopfront glass. Further, the unit can be dressed from inside the shop with the display when in the out position. This avoids having to go outside to view the effect. With a fixed refrigerated top, resting on a support structure, the best method is to have the type that is fitted to the shape of the window with watertight seals to sash and wall end and preferably finished with a stainless steel cover strip. This avoids the need for glass upstands at sealed edges. Should no refrigeration be desired, a display slab would be incorporated and may be of marble, pre-cast terrazzo or laminate bonded to blockboard on a support structure to set slab on an approximate 7·5 cm (3″) slope. Here again, edges to the window and wall must be effectively sealed. The use of trays, preferably stainless steel, for display is most effective and simplifies operating service. It is suggested that spare matching trays should be readily available, dressed in readiness to replace those in the window when they begin to look untidy. This of course would not apply to every tray but it is well to remember that a couple of disarranged ones could mar the entire window display.

For customer service a refrigerated display counter with high front glass, normally to a height of 1·25 m (4′) above floor, with stainless trays, back shelf for cash, scales and wrapping, is most attractive. It is efficient and expedites service. Again, spare trays are advisable.

459

Fig. 10. *Metal framed counter. White cleanoid panels secured with polished stainless steel angles and strips to heavy angle iron frame.*

The alternative is a conventional counter with marble, terrazzo or laminate top, preferably with a glass screen to front edge between scale and till, for hygiene and protection. Should a block at the end of the counter be dispensed with, the counter could be 3 m to 3·5 m (10′ to 12′) long depending upon the decision to form an "L" shape serving arrangement, or not. If decided upon, the return section of counter across the rear of the shop would be 1·5 m (6′) in length, and this could be a frozen food display cabinet if desired. It is advisable to position the cash register at the far end of the counter to draw customers away from the entrance end.

Equipment required is governed by the scope of the business envisaged. This may encompass sausage making and the preparation of hams, cold meats, or even a range of delicatessen and pies, etc. Such equipment may partly be positioned in the sales area, a slicer and possibly a mincer, for example, but for obvious reasons this should be within the preparation area at the rear, which would be laid out by the shopfitter to accommodate the machinery etc. in the most suitable sequence and position within the area available. A bandsaw is a most useful item of equipment, and this, together with mixers, mincers, fillers, etc., etc. of various makes and capacity may be examined at a reputable butchery equipment suppliers, who will give helpful advice.

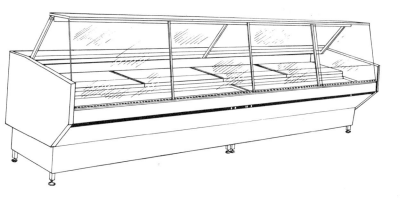

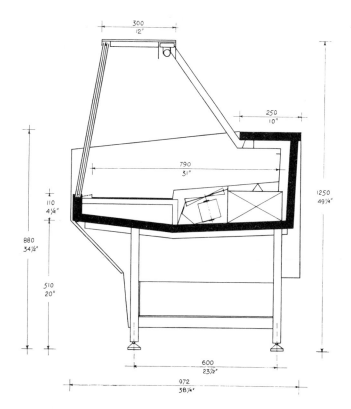

Figs. 11 and 12.   Forced air cooled display cabinet for fresh and cooked meats, thermostatically controlled.

461

Another facet of trade might be poultry and game to varying degrees, when additional stainless steel display rails would be needed, possibly ceiling type across the rear of the sales area. Sundry equipment such as knives, cleavers, saws and hooks, would be obtained from a butchers' supplier and chosen from the wide range available.

ELECTRICAL

A well illuminated window and sales area must enhance the displays and general attraction of a shop, and for this purpose tubular fluorescent lighting is recommended with the possible addition of some spot lights in the window to emphasise a particular commodity. Manufacturers have developed special tubes, such as warm white, for the trade. These do not affect the natural colour of the meat, and here again the shopfitter in conjunction with the electrician will advise upon the number and arrangement suitable to the shop layout, also upon the number and positions of power outlets to serve the refrigerated components, machinery and equipment.

VENTILATION

The need for adequate ventilation is paramount: a cool shop helps to provide a hygienic atmosphere, apart from the necessity of maintaining the condition of goods. Ventilation can normally be effected by incorporating vents over the window sash, entrance doors and to the extreme rear of premises where an extractor fan should be positioned. To ensure a free passage of air, any cross screens must therefore have vent gaps at ceiling level. This is particularly important when the shop is closed and heat tends to build up. This could affect the refrigeration unit. To supplement this in difficult circumstances, a ceiling fan of slow rotation type could be installed to circulate the air when conditions demand. The most efficient method of course would be an air conditioning plant which ensures a constant change of air at a controlled temperature. If particularly desired, the trunking for this would be concealed above the suspended ceiling to the sales area. Should the refrigeration equipment be positioned over cold-rooms, the trunking would be so arranged as to incorporate the extraction of their heat externally.

The intention of the foregoing is to describe in some detail the general factors in planning a new establishment of the dimensions given, with possible alternative materials, and to provide basic guide lines relative to a town type shop.

Design and layouts will vary to some extent according to location and area of premises. A village shop would differ from a town shop or one within a new shopping development, where trade and customer relationship may be quite different. This again could affect the extent of display and refrigeration, and even staff clothing. Other elements, some minor but important, to be considered, are printing of bags and wrapping materials, string and skewers, hooks, clean clothing, provision of ring externally for dogs, possibility of delivery service, need of internal blinds—with an attractive and clean shop why not show it—and of course security. Another factor worthy of consideration is the provision of mirrors, for the ladies and to enhance displays. Mirrors about 45·72 cm (18″) high on tilt, about 1·524 m (5′) above the floor to length of the wall above the shelf behind the counter are most effective as they reflect the whole serving counter display from the customer's position. This is also a deterrent to pilfering.

It will be appreciated that to cover every facet of differing types of trade layouts would be a formidable undertaking, and it is therefore hoped that the information and details provided give a sound basis for the preliminary consideration and ultimate decisions when venturing upon a new establishment. First, make quite sure of and decide upon the basic needs and requirements desired, then discuss them with the shopfitting and refrigeration specialists, so that all have a clear picture of the ultimate aims.

Refrigeration
of Meat

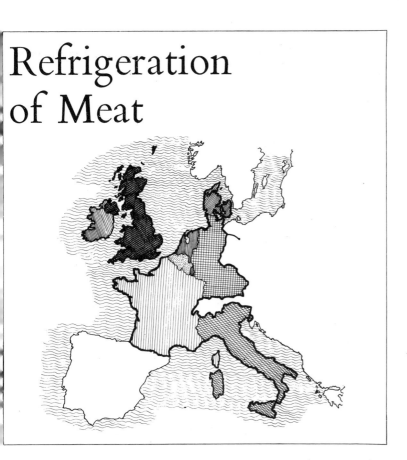

The proper refrigeration of meat is vital for its economic handling with minimum losses of quality and value.

This chapter does not claim to tell all there is to know about the refrigerating of meat. Engineering aspects can only be lightly touched on. Furthermore, many factors influence the sales value of meat quite apart from its refrigeration. However, over the past few years experimental laboratory investigations have been carried out alongside measurements of industrial meat refrigeration practice in the UK and to a lesser extent in other EEC countries, involving abattoirs, markets, shops, transport, freezing plant, cold stores and other meat handling procedures such as at the docks on imported meat. It is therefore now possible to indicate a number of points in the refrigeration chain where performance could be improved. Some of these conclusions, and the results on which they are based, have already been presented in greater detail than is possible here in the collected papers of two symposia published by the Meat Research Institute on "Meat Chilling" (1972) and "Meat Freezing" (1974).

Basic Principles of Refrigeration

Refrigeration is the removal of heat from a substance which results either in a reduction in its temperature, e.g. cooling, or a "change of state" for example, water to ice. A measure of heat is the British Thermal Unit (Btu). One Btu of heat if given to one pound of water raises its temperature by one Fahrenheit degree, or, taken from one pound of water lowers its temperature by one Fahrenheit degree. This amount of heat corresponds to the "specific heat" of water which by definition is 1. To change the temperature of one pound of ice by one Fahrenheit degree requires approximately 0·5 Btu. Thus the "specific heat" of ice is 0·5. Now at the change of state from solid to liquid (e.g. ice to water) or liquid to vapour (e.g. water to water vapour or steam) relatively large amounts of heat are absorbed— at a steady temperature in the case of pure substances. Pure water requires the addition of 144 Btu per lb in the change from ice to water and 972 Btu per lb to change from water to steam. These quantities are known as the "latent heat of fusion" and the "latent heat of vaporisation" respectively. Of course the reverse processes require the extraction of the same amounts of "latent heat".

REFRIGERANTS

In a refrigeration system this heat is disposed of by a refrigerant liquid which absorbs the heat in order to boil. Examples of liquids used as refrigerants are ammonia which boils, at normal atmospheric pressure, at $-33°C$ ($-28°F$) and R12 which boils at $-30°C$ ($-21·5°F$). In its simplest form a refrigeration plant can be represented by a container of a liquid, exposed to the atmosphere, which boils at the required temperature. Provided the boiling liquid is replenished as it evaporates then the refrigeration temperature can be maintained. The limitations of this simplification obviously are that the refrigerant is lost and must be replaced and that the minimum evaporation temperature is the boiling point of the liquid.

However by restricting the flow of vapour, thus increasing the pressure, higher evaporation temperatures are produced whilst lower evaporating temperatures can be achieved by reducing the pressure by pumping away the vapour at a faster rate. In order to make the system economical the vapour is recycled by condensing to liquid for re-use in the evaporator. This is done by compressing the refrigerant vapour which is then discharged into a condenser which is usually cooled by available air or water, and the liquid is condensed at high pressure.

The liquid formed in the condenser may be a few degrees lower than its condensation temperature, that is sub-cooled, before it leaves the condenser.

To complete the cycle the high pressure liquid is returned to the evaporator under control effected by an expansion valve. When the liquid passes through the expansion valve the pressure drops and a proportion evaporates, taking heat from the remainder of the liquid. Thus a mixture of liquid and vapour is in fact returned to the evaporator, the vapour passing on to the compressor, the liquid being evaporated effecting further refrigeration. To minimise the proportion of the vapour entering the evaporator at this stage sub-cooling of the liquid is arranged before the expansion valves.

MECHANICAL REFRIGERATION SYSTEM

The simple mechanical refrigeration system then is made up of the four basic components, evaporator, compressor, condenser and expansion valve. Figure 1 diagrammatically illustrates the connection of these. Such is the basis of the mechanical refrigeration plant used to remove heat from meat in order to lower its temperature and increase its storage life.

The heat absorbed from the meat, via the air in a chill-room or blast freezer, results in the liquid refrigerant contained in the evaporator

467

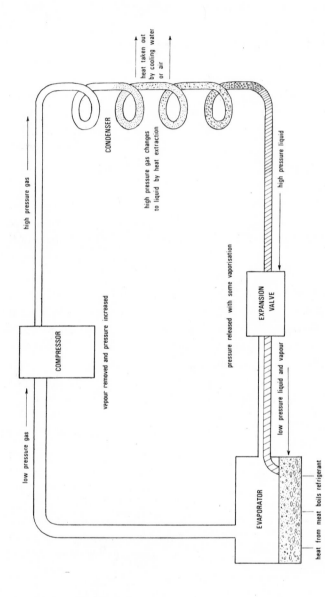

Fig. 1. *Diagram of the components of the basic mechanical refrigeration cycle.*

468

boiling and the vapour produced is removed from the evaporator by the compressor. The compressor increases the pressure on the refrigerant vapour to such a level that it will condense back to a liquid in the condenser giving up heat to air or water coolers. The high pressure liquid is reduced to evaporator pressure and returned to the evaporator by the expansion valve when more heat is absorbed and so the cycle continues.

It is a simple matter for a refrigeration engineer to calculate how much heat has to be extracted from a load of meat in order to cool it through a certain temperature change, and therefore the capacity of the refrigeration plant required to perform this operation in a given time. However, the heat has first to be got out of the meat. Cooling or freezing time of meat is influenced by the time it takes to shift the heat from the deep parts of the meat to the surface, its shape and size and the speed of removal of heat from the surface to the refrigerant in the evaporator. Still air is a poor conductor of heat and has a low heat capacity so that high air speeds are necessary to remove heat from the meat surface. However, high air speeds need powerful fans which create heat which also has to be removed by the refrigeration machinery. The output of a chill-room or blast freezer cannot be described in tons of meat per day unless the package size is specified.

The rate of conduction of heat through meat and fat (which is the better insulator of the two) and the heat transfer from meat to air at the meat surface set practical limitations to the rate which is either physically possible or reasonably economic. Consequently, impracticable schedules are sometimes proposed in ignorance of these physical properties of meat. It is therefore useful for meat operatives and refrigeration engineers to have a set of tables or graphs for cooling times of carcases and other meat of different types and weights under various conditions of cooling and freezing. In the recent experimental results given in Table 1 we have a typical example of the production of such design information at the UK Meat Research Institute.

469

TABLE I

Chilling Rates of Beef Sides, 145–171 kg (319–371 lb)
(experimental data of C. Bailey, 1973)

Chilling temperature °C (°F)	Air speed m/s (ft/sec)	Time to 10°C (50°F) deep loin* hours	Time to 10°C (50°F) deep leg hours
0 (32)	0·5 (1½)	9	26
	1 (3¼)	9	24
	2 (6½)	7	23
	3 (10)	6	22
4 (39)	0·5 (1½)	13	37
	1 (3¼)	12	32
	2 (6½)	9	32
	3 (10¼)	9	29

* Longissimus dorsi ("eye muscle")

RELATIVE HUMIDITY

In the refrigeration of meat it is important to consider the high humidity which develops as moisture evaporates from the meat. Humidity is usually expressed as "relative humidity" which is the percentage of moisture in the air compared to fully saturated air at the same temperature, which would be 100%. Relative humidity around or just below 90% of saturation would seem to be a suitable compromise level, lower values encourage excessive evaporation from the meat with unnecessary weight loss, whilst higher values approaching 100% saturation, incur the hazard of condensation.

CONDENSATION

In the change from water to water vapour heat is absorbed. Therefore when water evaporates from meat surfaces some cooling of the surfaces occurs. In the reverse process when water vapour condenses on a meat surface the surface is warmed up by the heat released by the change from vapour to liquid. When moist warm air enters a chill-room it will be cooled whilst passing over cold meat surfaces and water will condense out, thus warming and wetting the meat. Also if warm carcases are put in a chill-room already containing cold meat then the warm meat will heat the room air and give up water vapour to the refrigerated air which will condense not

470

only on the evaporator coils but also on the cold meat, with adverse microbiological effects.

Some of the moisture collected from the room is given up to the cold evaporator fins. The frost built up on the cooling pipes does not transfer heat as quickly as the bare metal and the efficiency of the cooler consequently diminishes. It is important therefore to defrost the cooler regularly. This is done quickly and usually automatically by pumping hot gas through the pipes or by electric heating. Such automatic periodic defrosts must be so regulated as to keep the warming up of the product to a minimum and it is important that fans do not come on during defrost and thus circulate warm air. The water from defrosted coils must be effectively drained away or else this can cause wetting of the meat by spray distribution when the fans come on. Of course, defrosting can be achieved, albeit more slowly, by switching off until the ice thaws and drips off.

Cooling

Immediately after an animal is slaughtered for meat its temperature is, of course, at blood heat, normally around 38°C (100°F). As a result of post mortem biochemical activity in the muscle there is in fact a slight rise in temperature after death. In consequence it can be as much as 8 hours before the deepest part of a side of beef falls below its original temperature, even in a chill-room.

Although meat that is allowed to cool down naturally to ambient conditions can be satisfactorily handled and distributed for local consumption it is better to chill artificially by means of mechanical refrigeration. This is because the bacteria that contaminate meat multiply faster the higher its temperature. The time it takes a single bacterium to grow and split into two new cells, which then repeat the cycle, becoming four, is determined largely by temperature. The magnitude of this effect is shown by Table 2, which brings together the most reliable published results as well as unpublished results of the UK Meat Research Institute.

TABLE 2

Times (in days) to Bacterial Spoilage of Meat
(at 100% relative humidity)

°C	°F	Initial bacterial numbers per cm		
		0–100 (superb hygiene)	100–10000 (very clean)	10000–1000000 (typical UK)
0	32	16	11	6
5	41	9	8	4
10	50	5	3	2
15	59	3	2	about 1
20	68	2	2	about 1

The times given in Table 2 represent the periods elapsing before the surface of the meat becomes covered with bacterial slime, the condition which usually corresponds with a bacterial count of about 10 million bacteria per cm. These times are prolonged to some extent by some surface drying of the meat, which takes place when the relative humidity is less than 100%. The organisms that live and thrive deep in the meat likewise multiply faster the higher the temperature and this can cause "bone taint", unless the meat is cooled right through and reasonably soon after slaughter. It has often been said, quite erroneously, that this condition could somehow arise from the so-called "animal heat" becoming "sealed in" by artificial refrigeration.

The types of micro-organisms that can transmit diseases to man do not normally multiply at temperatures much below 10–15°C (50–60°F) so that health precautions similarly demand chilling as rapidly as possible.

Two conclusions can therefore be drawn:

1. The faster a temperature close to 0°C (32°F) is attained throughout the meat the better for keeping quality.
2. The less the post-slaughter contamination of the exterior of the carcase the better, so that any washing procedure (e.g. by high-pressure spray) to be effective should markedly reduce the surface bacterial count.

Chilling is therefore the practical technique on a commercial scale of lowering the temperature of meat by extracting its heat. Of course, once chilled, meat has to be kept under refrigeration otherwise it would warm up again to atmospheric temperature and accelerated spoilage would result.

During the cooling process, the surface of the thinnest portions of the belly cool first and the centre of the thickest part of the meat takes longest to cool. Figure 2 gives a typical example of how these temperatures fall during chilling and how the weight loss begins to tail off. In this experiment the air temperature and air speed have been kept constant and the relative humidity of the air has risen to a fairly steady value. Under commercial conditions of operation there are various reasons why all these factors tend to vary and cooling times are subject to considerable variation.

CARCASE COOLING

Chilling is inevitably accompanied by some loss of weight due to evaporation of moisture from the surface of the meat. The warmer the surface of the meat, the greater is the tendency of moisture to escape into the atmosphere. Therefore the slower the cooling, the greater the loss of weight. Hence various systems of rapid chilling have been devised by refrigeration engineers, particularly for pigs, which result in minimal weight loss. For example, pigs cooled for 24 hours in ambient air followed by 24 hours at 4·5°C (40°F) are liable to lose between 2·5 and 3·5% by weight. This can be reduced to about 2% by refrigeration from the outset. Cooling down the surface of the hot pig quickly, for example by means of air at say −20°C (−4°F) and 3 m (10 ft) per sec for 4 hours followed by 10 hours in air at −1°C (30°F) and 0·5 m (1·5 ft) per sec, can reduce this loss to about 1·3%. Even 1% has been claimed for certain carefully calculated time/temperature/air speed schedules. The weight loss from beef sides similarly, can be reduced from about 2·5% to 4% without refrigeration to 1·5% to 2% with slow refrigeration in the traditional chill-room and as little as 1% or less with low temperature schedules. Lambs can lose as much as 5% when allowed to cool naturally, compared with 2% in a traditional chill-room and it is difficult to reduce this figure much further.

COOLING WEIGHT LOSS

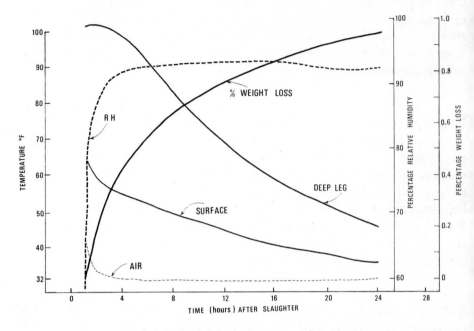

Fig. 2.   Weight loss of 231 lb beef side in air at 32°F and 3 ft/second.

DRIP

Although "drip" is unavoidable, rapid carcase cooling results in less drip loss during subsequent cutting and in packages than occurs in the case of slow cooling. For example, joints cut from pigs chilled promptly after slaughter lose less than those from pigs held at ambient temperature for 6 hours before chilling, as shown by some experimental results of A. A. Taylor, UK Meat Research Institute, given in the following table:

TABLE 3

% Drip Loss after 2 Days at 0°C (32°F) from Pork Legs
Cooled at Different Rates

Breed	Slowly cooled	Quickly cooled
Landrace	0·47	0·24
Large White	0·73	0·42
Wessex × Large White	0·97	0·61
Pietrain	1·14	0·62

474

The same type of result is obtained for beef, slices of "eye muscle" losing 3% drip when slowly cooled.

Certain types of pig exhibit a condition described as "watery pork" which cannot be eliminated by rapid cooling, although its incidence and the amount of drip loss can be reduced thereby.

COLD SHORTENING

However, recent research results indicate that it is possible to toughen beef and lamb by cooling them rapidly soon after slaughter. If chilling is carried out too quickly after slaughter and the temperature of the meat becomes lower than 11°C (52°F) before rigor mortis sets in then the muscles shrink. This is known as cold shortening. The muscular shrinkage remains after rigor mortis is resolved and the meat is extremely tough to eat. The extent of this toughening is greater the lower the pre-rigor temperature. "Cold shortening" toughening is not normally encountered with large or very fat carcases when cooling is slower.

It appears therefore that it may be unsafe to use air temperatures much below about 10°C (50°F) in the first 10 hours of cooling, although after this, lower temperatures may be safely applied to complete the chilling process. In the case of lambs a delay of 6 hours holding at ambient temperature before putting into the chiller has about the same effect.

In the case of beef attention must be paid to the cooling time to 11°C (52°F) of the "eye muscle" because if it is below 10 hours then there is some risk of toughening this valuable muscle by "cold shortening" (see Table 1). Pigmeat is not so markedly susceptible to toughening due to cold shortening. The quick chilling methods adopted to minimise weight loss in pigmeat cannot be applied to beef and lamb without risk of causing toughness.

HOT DEBONING

A saving in refrigeration cost is one of several advantages claimed for hot deboning, which is the process whereby the meat is removed from the carcase and vacuum packed before chilling. The smaller bulk of the packed meat and the reduced thickness of the portions

475

makes chilling easier. However, the rate of cooling the hot de-boned meat, which may be around 35°C (95°F) after packing, must be controlled to satisfy three requirements:

1. To give maximum shelf life by slowing down bacterial growth, the temperature of the meat must be lowered quickly to around 0°C (32°F).
2. To reduce subsequent drip loss the initial stages of cooling must be quick.
3. To avoid toughening by cold shortening the temperature of the meat must not be reduced to below 10°C in the first 10 hours after slaughter.

A compromise of careful chilling in two stages may be used; a period when the temperature does not go below 10°C (50°F) followed by the completion of cooling at a lower temperature. However, a two stage system may be inconvenient and techniques using the conditions found in conventional carcase chillers may be established to produce meat in good bacteriological condition with the minimum of drip and without cold shortening. For example, in one abattoir chiller where air at 0°C (32°F) and 0·5 m/sec (1½ ft/sec) was applied to 5″ thick vacuum packs of beef in plastic stacking trays, meat temperatures were reduced to 17°C (63°F) in the centre and 10°C (50°F) at the surface after 10 hours from slaughter. (See also p. 167.)

Electrical Stimulation

The problem of cold shortening can be overcome by increasing the rate at which the carcase goes into rigor mortis. The technique now being developed is called electrical stimulation which involves the application of a high voltage oscillating electrical current to the carcase for up to two minutes. After electrical stimulation carcases may be cooled rapidly without danger of cold shortening.

COMMERCIAL COOLING

A survey of abattoir cooling performance for beef in the meat industry in the UK and a smaller sample of abattoirs in three other EEC countries suggests that there is at present little incidence of exposure to chilling conditions, that could incur risk of cold shortening. See following table:

TABLE 4

Commercial Cooling Times of Beef Sides

Deep leg temperatures	UK			EEC		
	to 15°C (59°F)	to 10°C (50°F)	to 7°C (45°F)	to 15°C (59°F)	to 10°C (50°F)	to 7°C (45°F)
Number of sides investigated	64	36	32	14	8	7
Weight kg (lb) average	129·7 (286)	128·8 (284)	128·37 (283)	127·92 (282)	124·74 (275)	127·74 (275)
kg (lb) range	79·38–184·2 (175–406)	100·25–184·2 (221–406)	100·25–184·2 (221–406)	66·68–214·55 (147–473)	92·53–174·64 (204–385)	92·53–174·64 (204–385)
Cooling times (hours) average	25	34	40	19	27	33
range	(17–48)	(23–62)	(24–86)	(12–27)	(21–34)	(28–45)

However, as faster cooling times are achieved by improved refrigeration so the risk of toughening parts of carcases increases.

CHILLED MEAT STORAGE

An ideal storage temperature for fresh meat is at just above its freezing point of about −1°C (30°F). The expected storage lives given by the International Institute of Refrigeration of various types of meat at this optimum temperature are presented in the following table:

TABLE 5

Expected Storage Life of Chilled Meat at −1° to 0°C (30° to 32°F)
(Recommended Relative Humidity around 90%)

	Expected storage life
Beef	up to 3 weeks (4–5 weeks with strict hygiene)
Veal	1–3 weeks
Lamb	10–15 days
Pork	1–2 weeks
Edible offals	7 days
Bacon★	4 weeks
Rabbits	5 days

★ −3°C to −1°C (27° to 30°F)

However, for numerous reasons it is rarely possible to maintain meat temperatures as low as $-1°C$ to $0°C$ ($30°$ to $32°F$) under commercial conditions so that in practice storage life is usually less than given in Table 5. The times quoted for beef are also much longer than those given in Table 2 for 100% relative humidity because at lower relative humidity levels, even 90%, some dehydration occurs on the surface with the result that bacteria do not grow as well.

The temperature at which bacon is stored without freezing is somewhat lower than for fresh meat owing to the presence of salt which lowers its freezing point.

COMMERCIAL CONDITIONS

Surveys from the Meat Research Institute of temperature in commercial meat handling premises have probed the difficulties of maintaining low temperatures under factory conditions. In fact, the recommended temperatures are rarely achieved. Figure 3 illustrates the variation in beef temperature during cooling and in various distribution stages.

The relative humidity in commercial chilled meat storage rooms is subject to considerable variation from time to time and between one store and another. For example from continuous recordings of relative humidity taken over periods of up to 3 months in sixty-seven chilled meat stores, the results showed variations ranging between 55% and 100% RH with about one-third of the recorded time at above 90% and about a quarter of the time below 85% relative humidity.

EEC REGULATIONS ON CHILLED MEAT

The EEC Directive on intra Community Trade in Fresh Meat, which also broadly governs imports of meat from countries outside the Common Market, requires that:

1. Meat must be chilled immediately after post mortem inspection and kept permanently at an internal temperature not exceeding $7°C$ ($45°F$) for carcases and cuts and $3°C$ ($37°F$) for offal. These temperatures apply also to storage and transport.
2. Cutting may take place only if the meat has reached an internal

478

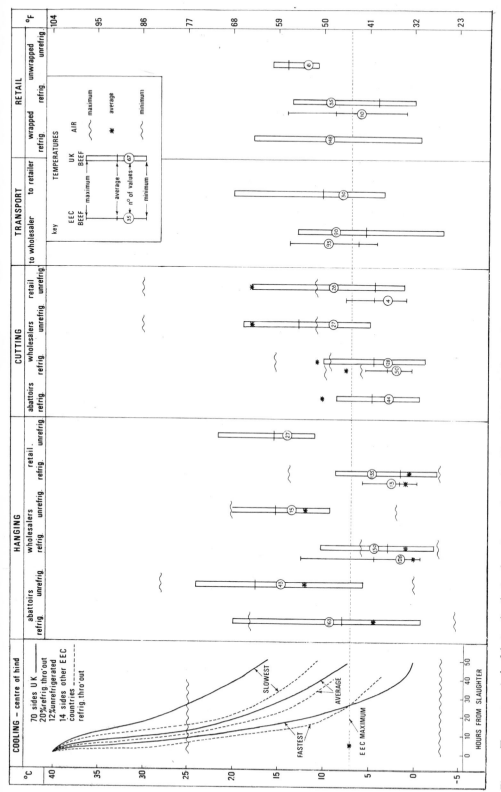

Fig. 3. Temperatures in beef during distribution from Survey of meat handling in U.K. and other EEC countries.

temperature not exceeding 7°C, except in the case of meat that is hot boned.

3. In order to ensure this, during cutting the temperature in the premises must not exceed 10°C (50°F).

Under 1 no time of cooling is proposed and it is therefore important for operatives wishing to comply with the regulations to make sure at the same time that beef or lamb is not cooled so fast as to toughen it.

Obviously offal is more easily perishable than flesh but in practice special measures have not been seen to be adopted on the Continent where the EEC directives originated, to keep it at a specially low temperature.

In any case, for storage and transport 7°C (45°F) is generally too high for a reasonably long shelf life (see Table 2). Transport operators, trading meat between EEC countries therefore endeavour to maintain air temperatures below 2°C (36°F).

Provision 2 taken in conjunction with Table 2 implies that meat cannot be boned within 24 hours of slaughter, and in cases where it has been cooled as quickly as this, there would be risk of toughening. About 3 it can be said that although boning rooms can and are maintained at 10°C (50°F), not all EEC countries appear to accept that there is any need for this temperature which can be rather uncomfortable for personnel, provided that meat for cutting is not kept out of the chill-room for more than $\frac{1}{2}$ hour or so before returning after completing the operation.

On the other hand cutting rooms in Holland are by regulation maintained at around 8°C (47°F), and in Denmark one has been observed running at 0°C (32°F).

Freezing

WHY FREEZE?

The prime cause of meat spoilage is the activity of bacteria and moulds. As has already been discussed, cooling of fresh meat will

480

slow down the growth and multiplication of bacteria and other micro-organisms and the closer meat is to its freezing point the more slowly it will spoil. To take this a stage further, by reducing the meat temperature to below its freezing point microbial spoilage is reduced considerably further until at around −8°C (17°F) bacteria and moulds cease to develop—they become dormant. Even at temperatures lower than −8°C (17°F), however, other spoilage proceeds, as a consequence of chemical and physical changes, again more slowly the lower the temperature. Even at −30°C (−22°F) meat will not keep indefinitely; enzymes within the meat induce chemical changes which gradually alter flavour and odour; oxygen reacts with certain fats to give rancid flavours, whilst physical changes result principally in dehydration at surfaces and movement of water within the meat. The surface dehydration of meat is irreversible, the water lost cannot be put back, and also as dehydration proceeds oxygen penetration is improved, speeding up rancidity development. The fat of pigmeat is particularly vulnerable to this type of spoilage giving the meat a shorter shelf life than beef and lamb. Cured pigmeat suffers even faster development of rancid flavour in cold store, attributable to the presence of salt. So, from these considerations, meat spoilage can be considerably slowed down by freezing, but the most important factor in maintaining the quality of frozen meat is the temperature at which it is stored.

THE FREEZING PROCESS: In order to freeze meat, large quantities of "latent heat" must be removed by a refrigerator to convert the water in the meat to ice. However, as meat is only partly water, for example lean beef is about three-quarters water, less heat has to be removed to freeze it than to freeze the same weight of water. The leaner the meat the higher the water content, fat meat contains less water and requires less heat extraction to freeze. Unlike pure water which freezes completely at 0°C (32°F) the water in meat which can be compared to a solution of salts freezes over a range of temperature beginning at about −1·5°C (29·5°F). About half of the water is frozen at −2·5°C (28°F), three-quarters at −7°C (20°F) and some water is still unfrozen at −18°C (0°F). In freezing meat then, it follows that half of the heat extraction is required to lower the temperature to 28°F, a further quarter to 20°F and only 15% to lower the temperature by another 20 degrees to 0°F. Figure 4 diagrammatically shows the relationship of heat taken in or given out with temperature change in freezing or thawing.

A piece of meat does not freeze uniformly, because heat is removed only from the surface by the applied refrigeration. The heat from the centre of the meat has to travel to the surface before it can be removed. Meat is a relatively poor conductor of heat although

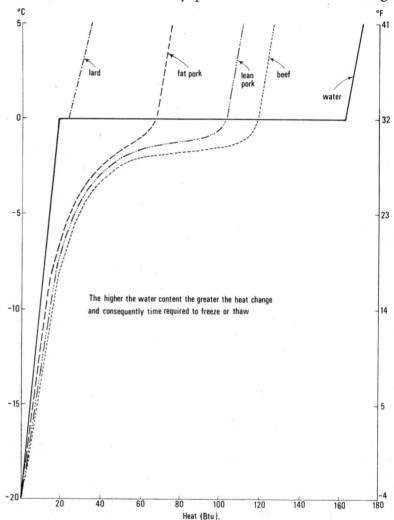

Fig. 4. *Heat extracted during freezing or absorbed during thawing of* 1 *lb meat or water related to temperature change.*

frozen meat conducts heat better than unfrozen. The meat freezes from the outside inwards. The time taken to move the heat from the centre to the surface depends on the distance; the thicker the piece the longer it takes to freeze.

Under ideal conditions freezing time varies as the square of the thickness. Under practical conditions a 150 mm (6″) thick meat block will take more than three times as long to freeze as a 75 mm (3″) thick block in the same freezer conditions. Thus the output of a freezer for meat blocks would be considerably increased by introducing thinner meat blocks, assuming that there were sufficient refrigeration capacity. Also the rate of freezing is increased when wrappings are left off—heavy packaging and enclosed air pockets have an effect equivalent to an increase in thickness.

If freezing takes place equally from both sides of a block, then the centre will be the slowest to freeze. Thus, in judging whether a product is frozen it is the temperature at the centre that should be measured. Although the outside may be cold and hard frozen, the centre may still be soft. It is important to ensure that enough time is allowed for the freezing of the centre, as the inspection of the surface will not show whether or not the centre is frozen. The freezing of meat has to be planned and the meat for freezing has to be selected in some way. For example meat should not be frozen too soon after slaughter, before rigor has set in, otherwise the meat is liable to go into rigor on thawing (so-called "thaw rigor") with consequent toughening and drip exudation. Then special equipment is required for the actual freezing process. Carcase meat and joints are usually frozen in moving air, some in efficient blast freezers operating at $-40°C$ ($-40°F$) and 3–6 m (10–20 ft) per second. Regularly shaped packages, whether the standard 12·5–15 cm (5–6 in) thick, 60 lb cartons or so-called quick-frozen consumer packs, may be frozen between refrigerated plates in multiple freezers.

Irregularly shaped small products, such as chicken, steaks and chops, can be effectively frozen by immersion techniques or by so-called cryogenic methods using evaporating liquid freezers. These methods in general are more expensive to operate than mechanical refrigeration units which cycle the refrigerant without allowing it to escape.

IMMERSION FREEZERS: The product is protected by sealed high quality packaging material and dipped in the freezing liquid, for example, salt or propylene glycol solutions, as it is conveyed through the tank.

EVAPORATING LIQUID FREEZERS: The product is sprayed with liquid nitrogen at −196°C (−320°F) resulting in rapid freezing rates. The nitrogen evaporated passes over the product entering the freezer chamber, effectively precooling it, and escapes to the atmosphere. Using a specially purified liquid fluorocarbon, dichloro di fluoro methane, which boils at −30°C (−22°F) at atmospheric pressure, the product is first dipped and then stacked and sprayed until frozen. It is claimed that only small traces of fluorocarbon remain on most products. The equipment is so designed that as the freezant evaporates, the vapours are condensed on refrigerated coils and thus recycled with minimal losses to the atmosphere.

BLAST FREEZERS: Although blast freezers, because of the fans, consume more power than an equivalent contact plate freezer, they are generally more suitable for meat because of flexibility and convenience in accommodating all shapes and sizes of product. The best performance from blast freezers can only be approached by presenting meat for freezing without wrappings, for example pork sides without stockinet or polyethylene or cuts of meat on open trays, metal for example, with efficient heat transfer properties. Not only do most wrappings insulate the product from the refrigeration, but also the air which is usually trapped between packaging materials and the product affords even greater insulation.

Most commonly blast freezers are misused by incorrect loading. The blast of refrigerated air must pass uniformly over the product, any large air gaps beneath or alongside racks of product or within racks due to part loading will allow the air flow easy passage, bypassing the products. Wrongly positioned racks or trolleys can severely restrict incoming air flow and prevent the efficient refrigeration of the remainder of the freezer load.

RATE OF FREEZING

When satisfactory freezing techniques were first being developed

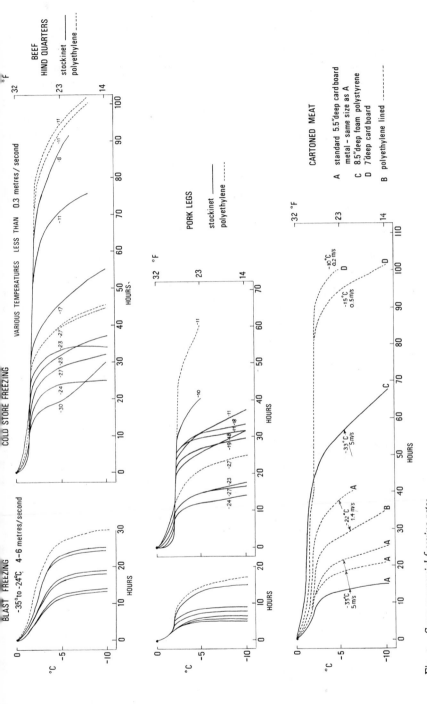

Fig. 5. Some commercial freezing rates.

485

forty years ago it was thought that it was essential for good quality to freeze "rapidly", which was defined usually as reducing the temperature at the centre of the piece of meat from $0°$ to $-5°C$ ($32°$ to $23°F$) in a period varying with different estimates from $\frac{1}{2}$ hour to 2 hours. However, it was realised that this was impossibly fast for carcase meat. Nevertheless, frozen New Zealand lamb came to be recognised by the British consumer as an excellent product. Subsequent work has led to the realisation that whatever the situation with some food commodities such as fish and soft fruits, a wide range of freezing rates has very little effect on quality or acceptability of frozen meat. Consequently, choice of freezing rate and therefore of freezing method has to be decided on the basis of economics or production layout rather than on grounds of quality.

Of all freezing processes, however, air freezing, either in blast freezers specially designed for the job, or more slowly in cold stores, is generally used for meat. Although carcase meat is frozen, the most generally acceptable frozen product is the 14 cm ($5\frac{1}{2}''$) thick 28 kg (60 lb) carton of boned-out meat. Some freezing times determined in commercial freezing plants for various carcases and cartons of meat under various conditions of temperature and air velocity are shown in Fig. 5.

FREEZING WEIGHT LOSS

Losses of weight in freezing are less the quicker the freezing, for example for a beef quarter about 0.6% in a blast freezer, at $-30°C$ ($-22°F$) and high air speed compared with around 2% in slow air freezing at about $-10°C$ ($14°F$). Blast freezing lambs results in a loss of less than 1% at $-30°C$ ($-22°F$) but in still air at $-10°C$ ($14°F$) the losses are around 2.5%. Even if packaged in polyethylene before freezing there is some evaporation inside the film which is manifested as frost on the inside of the wrapper, amounting in the case of pig sides to between 0.7% and 1.3% depending on the weight of the side.

Serious temperature increases can take place during the transfer of frozen meat from freezer to cold store, particularly if there is additional handling, for packaging for example. For small packages of frozen meat exposed in a factory room the temperature rises very quickly (see Fig. 6). After packing in master cartons, which in cold store become part of a large volume of cartons, the temperature of such warmed up packages may take a considerable time to fall again to store temperature, days or even weeks rather than hours.

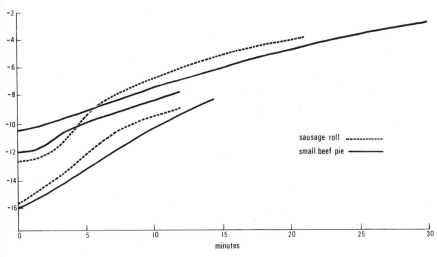

Fig. 6.   *The rate of temperature rise in frozen meat products during packing.*

FROZEN MEAT STORAGE

The UK frozen meat industry has largely grown up around imports from meat producing countries in the southern hemisphere. Cold storage for meat in the UK has therefore tended to approximate to the same level as that which was maintained on the transport vessels, namely about $-10°C$ to $-9°C$ ($14°$ to $16°F$), which just prevents spoilage due to micro-organisms. In fact, this temperature over the years has become accepted as ideal for frozen meat. Nevertheless it has been realised in commercial circles in the past few years that quality is maintained better, the lower the temperature. Table 6 reproduces the best advice on the storage life of various types of frozen meat at various cold storage temperatures, collected by the International Institute of Refrigeration in the first and second editions of "Recommendations for the processing and handling of frozen foods".

Some of the figures from Table 6 have been plotted in Fig. 7, p. 489. These times can only be used as an approximate guide, generally

487

TABLE 6

IIR Recommendations for Storage Life of Frozen Meat
(months)

°C (°F)	−12 (10)	−15 (5)	−18 (0)	−25 (−13)	−30 (−22)
Beef	5–8	6–9	12	18	24
Lamb	3–6		9	12	24
Pork	2		6	12	15
Offals			4		
Bacon (unsmoked)			2–4	6	12
Rabbits				up to 6	

indicating the time after which noticeable differences from fresh meat are detectable to the typical consumer.

It must be remembered also that the initial condition of the meat, or even the animal from which the meat was taken, influences the keeping time. The temperature at which meat is stored is the major factor governing the time meat will keep. Also once storage life is used up at one temperature it cannot be restored by lowering the temperature. From the graph (Fig. 7) which shows storage life of beef, lamb and pork at temperatures between −10° and −30°C (14° to −22°F) it can be seen that around 200 days at −12°C (10°F) is equivalent to around 400 days at −20°C (−4°F) and over 700 days at −30°C (−22°F). If therefore meat was kept at −12°C (10°F) for 100 days and then its temperature reduced for further storage at −20°C (−4°F) it is most likely that its storage life at the lower temperature would be about half of the 400 days shown. Similarly the length of storage before freezing must be proportionately discounted from the frozen storage life. Meat frozen when fresh will keep longer in the frozen state than initially stale meat.

WEIGHT LOSS FROM FROZEN MEAT IN STORAGE

In addition to the improvement in quality brought about by lowering temperature, there is also a valuable reduction in evaporative losses in weight on storage. These losses become apparent as "freezer burn", when the meat surface dries out giving a white spongy appearance. In most modern stores at −10°C (14°F) there is a loss of

488

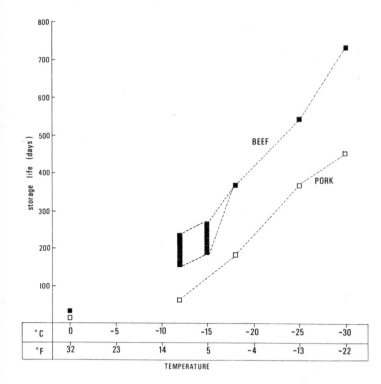

Fig. 7. Storage life of beef and pork (International Institute of Refrigeration Publication).

about 1% per month from stockinet wrapped carcase meat compared with about 0·2% at −30°C (−22°F). However, such stores are usually air duct cooled and it has been found that with old pipe cooled stores at −10°C (14°F) the rate of loss is also about 0·2% per month, about the same as the air cooled stores at −30°C (−22°F). In Fig. 8 are given the average loss of weight results for stockinet wrapped carcase meat from tests in thirteen different commercial stores. It is the general belief of the meat industry that lower temperatures than the customary −10°C (14°F) cause greater evaporation loss. This is of course contrary to the general scientific principles and to all experimental evidence (except where other factors, such as excessive radiation or abnormal temperature fluctuation interfere). It must therefore be concluded that the trade judgement on this issue is based on a comparison of the performance of old pipe-cooled stores at −10°C (14°F) with new air-cooled rooms at say −18°C (0°F), in which case the latter could well result in

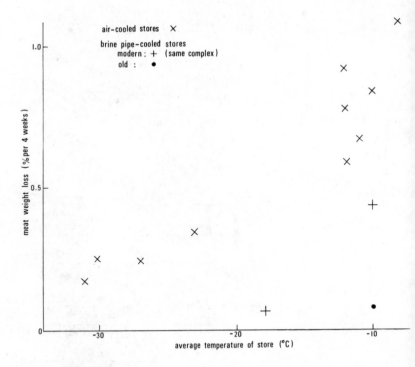

Fig. 8. Average loss of weight (per 4 weeks) from stockinet wrapped carcase meat in 13 U.K. Cold Stores.

greater dehydration not because of the lower temperature but because of the greater air movement and probably lower relative humidity in the lower temperature store.

To reduce dehydration of frozen meat in cold stores, the temperature of the product must be kept constant, both fluctuations and temperature rises increasing dehydration damage. It is important therefore that products should enter the cold store at the same temperature as the store. Whenever there is a difference in temperature between the product and some other part of the room there will be a difference in water vapour pressure and water, in the form of vapour, will migrate from the warmer surface to the colder. If the meat temperature is only slightly higher than cold store surfaces and coils, then water will evaporate from the meat. The greater the temperature difference, the faster water will be transferred and the quicker the meat will dehydrate. Even though the

490

water is in the form of ice at, say, $-30°C$ ($-22°F$), the transfer takes place, solid ice subliming from the meat to be condensed as frost on colder surfaces. Thus, refrigerated coils and pipes frost up and meat dries out. Meat dehydration is accelerated by movement of air and discouraged by good packaging of the product.

Good cold store design and operation will reduce temperature differences and restrict air movement. Drying can be reduced by restricting the heat entering the cold store, through the insulation, from lights and people working in the store, from fans and, probably most importantly, through the door. Heat can also get into a cold store by conduction through the insulation by supports and pipes. The proportion of heat leaking through the insulation would increase if the store temperature was reduced because of the greater temperature difference between the outside and the inside of the insulation. Provided this increased heat load is coped with by the refrigeration plant, the savings in evaporation loss from meat the lower temperature should be maintained. "Hot spots" near the walls and floors of cold stores are caused when heat coming through the insulation is trapped. The use of pallets or dunnage to keep meat clear of cold store walls and floors allows air to circulate and remove this incoming heat.

The only such regulations relate to meat frozen after "intervention" price support buying operations for beef and veal and for pigmeat, the details of which are listed in the following table:

EEC REGULATIONS ON FREEZING OF MEAT

TABLE 7

EEC Requirements for "Intervention Freezing" of Meat
Maximum temperatures °C (°F)

Maximum temperature	Beef and Veal	Pigmeat
of meat on delivery for freezing	<7 (45)	<4 (39)
of freezing air	<-30 (-22)	
of meat on removal from freezer	<-7(20)	<-15(5)
of storage air	<-17 (1·5)	<-20 (-4)

491

Although no time is specified for "intervention freezing" it i
obviously necessary, for planned operations, to know this. As w
have already seen in Fig. 5, freezing time is less in a blast freezer a
the specified maximum of $-30°C$ ($-22°F$) than it is in a low ai
flow cold store at the same temperature.

The EEC regulations also specify packaging after freezing ir
stockinet and polyethylene ("polythene") film of not less thar
0·05 mm thickness. Packaging before freezing would, of course
reduce the evaporative loss during freezing but would also increase
the freezing time through the specified temperature interval by 80%
for pig sides and 40% for beef hind quarters in a blast freezer and
40% and 25% respectively in an air-cooled store at $-30°C$ ($-22°F$)

Thawing

THAWING PROCESS

The heat change processes for the thawing of frozen meat in moving
air or water are the same as those discussed previously for freezing,
but of course in reverse. Heat has to be put into the meat through the
surface to change it from the frozen to the defrosted state. This is
most commonly achieved by passing air or water over the meat
surface or over the surface of the package containing the meat. Heat
can also be imparted by condensing water, releasing latent heat, on
to the surface in a vacuum chamber. The rate at which thawing can
be done depends on the thickness of the product, whether or not it is
packaged, and the speed at which heat can be put to the meat
surface and then moved from the surface to the centre of the meat.
Thawing by conventional means is inherently a slower process than
freezing. This is because of two related factors; one is that whilst
freezing can be achieved using very low temperatures, thawing
cannot be done at very high temperatures because of the limiting
hazards of accelerated spoilage and cooking; the other is that once
thawing starts and the meat surfaces soften, the speed of transfer of
heat to the centre is reduced by the increased insulating property of
wet as compared with frozen meat.

Thawing can also be achieved rapidly by heating the meat internally
by molecular vibration using certain radiations, for example micro-

492

waves. The disadvantages of this technique for meat become apparent when complete thawing is required. Because of the variation in water content and therefore the heat required to thaw the different parts of a meat piece, fat and lean for instance, plus the non-uniformity of shape of carcase or block, runaway heating occurs in the parts thawed first giving rise to cooking before the rest of the block is thawed.

EXPERIMENTAL THAWING TESTS

Experimental work on the thawing of meat by various methods is currently being undertaken at the UK Meat Research Institute. The figures given in Tables 8 and 9 give examples of these findings.

TABLE 8

Thawing Times in Hours from −40°C (−40°F) to 0°C (32°F) at the Centre of Pork Legs by Water, Vacuum and Air Methods (Experimental results of C. Bailey)

Thawing conditions	Leg weight		Thawing temperature			
			5°C (41°F)	10°C (50°F)	20°C (68°F)	30°C (86°F)
	kg	lb				
Water 0·34 metres/minute	3	6½	—	10	6	5
(1·1 feet/minute)	6	13	—	18	11	8
1·4 metres/minute	3	6½	13	8	6	4
(4·6 feet/minute)	6	13	30	17	11	8
Vacuum	3	6½		11	8	
	6	13		17	11	
Air (85% relative humidity) still	3	6½		13	7	6
	6	13		23	14	9½
5·5 metres/second	3	6½		12	6	5
(1070 feet/minute)	6	13		19	12	8

It would appear from these figures that water movement influences thawing time less at the higher temperatures and that increased air speed accelerates the thawing particularly of the heavier legs. Tests on a commercial example of water thawing of 9 kg (20 lb) pork legs using mains water at about 10°C (50°F) overflowing from tanks containing the meat, resulted in thawing times of around 16 hours

493

for deep meat from −7°C (20°F) to 0°C (32°F). Water thawing results in far less weight loss from meat than air thawing. Considerable drying out of meat surfaces occurs in air thawing at high temperatures. Beef blocks, usually about 27 kg (60 lb), dimensions 61 cm×40 cm×13 cm (24 in.×16 in.×5 in.) of boneless meat pieces, cannot satisfactorily be defrosted by immersion in water because the block breaks up and the meat becomes waterlogged. Some investigation has been made of air and vacuum thawing of beef blocks.

TABLE 9

Thawing Times in Hours from −40°C (−40°F) to 0°C (32°F)
at the Centre of Beef Blocks
(Experimental results of C. Bailey)

Thawing conditions	Thawing temperature		
	10°C (50°F)	20°C (68°F)	25°C (77°F)
Vacuum	20	13	9·5
Air (85% r.h.) still	32	22·5	13
5·5 metres/second (1070 feet/minute)	26	17	12

These few results show a decided advantage in thawing time for the vacuum thawing system over air thawing and also in air thawing the significant effect of high air speed over the meat.

COMMERCIAL THAWING TESTS

In the following example (Table 10) of thawing times and meat weight loss in a commercially used, programmed air thawing tunnel the overall loss of weight (mainly by evaporation from meat surfaces but also including some drip) from one particular schedule was markedly reduced by wrapping in stockinet even though thawing took somewhat longer.

In a somewhat different schedule in the same thawing tunnel the results in Table 11 show a 50% extension of thawing time when meat is wrapped, as specified by the EEC regulations, for storage of carcases frozen for "intervention" purposes, in polyethylene plus stockinet, compared with stockinet only.

TABLE 10

Weight Loss and Thawing Times in Hours from −10° to 0°C (14° to 32°F)
of Meat Cuts in a Commercial Programmed Thawing Tunnel in Air at:

1. 23°C (73°F), 72% rh, 4 m/s (780 ft/min) for first 4 hours
2. altered over next 4 hours to
3. 7°C (45°F), 85% rh, 1 m/s (200 ft/min)

| | Weight | | Thickness | | Unwrapped | | Stockinet | |
					Hours	% loss	Hours	% loss
	kg	lb	cm	in				
Beef hind top	20·2	44	28	11	51	1·7	59	1·4
Pork leg	13·5	28	18	7	40·5	3·2	42·5	1·8
Pork leg	5·6	14	14	5·5	17	1·0	19	0·3
Lamb leg	1·9	4	8	3·5	3·5	2·0	4·5	0·4
Boneless lean beef	3·7	8	10*	4*	4·75	2·3	4·75	0·05

* roughly cubic

TABLE 11

Thawing Time in Hours −10 to 0°C (14° to 32°F) at the
Centre of Pork and Beef Hind Legs at 15°C (59°F),
85% Relative Humidity and 1 m/s (200 feet/minute)

	Stockinet	Stockinet over polyethylene	Polyethylene over Stockinet
Beef 30 cm (12") thick	48	72	70
Pork 12·5 cm (5") thick	14	22	22

Weight loss during commercial air thawing due to drip alone has been observed to amount to around 1% in boxed beef and up to 8% from cartoned offals.

The heat necessary to thaw meat from very low temperatures is required in steadily increasing proportion for a given temperature interval as the temperature rises to about −5°C (23°F) when the bulk, 70 to 80% is needed to complete defrosting from −5° to −1°C (23° to 30°F). The cost of warming meat from say −18°C (0°F) to

$-5°C$ (23°F) is therefore low by comparison with the cost of completing the thawing through the $-5°$ to $1°C$ (23° to 30°F) range. It follows that if frozen meat can be further processed without completely thawing, at $-5°C$ (23°F) for instance, then considerable savings are effected in the energy consumed and the time involved.

Tempering

Commonly in manufacturing and meat canning, meat can be chopped to requirements without thawing. Block chopping machines working on low temperature meat are in use. Other machines chop frozen meat that has been "tempered", that is raised in temperature so that it becomes partly defrosted. This economises in thawing costs and reduces thawing time as well as largely eliminating drip losses and also obviates risk of spoilage due to high temperatures during processing.

Forming

Cuts of meat are sometimes frozen in order that they may be "formed" in presses before mechanical cleaving. The aim of this process is to produce uniform shape and weight of slice ("portion control"), for example with sirloin steaks. The temperature at which this pressing is carried out is rather critical. The ice content of meat is around 30% at about $-2°C$ (28·4°F) compared with 75% at $-5·5°C$ (22°F) and because of this the compressibility is much greater at higher temperatures. In fact to obtain a volume change at $-5·5°C$ (22°F) equivalent to that at $-2°C$ (28·4°F), approximately six times as much pressure would be needed.* The pressure required would be doubled for the change from $-2·5°C$ to $-3·5°C$ (27·5°F to 25·7°F).

Retailing

CHILL-ROOMS

The typical butcher's chill-room is by purpose and because of size somewhat different from the large carcase holding chill-rooms at abattoirs for instance. As they are small rooms, a large proportion of the cold air drifts out every time the door is opened; which may be 200 and more times a day in a typical, busy shop. This fall-out

* Calculations of Dr. C. A. Miles, Meat Research Institute

Beef chill room

Courtesy H.T.I. Engineering

Heavy duty wall and floor tiles in a beef and sheep slaughterhall

Side view of turkey

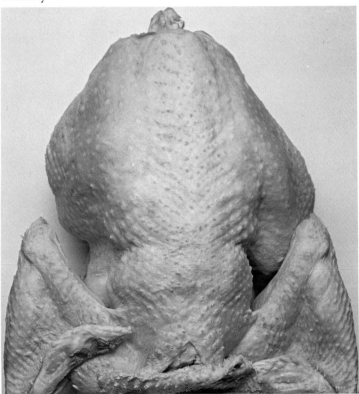

View of turkey from above

Courtesy Bernard Matthews Ltd.

Remove giblets and cut straps retaining legs. Remove neck and cut off mid-wings. Cut down the back, then cut round the oyster.

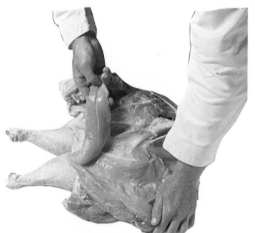

Break leg from ball socket joint and separate leg from carcase without cutting skin

Slit down shoulder to wing joint to expose the bone. Cut round the head of the bone and pull away from meat

Steps in portioning turkey

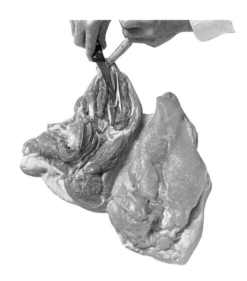

Trim away the breast muscle from the carcase, locate the two fillets by finding protruding shite tendon ends and simply pull away. Then slice off two thin escallops from each breast

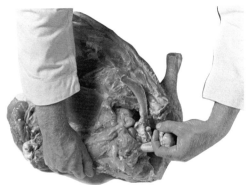

Debone thigh and drumstick keeping bones intact. Hold leg bones and separate all sinews from the drumstick meat

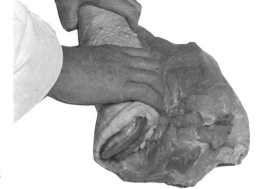

Cut the carcase away from the meat and divide the skin at the centre on the site of the keel bone. Roll firmly (starting with the dark meat). Cover with a thin layer of pork fat and string-tie

Courtesy Bernard Matthews Ltd.

Roasts

Courtesy Bernard Matthews Ltd.

Escallops

Presenting turkey cuts

Wings and neck

Fillets

Slices from a roast

Mid-wing in a casserole

Escallops

Ways of cooking turkey portions

Courtesy Bernard Matthews Ltd.

Retail Butcher's Shop

of cold air and its replacement by warm shop air can be limited to some extent by inner swing doors. These are effective only when they are well fitted and maintained and cannot easily be fastened back. Light alloy framed transparent inner doors, "made to measure", are possibly the most efficient. Air curtains formed by blowing a barrier of warm air across the chill-room entrance, usually vertically downwards from the top, when the door is opened are quite effective. Again, these must be accurately adjusted as any deviation of the air flow can effectively circulate outside air into the chill-room. When air curtains are not used the cooling fans should switch off when the door is opened. Heat from outside also leaks in through the insulation, through walls, floors and ceilings. For this reason, meat touching walls and floors in a chill-room absorbs the heat leakage and thereby warms up. Air must circulate between meat and walls to remove this incoming heat on the coils. Insulating materials used in chill-room construction must be kept dry, as the insulating properties and efficiency depend on air trapped in the material. Thus, if it becomes wet the insulation breaks down and heat passes through. This consideration is important when erecting a chill-room; care must be taken to ensure that the insulation cannot become wetted, for example the floor of a prefabricated chill-room is best raised above a shop floor which is frequently washed.

Ideally a retail butcher needs two chill-rooms; one for long-term storage of the stock of maturing carcase meat, when temperature changes affecting the meat surfaces by causing rapid dehydration and darkening, can be kept to a minimum, and the other for joints and prepared meat ready for sale.

Regulating the temperature in butchers' small chill-rooms presents problems. These rooms, because they are small and in continuous use by day, are difficult to maintain at low and steady temperatures although conditions do become steady by night. Snags can arise when refrigerators are made to work harder by day to maintain low temperatures but at the same time thereby creating the possible hazard of slowly freezing the meat by night. Meat suffering such cycling of temperature loses in both weight and appearance. Also the continuous intake of moist warm air into chill-rooms during the working day results in excessive ice deposits on the evaporator coils which consequently need frequent defrosting. As a result of automatic defrosting in the newer types of chill-room, periodic high temperatures around 15°C (59°F) can result, and moisture condenses on cold meat surfaces with consequential heating and detriment to appearance and keeping quality when cold conditions follow the defrost, and the meat surfaces wetted by condensation cool and dry out. This periodic wetting and drying leads to the development of hard and darkened cut meat surfaces. Drying loss on beef joints in this type of chill-room has been found to be as much as 5% per day where the air blows directly on to the meat and 2·5% elsewhere in the room. When this was checked experimentally in a particular butcher's chill-room, the meat in the worst affected position lost 29% after seven days' storage compared with only 8% in other places.

It is important in a butcher's chill-room to ensure air circulation through the meat in order to avoid rapid spoilage. Even at 0°C (32°F) meat can spoil rapidly if it remains wet in consequence of static pockets of 100% relative humidity which cause condensation on the surface.

Bacon which requires an even lower relative humidity for storage because of its salt content, has been observed in a modern chill-room to attract atmospheric moisture, which then drops off.

Ideally, the relative humidity should be around 90% to limit spoilage and minimise evaporation loss.

WEIGHT LOSSES IN RETAILING

Evaporation and trimming losses occur throughout retail handling, storage and display. Tests in one butcher's chill-room, for example, resulted in losses from pig carcases in the weight range 40–45 kg (90–100 lb) of 1·5% after 2 days and up to 2·6% in a week, whilst hindquarters of beef lost on average 2% in seven days. During hanging outside the chill-room, these evaporation losses are appreciably greater. Losses in cutting and trimming vary widely and are difficult to assess, in one case being 3·1% on a 43·55 kg (96 lb) pig. Also carcases and cuts of meat left for many hours in a cutting room suffer appreciable weight loss by evaporation, especially when the air temperature is high.

Meat taken from the butcher's chill-room to a warm preparation room will fairly quickly warm up to within a couple of degrees or so of the air temperature. The rate of this temperature increase is very variable, depending largely on the difference in temperature between the room and the meat, the size of the meat piece, and of course the time it is out of chill. Observations during cutting and

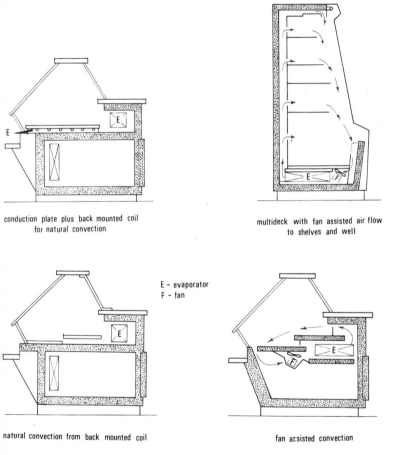

conduction plate plus back mounted coil
for natural convection

multideck with fan assisted air flow
to shelves and well

E - evaporator
F - fan

natural convection from back mounted coil

fan assisted convection

Fig. 9. Refrigerated display cabinets (permission of P. C. Moesman, Netherlands Centre for Meat Technology).

499

preparation of meat, taken from butchers' chills at 0 to 2°C (32 to 35°F), have revealed temperature rises, in the thickest parts, of 1C° (2F°), and in surface layers almost 4C° (7F°) per 10 minutes exposure to cutting room temperatures of around 14°C (57°F), and up to 2C° (4F°) in deep meat and 5C° (9F°) at the surface per 10 minutes in cutting rooms around 19°C (66°F).

REFRIGERATION DISPLAYS

Experiments at the UK Meat Research Institute have shown that unwrapped cuts of meat displayed for 6 hours in refrigerated cabinets lost 0·4% to 1·5% by evaporation in a conduction plate cooled display cabinet, and from 0·5% to 2·5% in a cabinet with forced air at about 0·3 metres per second (60 feet per minute) over the meat. Some of these results are given in Table 12. The beef and pork which was displayed in 1″ thick slices lost weight 50% faster, and was noticeably drier in the fan assisted cabinet than in the conduction cooled. Fig. 9 shows in cross-section some of the design features of different types of refrigerated display cabinets. The temperature of unwrapped meat was fairly controllable in either of the cabinet types, although there was a tendency for the bottom layer to freeze in the conduction plate type. Fig. 10 illustrates the instrumentation of one of these tests.

TABLE 12

Weight Loss (%) of Unwrapped Meat after Periods of Display
in Two Different Cabinets

	Conduction-cooled		Fan-assisted	
	6 hours	24 hours	6 hours	24 hours
Beef				
Pork	1·0	3·0	1·5	4·5
Mince	1·5	6·3	2·5	8·4
Liver	1·5	4·5	2·5	6·5
Bacon				
joints 1·5–3·5 kg (2½–8 lb)	0·4	1·9	0·5	1·8
sliced 1 kg (2 lb) heaps	0·9	4·2	1·3	5·2
sliced 1 kg (2 lb) layers spread out	1·3	5·5	1·5	6·3

Results of temperature measurements in various types of display cabinets in shops showed that meat temperatures were only close to

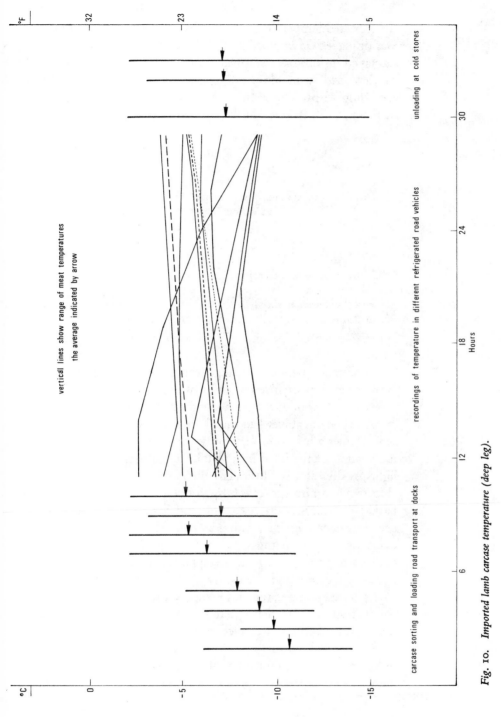

vertical lines show range of meat temperatures
the average indicated by arrow

°F
32
23
14
5

unloading at cold stores

recordings of temperature in different refrigerated road vehicles

Hours

carcase sorting and loading road transport at docks

°C
0
-5
-10
-15

Fig. 10. Imported lamb carcase temperature (deep leg).

the expected temperature, namely that of the cabinet air, when meat was displayed unwrapped or tightly wrapped. Meat slices or chops displayed in film overwrapped foam polystyrene trays were shown to vary widely in temperature in different types of cabinet and according to pack position in the cabinets. A summarised example of these results is given in the following table.

TABLE 13

Comparison of Average Meat Temperatures °C (°F) in Top and Bottom
Foam Polystyrene Film Overwrapped Trays Displayed in Two Layers
in Various Types of Cabinet

	Number of tests	Top pack	Bottom pack
Conduction plate + natural convection	13	9·9 (49·8)	3·3 (37·9)
Natural convection	3	17·4 (63·3)	14·2 (57·7)
Fan assisted convection (shelf)	18	7·8 (46·1)	6·3 (43·3)
Fan assisted convection (well)	21	6·8 (44·2)	4·7 (40·5)

The difference between average top pack meat temperature and average bottom pack temperature was greater in conduction plate type cabinets than in the fan assisted air cooled type. The top packs in each type of cabinet had the highest temperature, even in the fan assisted ones where cold air is blown over the packages. The effect of the heating of the meat by radiation from shop surfaces, including lights, is thus clearly illustrated; the fan assisted type cabinet is best able to minimise the effect by forcing refrigerated air over the package surfaces, thus removing much of the heat gained. The packages encourage the heating by radiation of the product, the film overwrap acting in the same way as glass in a greenhouse, and at the same time heat is further trapped in the meat by the insulating effect of the tray and the air trapped between meat and package. The greatest temperature difference shown in Table 13 is in the conduction plate cabinets where the top packs are exposed to radiation heating and are protected from the cold plate cooling surface by the tray and bottom pack. In Fig. 11 meat temperatures recorded in butcher display cabinets of the conduction plate type are shown. Figure 11a compares temperatures of unwrapped and overwrapped

502

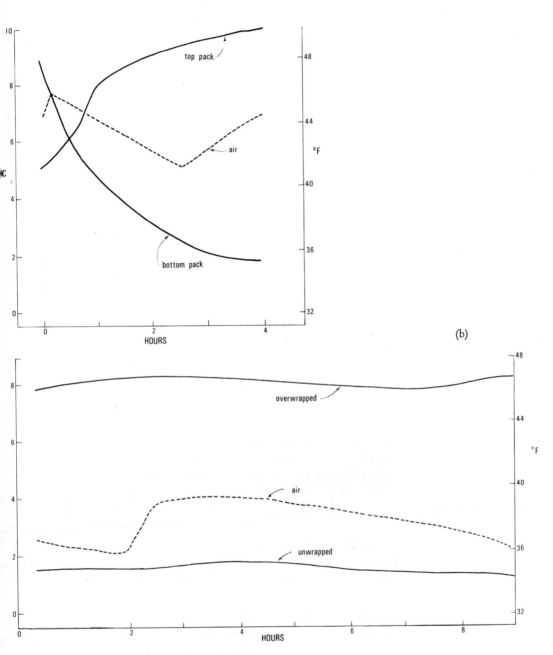

(a)

(b)

Fig. 11. Temperatures of air and meat in a conduction plate cooled display cabinet a. Compares top pack (exposed), and bottom pack (sheltered) b. Compares unwrapped and film overwrapped trays of beef slices.

503

meat slices displayed side by side, clearly showing the temperature elevation of the wrapped meat due to the "greenhouse effect". Figure 11b compares temperatures of overwrapped meat exposed to, and sheltered from, radiation during display.

Cabinets depending on gravity convection only, the convection induced by an evaporator coil at the back of the display area giving a well of cold air, do not seem very effective in maintaining low meat temperatures. They are more commonly and successfully used to display pies and cakes at around 13°C (55°F) (which is cooler than the usual shop or supermarket temperature of around 21°C (70°F)). Draughts around these displays easily displace the cold air and the display temperature is only slowly restored. Cabinets of the conduction plate type are often equipped with a back mounted coil to assist the convection of cold air over the displayed products.

Display cabinets of the multideck type (shelves + well) are similar in principle to the fan assisted well types. The cold air zone is achieved by a vertical cold air curtain down the front of the display in addition to ducted air over the display shelves. High air velocity has to be maintained and, because of great losses of refrigerated air to the shop, they are relatively expensive to run.

Although illumination in shops is a considerable source of radiation heating, the background radiation is also important. Protective covers over displays when the shop is shut help to reduce the heating of the contents. Such screens are being recommended as night covers for the protection of frozen food. Another attempt to reduce the radiation effect is the development of a special reflective ceiling material which reflects radiation from the walls and floor of a shop back in the same direction. This, however, has been shown to have little beneficial effect for chilled meat displays.

To draw conclusions, it is clear that fan assisted cabinets cause greater drying of unwrapped meat than do conduction plate types.

Meat temperature is more difficult to control in the latter, especially when the meat is loosely wrapped. Radiation heating of such

504

packages is increased with increased illumination and lessened by aluminium foil wrapping.

Where there is neither air space nor insulation, as with tightly wrapped meat or products such as sausages in polyethylene film, the heat is extracted practically as fast as it is absorbed.

When meat temperature is higher than anticipated due to the radiation effects discussed above, spoilage is accelerated and shelf life shortened. This spoilage is manifested firstly in the appearance of cut lean surfaces; the colour becomes less acceptable more quickly the higher the temperature.

TABLE 14

Time (days) for Meat Colour to Become Unacceptable
(Data of D. B. MacDougal, UK Meat Research Institute)

	$0°C$ ($32°F$)		$5°C$ ($41°F$)		Retail display cabinet $5°C$ ($41°F$) with defrost	
	Sirloin	Fillet	Sirloin	Fillet	Sirloin	Fillet
Time to change from very acceptable (red)					1 day* plus	
to acceptable (brownish-red) when not directly compared with fresh meat	4–5	2	1	<1	$\frac{1}{2}$	$<\frac{1}{6}$
to unacceptable (reddish-brown to brown)	6–9	4	2–3	$1\frac{1}{2}$	$<1\frac{1}{2}$	$\frac{1}{6}-\frac{1}{3}$

* These samples were held at $2°C$ for 24 hours before putting on display

Table 14 gives results comparing the effect on the colour of cut sirloin and beef fillet, the two beef muscles representing extremes of colour stability, of holding at two different chill-room temperatures with the air saturated at 100% relative humidity. For comparison, results are also given of the behaviour of meat in a supermarket display cabinet in which the temperature was variable, the humidity lower and drying greater.

**DISPLAY OF
FROZEN MEAT**

In-package dehydration is greatly increased by a wide temperature differential setting, causing extensive air temperature fluctuations in a display cabinet, and by heat radiated from the shop fittings and surfaces, especially from lights placed close to the meat display. This latter factor has less effect with tightly wrapped joints than with film overwrapped tray packs which act as miniature greenhouses. Tests in shops displaying frozen meat showed that top packs of film overwrapped foam polystyrene trays of chops in a poorly run cabinet at $-12°C$ ($10°F$) lost $0·5\%$ per day, four times as much as in a cabinet running at $-8°C$ ($18°F$) with hardly any fluctuation and ten times as much as in one at $-29°C$ ($-20°F$). Obviously a low and constant cabinet temperature minimises loss.

Transport

Meat transported by road in the UK involves four distinct types of vehicle:

1. Refrigerated insulated vans or trailers, the refrigeration being generally mechanically produced cold air. Some hauliers use liquid nitrogen or carbon dioxide, which is injected into the loaded van, or solid carbon dioxide chunks through which air is passed by a fan.
2. Insulated vans without built-in refrigeration. However, these are sometimes refrigerated by adding solid CO_2 blocks to the load.
3. Uninsulated vans which are often constructed without a bulkhead between the load and the driving compartment.
4. Flat open lorries.

It would seem immediately obvious to condemn the transport of any sort of meat by vehicles in groups 3 and 4 and even to aim for the ideal of preferring 1 to 2. Sound reasons for the meat and road haulage industries using all of groups 1 to 4 must exist; these are mainly based on economics. In some cases, however, it is a lack of realisation of the extent of temperature changes and the damage to product and subsequent shelf life occurring in transport that allows poor practice to continue.

506

Most cold or frozen carcase and boxed boneless meat is carried either in refrigerated lorries, especially on long journeys and when it is for export, or in insulated vehicles which are generally used for short journeys. Uninsulated vans are commonly used for daily deliveries of meat products such as pies. Bulk transport of sides of bacon in hessian sacks is still carried out by open flat lorries, particularly from Northern Ireland to English destinations. In warm weather ice is packed amongst such loads.

MEAT PRODUCTS

Typical examples of recent measurements of meat product temperature changes recorded during journeys from the same factory on warm summer days in mechanically refrigerated, insulated and uninsulated vans are shown in Fig. 12. In the case of the uninsulated van the product temperature rapidly rises to approach the air temperature in the van. These vehicles easily heat up to higher than ambient temperatures in summer weather and particularly when the driving compartment is not separated from the load, even winter van temperatures can be high due to the use of the cab heater-windscreen demist systems.

Insulated vehicles offer some protection from excessive heating of the product during transit, but obviously the inside of such vans and the insulation are initially at the ambient temperature of the loading bay, and it is the cold product itself that provides the refrigeration—perhaps cooling the air and the van a little, but at the same time becoming warmer.

The use of the refrigerated lorry (in Fig. 12) was at the insistence of the firm being supplied. Even for the short journey illustrated, its value is obvious in terms of maintaining low product temperature between factory and shop.

FROZEN IMPORTED MEAT

A recent investigation of temperature changes in imported frozen meat during transport from the docks to inland cold stores revealed a wide variation in the pattern of use of refrigerated lorries. Generally it was found that the product transported was allowed to warm up considerably at the docks because of delays in loading of up to 5

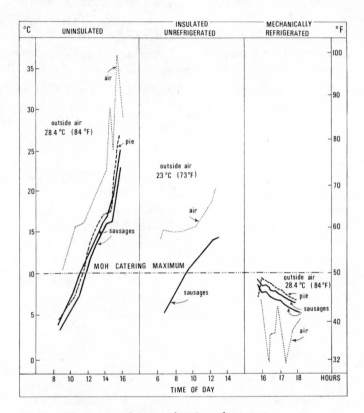

Fig. 12. Temperature of meat products in road transport.

hours in unrefrigerated conditions (see Fig. 13). Other investigations (in Holland) have shown that loading times for frozen carcases at 15°C (59°F) ambient temperature of longer than 20 minutes may result in detrimental increases in temperature, and for cartons 100 minutes; the edges of cartons are most affected. Palletised loading in 15 to 30 minutes brings loading and transfer times into safe limits, but this is not a practical proposition without altering the whole system of dockside sorting and handling. The shortening of the handling time by speeding up the handling process is most important, even more so for frozen meat than for chilled. The use of large units in the form of pallet loads not only shortens handling times but also reduces the surface to volume ratio and consequently the heat gain.

508

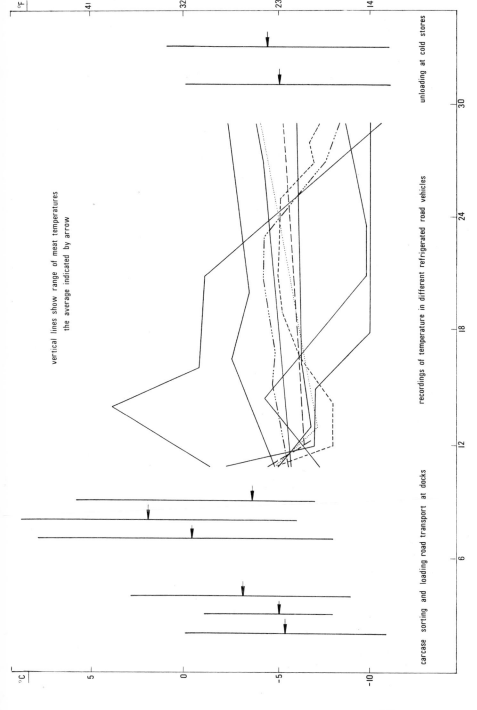

Fig. 13. Imported lamb carcase temperatures—flank.

Precooling a vehicle is time consuming and troublesome when condensation occurs on the inside on exposure to ambient conditions during loading.

On the journeys the operation of the refrigeration machinery was largely left to the driver's discretion, for example thermostat settings varied from lorry to lorry between −6°C (21°F) and −19°C (−2°F) and the period of operation from only 1 hour out of 39 to full time running.

Although a 2–3 day journey cannot be expected to produce in a transport vehicle a major cooling effect on a load, and a short journey can at least maintain the initial product temperature, nevertheless the investigation showed that refrigeration operated at low thermostat settings was in fact effective in cooling the load.

Fig. 13 shows some of the variation in meat temperature recorded in different lorry loads of frozen carcases in consequence of different usage of the refrigeration. Some warm loads are cooled during the journey, others actually warm up.

TABLE 15

Recommended Temperatures for Carrying Meat as Laid Down by a UK Meat Transport Firm

	°C	°F
Beef	1·1 to 2·2	34 to 36
Lamb	1·1 to 5·5	34 to 42
Mutton	0 to 1·1	32 to 34
Pork	2·2 to 4·4	36 to 40
Veal	2·2 to 4·4	36 to 40
Bacon	2·2 to 4·4	36 to 40
Chilled beef	−1·5	29
Frozen meat	−10 to −8·9	14 to 16

Air circulation around the load in road vehicles is important. As in the case of cold-rooms discussed earlier, heat leaks in through the insulation and the meat carried must be protected by ensuring air passage between walls, floor and ceiling and the meat. In long vehicles it is necessary to duct refrigerated air from the evaporator through the full length, or short circuiting will deprive part of the load of any refrigeration. The EEC Directive lays down 7°C (45°F) as the upper limit of meat temperature during transport. Generally road hauliers carrying export meat aim to achieve a lower temperature than this: one UK haulage firm set standards for meat transport as shown in Table 15.

It is curious that the temperatures recommended differ slightly for different species. The widest range of temperature 1·1° to 5·5°C (32° to 42°F) is recommended for lamb whilst mutton is distinguished by low temperatures 0° to 1·1°C (32° to 34°F), beef 1·1° to 2·2°C (34° to 36°F) is different from pork which is classified along with veal and bacon, 2·2° to 4·4°C (36° to 40°F).

It is worth noting the meat trade distinction between "fresh" and "chilled" meat. "Chilled" meat is usually distinguished as imported meat from southern hemisphere sources, preserved by careful control of temperature to −1·5°C (29°F), that is just short of freezing, in order to withstand the 3 weeks and more sea journey time, plus distribution time in the UK.

In the past, longer voyages, carrying chilled beef from Australia to the UK were made possible by maintaining the atmosphere in ships' holds at 10% carbon dioxide. The carbon dioxide suppresses microbial development on the meat with the effect of prolonging shelf life. Although in the past imported chilled meat was in carcase form, now it is presented as boneless joints, e.g. "Cryovac" packed in cartons. Total distribution times of "Cryovac" packed chilled beef from source to consumer of up to 14 weeks have been noted, the meat still apparently being acceptable, although a safer time limit of around 8 weeks should surely be considered. Any slight increase in temperature during transit of chilled meat for whatever reason would drastically reduce its potential shelf life.

The transport of frozen meat at −10°C to 8·9°C (14° to 16°F) in Table 15 relates to the traditional UK meat trade frozen meat handling temperature of −10°C (14°F) whereas the International

Institute of Refrigeration recommends −18°C (0°F) for the transport of "deep frozen meat and offals" which it distinguishes from "frozen meat" which can be transported at −12°C (10°F). The IIR also state a maximum acceptable temperature rise in the transport of frozen meat of 3°C (or 5°F).

Meat Spoilage

The Microbiology of Meat

Meat is a highly perishable commodity which is rarely treated as such by the trade. From the end of the slaughter line, bacteria begin to grow on meat at a rate dependent on its natural acidity, the amount of water available, and the temperature. When the surface bacteria reach high numbers, the chemical by-products of their growth reach concentrations which are regarded as objectionable, and the meat is rejected as "spoiled". Hence the shelf-life of meat, and one aspect of its quality, is limited by bacterial growth.

The microbiology of meat is complex and involves a wide variety of micro-organisms. It is beyond the scope of this chapter to describe

513

these microbes in detail. However, they can be grouped into categories meaningful with respect to their growth and significance on meat in terms of their ability to grow at different temperatures, their requirement for oxygen or its absence, and to some extent on their ability to cause food-poisoning.

MESOPHILES AND PSYCHROPHILES

Bacteria are notable to grow over an indefinite range of temperatures: those which grow in the warm temperature range 10–45°C (50–113°F) are termed "mesophiles" and those which grow over a low temperature range 0–28°C (32–81°F) "psychrophiles". An intermediate group which grows best over the warm temperature range and slowly over the lower is termed "psychrotrophs" (cold-tolerant). Almost all food-poisoning bacteria are mesophiles, therefore refrigeration below *c.* 10°C (50°F) offers good protection against food poisoning. Although many mesophiles are capable of causing food-spoilage, the most significant spoilage bacteria of meats are cold-tolerant or psychrophilic, since meat is normally stored at or below ambient temperature.

AEROBES AND ANAEROBES

Some bacteria, termed aerobes, have an absolute requirement for oxygen which restricts their growth to the surface of meat since aerobic conditions exist only in the surface few millimetres of meat. A second group, termed anaerobes, require the absence of oxygen and, therefore, grow only in the interior of meat. Others grow better in oxygen, but are able to grow slowly in its absence (facultative anaerobes). Food-poisoning bacteria fall into the anaerobe and facultative anaerobe categories, but the most important spoilage bacteria (Pseudomonas spp) are aerobic.

The groups of bacteria most important in meat microbiology are listed according to their temperature and oxygen relationships in Table 1, p. 532. This chapter describes the ways by which these bacteria contaminate meat. It describes the means by which such contamination may be minimised, the significance to public health and meat quality, of bacterial growth on meat during storage and transport, with particular reference to methods used to control this growth.

514

Factors Affecting the Growth of Bacteria

The growth of microbes in and on carcases is controlled by the temperature, the availability of water, and the acidity of the tissues. These three factors may interact and should not be considered in isolation unless it is assumed that the other two factors are at their optimum value.

TEMPERATURE

The control of temperature is normally considered the most important factor in the control of microbial spoilage. Microbes grow fastest at a particular "optimum" temperature assuming a_w—1·00 (see "water" below), and pH near neutrality (see p. 516). Reduction of the incubation temperature from the optimum progressively reduces the rate of growth until growth ceases just below the "minimum growth temperature". If the a_w is below 1·00, or the pH is below 7·0 the minimum growth temperatures may change.

WATER

In addition to nutrients, which meat is readily able to supply, and a suitable temperature, microbes require water for growth. The term used by biologists for physiologically available water is "water activity" (a_w) equalling equilibrium relative humidity (e.r.h.)/100. Bacteria, yeasts and moulds grow best at an a_w near 1·00. As the a_w falls, microbial growth becomes slower, until it eventually stops. The a_w of lean meat is c. 0·993, and offers near-ideal growth conditions for bacteria. If the e.r.h. of the atmosphere is below the a_w of the meat, water will be lost from the meat and the a_w at the meat surface, where most bacterial growth occurs, will fall. In practice the a_w at the meat surface does not fall to the corresponding atmospheric e.r.h. because water lost from the surface is partially replaced by diffusion from the interior of the bulk. If sufficient water is removed, or lost, the a_w falls and growth of certain microbes cannot occur. Moulds are able to grow at much lower a_w values than bacteria (see Table 2, p. 533), hence reducing a_w is selective with respect to microbial growth.

pH

The pH of muscle at slaughter is near 7·0 but continued enzymic activity within the muscle after death converts glycogen to lactic acid which normally reduces the pH to an ultimate value of approximately 5·7. This is important because the majority of meat spoilage bacteria and all food poisoning bacteria grow most rapidly near pH 7·0 and their growth becomes progressively slower as the pH is lowered. It is advisable, therefore, from a microbiological point of view that the normal low ultimate pH be obtained in the muscle and any practice which causes high ultimate pH (over 6·0) should be avoided. It is thought that the most common causes of high ultimate pH meat are stress and fatigue in the animal just prior to slaughter, both of which deplete the muscles of glycogen resulting in less lactic acid formation after death.

Some meats have higher than normal pH values (e.g. some muscles in the collar of pigs; and beef which is "dark cutting") and in these bacteria grow more rapidly, thereby rendering them more liable to bacterial spoilage. In manufactured meat products it is common practice to add alkaline polyphosphates which, in addition to increasing the water-binding of the muscle proteins, increase the pH value. There are insufficient experimental data on the antimicrobial effects of polyphosphates. Small increases in pH may change markedly the rate of microbial growth on cured meats, or meat in vacuum packs, but the effects appear to be less important in fresh meat. Practices which deliberately increase the pH value of meat for technological reasons should not be encouraged until it has been established clearly that the meat is as stable and as safe bacteriologically as meat of normal pH.

CONTAMINATION

The musculature of live and healthy animals contains relatively few microbes. Internal, or deep, contamination of muscles may, however, occur if animals are stressed just prior to slaughter, or if dirty instruments are used for bleeding out. Such internal contamination should be minimised as it increases the possibility of bone-taint, and can introduce anaerobic food-poisoning bacteria (e.g. Clostridium perfringens) which are capable of growth in the deep muscle. External or superficial contamination of beef and lamb carcases will occur during flaying, evisceration and dressing, and of pig carcases at all stages of slaughter following scraping. The number of bacteria on the finished carcase differs from one animal to another, and from one slaughterhouse to another (Table 3, p. 533) due to differing slaughter practice. It is important to minimise the external microbial contamination since the storage life of a carcase is dependent on the number of bacteria initially present. There is evidence in a recent survey that the differences in bacterial numbers on carcases between well-run and indifferently run abattoirs is of the order to make a noticeable difference to the storage life of the carcase.

"Soil" carried into lairage will contain 10^6 to 10^8 bacteria/g, and faecal material even larger numbers. Fodder and bedding also contain large numbers of moulds (fungi) and the related actinomycetes. The lairage should be kept clean in an attempt to minimise the numbers of bacteria on the live animal. Bearing in mind the high level of bacterial contamination on the hide it is relatively common to wash the live animals with water sprays, thereby hoping to reduce the numbers of bacteria on the skin. Gross soil, including some faecal matter, is certainly removed by this treatment, but there is no evidence that reduced bacterial contamination of the carcases results. It is also claimed that washing pacifies the animals, thereby minimising stress, but after such washing it is not uncommon practice to use mild electrical shocks to drive the animals to slaughter, and the overall reduction of stress seems questionable. Soil on the hooves has also been removed satisfactorily by walking the animals to slaughter through water-baths *c.* 1 foot deep.

Systematic attempts to reduce the bacterial contamination of lamb carcases by shearing the crutch or washing the live animal immediately before slaughter with water or a detergent solution have failed to effect a reduction in bacterial numbers on the carcases of any commercial significance, but improve their appearance by eliminating visible traces of faecal material.

During the stress associated with pre-slaughter handling and slaughter itself, and particularly in fatigued animals, the musculature may become more contaminated with bacteria. If the stress is minimal, the contamination within the musculature is small, but if the stress is severe, serious bacterial contamination can occur. This is

THE LAIRAGE AND HANDLING LIVE ANIMALS

517

believed to result from some change in the properties of the gut wall, permitting those bacteria normally present in the gut to invade the tissues, but convincing experimental proof that this happens under commercial holding conditions is lacking.

Clostridia

In one of the few conclusive experiments, the musculature and lymph nodes of cattle stressed just before slaughter were shown to contain larger numbers of clostridia than those of animals rested before slaughter. Increased numbers of clostridia are particularly undesirable since they render the meat and subsequent meat products more liable to putrefaction, particularly if stored under inadequate refrigeration. The genus Clostridium contains species (Cl.botulinum, Cl. perfringens (welchii)) which occur normally in the gut of healthy animals, but which are capable of causing food-poisoning if they are eaten after they have grown. In addition Salmonella spp and Escherichia coli, both of which are a hazard to human health, may be present in the gut of apparently healthy animals, and would likewise be expected to be present more frequently in the musculature of animals slaughtered under stress. Pre-slaughter stress also results in meat of higher than normal pH, at which the microbes are able to grow more rapidly, thereby increasing the rate of spoilage and the possibility of growth of food-poisoning bacteria.

Salmonella

In the lairage area microbial numbers are large, and such practices as are recommended are to protect the public health rather than to influence numbers of spoilage bacteria on the carcase. Animals obviously suffering from salmonellosis, and therefore excreting large numbers of salmonellae, are readily detected by inspection, but it is not uncommon for animals which show no such symptoms to excrete salmonellae. These animals are termed "symptomless excretors". Salmonella are easily transmitted from one animal to another in the lairage, and if this occurs they will be present on the carcase. It is increasingly common practice to starve animals in lairage since there is good evidence that if Salmonella spp are present in the gut of the animal certain feeds may promote their rapid growth resulting in very large numbers at slaughter, thereby increasing the chances of their occurrence on the carcase. If an animal is starved the gut will be essentially empty at slaughter, thereby reducing the likelihood of contamination of the carcase with gut contents.

During slaughter, flaying, evisceration and dressing, the carcase acquires a surface microbial flora comprising bacteria, yeasts and moulds (fungi). This flora is of broadly similar microbial composition for beef, pork and lamb, although differences become apparent when data of workers in different countries are compared. The most thorough investigations of the sources of this contamination were made pre-1940 in Australia when it was concluded that over 30% of the contamination of cattle carcases originated in dirt and on the skins of animals, about 50% from transport and storage, other sources including the abattoir atmosphere (c. 5%), the visceral contents (c. 3%), cutting and packaging (c. 2%) and miscellaneous sources, including personnel (c. 3%). These conclusions have never seriously been challenged by more thorough experiments, but do not explain the relatively uniform contamination of the carcase surface, even on areas untouched by personnel or equipment during flaying. Some areas, e.g. the crutch on lambs, are generally heavily contaminated, but no area appears free from bacteria.

THE SLAUGHTER LINE

Stunning

Differences in methods of stunning and sticking make general conclusions difficult, but the claim that bacteria introduced into the blood during stunning (e.g. by a captive bolt) are carried in the blood to the deep tissues is widely held but not well substantiated. Certainly use of a captive-bolt will often carry microbes from the hide or fleece into the brain, thereby reducing its value.

Pithing

The severing or destruction of the spinal cord is not practised in all countries, and it must be questioned whether pithing is essential. The practice in Australia is simply to sever the spinal cord at the atlanto-occipital joint, i.e. between the skull and the first vertebra. In the UK a "pither", commonly a coil wire spring, is inserted into the stunning wound and down the spinal column. The pither is frequently placed on the floor of the slaughter hall when not in use and commonly becomes very dirty. In principle it is bad practice to introduce into a carcase such a heavily contaminated object, although, again, direct evidence that serious bacterial contamination of the carcase occurs through the use of heavily contaminated pithers is lacking.

Bleeding

The skin of an animal is usually heavily contaminated with bacteria, and sticking with a dirty knife, or through such contaminated skin would certainly introduce bacteria into the blood system. The skin is certain to carry bacteria of faecal origin, many potentially pathogenic, and it is obviously imprudent to introduce such bacteria via the blood into a carcase, whether or not they reach the deep tissues. It is often claimed that the use of a dirty sticking knife increases the frequency of deep muscle- or bone-taints, but again the data are poor. In some countries slaughtermen use two knives, the first to cut the skin and to expose the carotid artery/jugular vein, and the second to sever the blood vessels, leaving each knife in a "steriliser" while not in use. Such practice is superior to using an improperly cleaned knife carried in a scabbard which is rarely cleaned. The use of clean sticking knives is simply achieved, and their thorough cleansing between each animal is strongly advocated. There are equally strong arguments for sterilising the skin before slaughter, but no suitable method has been found. In Britain sticking and bleeding are poorly controlled processes, and the blood is usually run to waste or for use in feed. This area is often grossly contaminated with bacteria, the bleeding trough and sticking knife commonly carrying 10^4 to $10^6/cm^2$. In Scandinavia blood is collected essentially free from bacteria, so it is a much more valuable product.

Beheading

The carcase is normally dehorned using a cleaver, and the head severed from the body using a knife. The scabbard used to hold these knives is commonly dirty and frequently contaminated with faecal coliforms. The severed head is then normally washed with mains water and hung from a hook, which is frequently heavily contaminated with bacteria.

Flaying

During the flaying process microbiology is rarely considered, and the design of flaying equipment often leaves much to be desired in this respect. The external surface of the hide (or fleece) is an obvious source of microbes and it has been shown important that removal of the hide should be in a downward direction. Removal upwards or sideways are both more likely to result in the deposition of microbes on the carcase. In the case of sheep, the fleece is believed to be a heavy source of contamination and it has become common practice in some countries to shear the crutch area of the live animal to prevent "blow-fly strike", that is, the laying of eggs in the wool, and because this area is often heavily contaminated with faecal matter, and therefore microbes. While this procedure has resulted in a reduction in gross faecal contamination of the carcase, it has not resulted in an overall reduction in the numbers of bacteria present on the carcase. Pigs are not skinned after slaughter but scalding, singeing and scraping can be arranged to reduce the population of bacteria on their surface.

Evisceration poses numerous problems since the contents of the gut and rumen contain from 10^6 to 10^{10} bacteria per gram (although the small intestines contain lower numbers). Hence if the gut contents spill on to the carcase, it will be virtually impossible to remove all these bacteria, even by careful washing. It is therefore imperative that no spillage from the gut occurs either from the oesophagus, or when the anus is removed. The most effective way of preventing this is to tie-off both ends of the alimentary tract, and this has been shown in Australia to contribute to an overall reduction in the bacterial numbers on the carcase surface. Large numbers of coli-forms on the carcase after evisceration is indicative of contamination by gut contents. The viscera of all animals should be removed intact since its contents may include the food-poisoning bacteria Cl.per-fringens, Salmonella spp and some Escherichia coli. Bacterial contamination may be transferred from one carcase to another by the abattoir equipment. This is reduced by frequent washing of the equipment with hot water, ideally after use on each carcase.

Evisceration

The carcase is normally split longitudinally with a reciprocating saw, and such equipment is difficult to clean thoroughly and hence may be a source of cross contamination.

Cleaving

A recently slaughtered and dressed carcase will be superficially contaminated with 10^2 to 10^6 bacteria per sq. cm. The actual number is important since the storage life of the meat will depend upon the initial level of contamination. Good hygienic practice is therefore essential during slaughtering both to minimise contamination by food-poisoning bacteria and to provide a reasonable storage life.

The overall impression from a consideration of accumulated data is that much of the surface contamination of carcases could be avoided, without a significant reduction in throughput, if the hands and aprons of operatives and the knives they used were more thoroughly cleaned. Knife "sterilisers" must be situated at convenient points on the slaughter line, and the knives must be cleansed frequently. One would expect a cleaner carcase to result if the animals had been stunned with a clean captive bolt, had been cut with one knife and stuck with a second, both of which were thoroughly cleaned, had been flayed in a downward direction, had been eviscerated with clean knives after having tied off the oesophagus and anus without puncturing the gut.

Spray washing

At this stage in past years it was common practice to wipe the carcase with a cloth. The recent prohibition of such wiping cloths can readily be shown to be desirable since cloths generally added bacteria to the surface of the carcase rather than removed them. With the prohibition of wiping cloths the trade turned to using water sprays to clean carcases. These were often commissioned before adequate data were available on the effects of nozzle design, water pressure and water temperature. These variables have been evaluated at the MRI. A fan jet used at a dynamic line pressure of 100 p.s.i., with a surface water temperature of 60°C (140°F), surface impact of 0·15 p.s.i. and a flow-rate of 8·5 litres/min (1·9 gall min) has been shown to yield consistently clean carcases with good bloom and a relatively low number of residual bacteria. Hot water gave carcases with better bloom and fewer bacteria than cold water. Spraying time had little effect on the appearance of the carcase, but carcases sprayed for 1 min had fewer bacteria than those sprayed for $\frac{1}{2}$ min. The addition of undesirably large amounts of hypochlorite o the spray water did not produce a large enough reduction in bacterial numbers to warrant its use.

t must be emphasised that the reduction in bacterial numbers by pray-washing is small and probably insignificant in terms of extending the storage life of the carcase but it is certainly preferable s a cleaning agent to using muslin wiping cloths. Having used a pray, it is important that the carcase be thoroughly dried, since there s some evidence that bacteria grow better on an inadequately dried arcase than on one which is well dried.

The overall impression from these studies and from US experiments n which acid washes at pH 2 failed to reduce significantly the pacterial numbers on carcases, is that once the bacteria are on the carcase it is very difficult to reduce their numbers, and that effort hould be applied to prevent contamination rather than remove it.

t is a common belief that the numbers of live bacteria on a carcase fall during the first 24–36 hours of storage, but this cannot be relied upon, and it is equally common for the bacterial count to remain constant during this time. Immediate chilling of the carcase is necessary to restrict bacterial growth on its surface and within the deep muscle. The surface of the carcase will cool quickly during chilling and the main problem is to cool the deep muscle sufficiently rapidly to prevent growth of the anaerobic mesophilic clostridia see Table 1, p. 532). The initial temperature of the deep muscle is approximately 39°C (102°F) which is near the optimum for growth of Clostridium spp, but in an unstressed animal sufficient residual oxygen is present in the muscle after slaughter to prevent their growth. However, cellular respiration continues after death and gradually uses up this oxygen until conditions become sufficiently anaerobic for clostridial growth. There is then the risk of internal putrefaction, which, apart from being objectionable, will also be dangerous if growth of Cl.botulinum or Cl.perfringens occurs. Chilling should therefore reduce the internal temperature of the muscle relatively rapidly to a level at which clostridial growth is severely limited (c. 15°C (59°F)) and finally to a level below 5°C (41°F). This entirely prevents growth of the common clostridia.

MEAT HANDLING
Cooling hall

If carcases are hung too close together, even touching one another the heat loss is reduced and bacteria grow rapidly, particularly on the surfaces in contact. For example, if carcases are taken directly from the slaughter line into chill, and hung touching, the temperature of the touching surfaces could be as high as 20°C (68°F) after 24 hours in chill, and bacterial numbers on those surfaces as high as $10^6/cm^2$. The storage life of such meat is very short, even if it is subsequently stored under ideal conditions. Carcases should be hung at least 2 in. apart to allow good air circulation. The thermal capacity of the system must be adequate (see p. 473).

If meat is cooled too rapidly before rigor mortis is complete "cold-shortening" of the muscle may occur, and, since it is not a reversible process will result in extremely tough meat. Cold-shortening is more likely in small carcases than large, hence the risk is greater in lamb than beef. There is no corresponding problem in pigs for biochemical reasons. Requirements such as those in EEC directives that a carcase shall be cooled to below 7°C (45°F) "as rapidly as possible" safeguard against the growth of food-poisoning Salmonella, but increase the chance of cold-shortening.

The r.h. of the chill-room atmosphere influences the rate of microbial growth on the carcase surface. Partial drying of the surface of a carcase will reduce the rate of growth of spoilage bacteria, but such drying is rarely uniform over the whole carcase, and some areas become "sticky" from bacterial growth before others. The control of r.h. is difficult, and, with presently available equipment, rarely attempted on a commercial scale. In selecting a chill-room r.h. the commercial operator is left with a difficult choice between increasing weight loss and attempting to limit microbial growth. The usual recommendation is to operate chill-room between 85% and 95% r.h., but intentions are sometimes thwarted by overloading chill-rooms, and commonly by hanging carcases in contact with one another leading to areas on the carcase which are improperly cooled and on which no drying has occurred, both promoting microbial growth.

At this stage the meat carries its characteristic cold-tolerant microbial flora and it is important that pathogenic bacteria are not introduced by persons handling and cutting the meat. The temperature of the meat in cutting rooms should not exceed 7°C (45°F), nor the air 10°C (50°F). These restrictions to limit the growth of pathogens will incidentally limit the growth of spoilage bacteria. Strict hygienic precautions must be taken to ensure that Salmonella are not introduced from handlers (e.g. suffering from salmonellosis) nor staphylococci (e.g. from septic wounds).

Boning and cutting rooms

Packaging materials which have been stored under hygienic and dust-free conditions are unlikely to add significant numbers of bacteria to meat. Plastic laminates carry very low numbers of microbes, but paper/card trays are occasionally heavily contaminated by moulds. Cardboard cartons, provided they have been stored under dust-free conditions, carry only low numbers of microbes, but the common storage of such cartons under poor conditions has led to proposals that cartoning be prohibited in meat cutting rooms. Convincing evidence that cardboard cartons are a major source of bacterial contamination is lacking.

Packaging

Chilled storage below 10°C (50°F) inhibits growth of mesophiles, including almost all food-poisoning bacteria (Table 4). Further reduction in temperature greatly reduces the growth rate of spoilage bacteria, but even at temperatures close to the freezing point of meat bacterial growth proceeds slowly. Many psychrotrophic spoilage bacteria continue to multiply slowly at −1°C (30°F), but the lower the temperature, the slower is microbial growth. Growth is roughly half as fast at 5°C (41°F) as at 10°C (50°F), and the rate is more than halved again at 0°C (32°F). Storage at −1°C (30°F) gives the maximum storage life without freezing.

Chilled and Frozen Storage

Shelf life will be extended even more by freezing, which is equivalent to physiological drying, but this is considered detrimental to quality and some microbial growth is possible even on frozen meat. There are reports of bacterial growth at −6°C (21°F), while in high humidity, surface discoloration has been reported on meat stored at

−8°C (18°F) due to mould growth. Cladosporium herbarum (causing "black spot") and Sporotrichum carnis (causing "white spot") have been common causes of defects on carcases stored under refrigeration, and growth of other moulds including Thamnidium and Mucor spp has been reported on frozen meat at −10°C (14°F). Although some of these reports must be viewed with suspicion, perhaps having been recorded in the absence of adequate temperature monitoring during storage, temperatures below −10°C (14°F) are advisable if prolonged storage is intended and microbial growth is to be completely precluded.

While it is relatively simple to freeze meat rapidly to a sufficiently low temperature to prevent microbial growth, thawing may result in microbial growth and this has received little attention experimentally. Commercial practice varies with the dimensions of the frozen meat, and may thaw in air, water or vacuum. The thawing temperature would be expected to be important, and has been shown to be so, but the velocity of air or water is also important in determing the thawing time and whether or not extensive microbial growth occurs at the meat surface. A compromise must be achieved between a high temperature with a desirably rapid thawing rate and a lower temperature with a slower thawing rate. This balance depends upon the size and weight of the article to be thawed.

Preservation and Spoilage

**FRESH MEAT
Storage in air or
gas-permeable wraps**

Joints or minced meat taken from a beef carcase and stored under refrigeration in air or in a gas-permeable wrapper will spoil due to the growth of Pseudonomas spp which produce off-odours when their numbers on the meat reach $c.$ $10^7/cm^2$ and slime and discolouration at $c.$ $10^8/cm^2$. Growth of other bacteria, for example Brocothrix thermosphacta and lactic acid bacteria occurs on beef stored in air but is not believed to play an important role in spoilage. Superficial growth of fungi is sometimes observed on beef joints after prolonged storage but by this time the meat has usually been spoiled by Pseudomonas spp. The spoilage of pork is similar to that of beef, but lamb is often observed to spoil through the growth of Brocothrix thermosphacta rather than Pseudomonas spp. The shelf-life of meat with an initial contamination level of 10^2–10^4 bacteria/ cm^2 in air at 100% r.h. and at 0°C (32°F) is approximately 12 days. If the initial contamination level is higher, or the temperature of storage increased, the shelf-life is reduced, but a reduction in r.h. below 95% will increase shelf-life in air.

The main purpose of vacuum-packing fresh meat is to extend the storage-life. If good conditions of slaughtering and cutting have been used, and low temperatures (-1°C) have been used for storage, beef may be stored in vacuum-packs for 10 weeks and remain in good condition. Plastic laminates used in vacuum-packing have a low permeability to gases, and therefore the escape of carbon dioxide produced by the metabolism of the meat or by microbial growth is prevented and it accumulates round the surface of the meat to concentrations which inhibit the growth of Pseudomonas spp. Gram-positive bacteria are more resistant to carbon dioxide and continue to grow. After prolonged storage, spoilage is due to "souring" caused by lactic acid bacteria. When beef of a high terminal pH (above 6·0) is vacuum-packed, growth of some Pseudo-monas spp may occur. Hydrogen sulphide produced by certain of these bacteria combines with myoglobin in the drip to form the green pigment sulphmyoglobin producing the characteristic "greening" which is sufficiently intense to warrant immediate rejection of the packs. Its common occurrence in Australia has led to a recommendation that the industry vacuum-packages beef only if the pH value is below 6·0. The storage-life of lamb is extended by vacuum-packing, and its spoilage is commonly due to Brocothrix thermosphacta. Fresh pork is rarely vacuum-packed and there is little information on the effectiveness of this technique in extending its storage-life. As with gas-permeable packs, the storage-life of vacuum-packed fresh meat is dependent upon the initial bacterial contamination level and on the storage temperature, but is independent of the r.h. It is important to refrigerate vacuum-packed fresh meat because anaerobic bacteria are able to grow at its surface.

Storage in Vacuum-packs

Carbon dioxide has been used in the past to increase the storage-life of carcase meat shipped from Australia and New Zealand but in

Gas Storage and Gas Packaging

527

recent years this trade has been replaced by better temperature control and by vacuum-packing. Packaging in gas-impermeable films with carbon dioxide has been used to some extent for bacon, but it is not used for fresh meat, having no obvious advantages over vacuum-packing. Both vacuum-packing and carbon dioxide packing adversely affect the colour of fresh meat.

Pre-packaging in gas-permeable films and storage of the packs in gas mixtures of oxygen and carbon dioxide will extend shelf-life and retain better colour so that meat may be pre-packed and distributed directly to retail stores over relatively long distances, but this has been little used.

BRITISH SAUSAGES

British sausages are traditionally prepared from a mixture of uncooked cereal, salted seasonings, diced meat and iced water. Spices are often heavily contaminated with bacteria but the quantity of spice included in the sausage is so small that it may have little effect on the total bacterial count of the final fresh sausage. The bacteriological condition of the casing rarely affects the bacteriological condition of the product, and the quality of the meat is the most important factor influencing the bacteriological condition of the fresh sausage. Sulphur dioxide (450 ppm) is present as a preservative in British sausages and extends the shelf-life by inhibiting the growth during storage of Gram-negative bacteria including Pseudomonas spp. The flora developing on sausage during storage is dominated by Gram-positive bacteria, in particular M.thermosphactum, and spoilage is accompanied by a sour odour. Yeasts are common in British fresh sausages and their growth at the surface can cause sliminess.

CURED MEATS

The main curing ingredients used in the UK for bacon production are sodium chloride, sodium nitrite, and sodium nitrate. Of these sodium chloride and sodium nitrite have the major preservative function. Nitrate probably has only a minor effect on microbial growth in bacon, other than as a precursor of nitrite.

Microbes differ in their sensitivity to sodium chloride: Pseudomonas

528

spp are particularly sensitive and their growth is rarely detected on bacon. However, several other groups of micro-organisms are salt-tolerant and capable of growth on bacon which must therefore be treated as a perishable food.

Wiltshire bacon is the most common cured meat on sale in the UK and is produced by injecting pig sides with a fresh brine followed by immersion in a cover-brine for 4–5 days followed by a final stacking of sides in cool conditions (c. 5°C (41°F) for a 7-day maturation period. The bacterial contamination level of the carcases going into cure probably affects the subsequent storage-life but there is little bacterial growth during immersion. An increase in microbial numbers occurs on the sides during maturation after which Micrococcus spp, Vibrio spp, Acinetobacter spp and yeasts are the most common organisms detected. High temperatures ($>7°C$) and high humidities ($>85\%$) in the maturation cellar increase bacterial growth. Subsequent storage in air of bacon sides or joints may result in surface sliming and spoilage due to continued growth of those bacteria detected at the end of maturation but which is putrefactive only if the pork has been improperly cured. Smoking considerably reduces the number of bacteria on bacon sides and this, combined with the inhibitory effect of chemical components of the smoke and the reduction in pH of the bacon which smoking causes, improves storage stability. Gram-negative bacteria are more sensitive to the effect of smoking than Gram-positive bacteria.

Apart from superficial spoilage internal tainting or souring is sometimes encountered, particularly of hams. A variety of micro-organisms including clostridia, Vibrio spp, Lactobacillaceae, Micro coccaceae, Enterobacteriaceae and Bacillaceae have been detected in association with these taints but their relative importance is not understood. Good chilling and refrigeration practice significantly reduce the incidence of internal tainting, and the penetration of salt into the deep musculature is also an important factor.

Vacuum-packing, as with fresh meat, markedly extends the storage life of bacon. The growth of yeasts and Gram-negative bacteria is usually inhibited in vacuum-packed back bacon and although micrococci grow their growth rate is slower than in air. Streptococci and lactobacilli (a typical streptobacteria) are the eventual cause of souring of vacuum-packed back bacon. Vacuum-packed collar bacon is less stable bacteriologically than back bacon and high numbers of Gram-negative bacteria are sometimes present on this product which should be stored under refrigeration.

COOKED MEATS

Most food spoilage and food-poisoning bacteria grow best at pH values near neutrality (pH 7). Hence acidification is an effective means of preservation because most of these bacteria do not grow in acid conditions. In addition, at low pH values the resistance of bacteria to heat is reduced. Hence in canning, foods are subdivided according to the pH value, since those of low pH require less heating to render them shelf-stable in the absence of refrigeration. In canning terminology meat is a low acid food, where the pH value plays essentially no role.

If meat is heated slightly some of the bacteria are killed. The cold-tolerant spoilage bacteria are most sensitive to low heat treatments (at 60°–70°C) (140°–158°F) which are best carried out in the final packaging to avoid bacterial contamination after completion of the process. Hence the shelf-life of the product under chill is extended. If the heat process is relatively severe (above 100°C (212°F)) and designed to kill more heat resistant bacteria, including the spores of Clostridium and Bacillus spp, the product will be shelf-stable without refrigeration. Such products may still contain small numbers of viable spores, but they are unable to grow in the product at the normal storage temperatures, and no spoilage occurs. This is termed "commercial sterility". This relatively severe heat process may be reduced and still yield a stable product, if the product is of reduced pH or contains curing salts. Large hams are heated only lightly, yet are relatively stable in the absence of refrigeration because they contain salt (sodium chloride) and sodium nitrite which prevent bacterial growth.

At the retail and domestic level it is important to remember that cooked meats are still susceptible to spoilage by bacteria which are permitted to recontaminate it, i.e. the cooking does not render the meat unsuitable for bacteria. If cooked meat is not eaten, it should be stored under refrigeration. Even "roasting" in an oven may not be sufficient to kill all the bacteria present in a meat joint, since the centre temperature rarely exceeds 75°C (167°F), and this temperature has little or no effect on spores. If, after cooking, a joint is held in warm conditions, the spores which have survived heating, particularly those of Cl.perfringens, may multiply and if the cells are not killed by subsequent heating may cause food-poisoning when the meat is eaten.

Food Poisoning

There are two basic types of food poisoning termed infection and intoxication. In an infection, live bacteria are eaten and multiply in the body. Salmonella cause infections, though the different types of Salmonella have different "infective doses". Some, like S.typhi, which causes typhoid, have very low infective doses, and even the ingestion of ten to twenty bacterial cells is sufficient to cause typhoid. Other types may require 10,000 or 100,000 bacterial cells to be eaten, before an infection is established in the host. In addition some are "host-specific", that is they may e.g. affect chicken, but not man. Salmonellosis is common, though mortalities are relatively rare. Each year in the UK there are about 6000 cases of salmonellosis, approximately half of which are due to S.typhimurium. The symptoms are diarrhoea, fever, and vomiting, and the illness may last 1–14 days after a 12–24 hour incubation period.

In an intoxication, the bacteria grow in the food before it is eaten, and produce a toxic chemical. This chemical produces the characteristic symptoms of food-poisoning after it has been ingested. Certain Staphylococcus spp produce a toxin which causes severe vomiting, violent headaches and prostration within 2–6 hours of being eaten. The duration is normally only 6–24 hours. This toxin is heat-resistant, and would not be inactivated by normal cooking. It is particularly troublesome in cooked cured meats, normally by recontamination after the curing process, and often through handling during slicing.

Clostridium botulinum produces a toxin which is one of the most poisonous natural substances known to man, and causes death by

TABLE I

Bacteria Commonly Present in Meats

	Aerobes	Facultative anaerobes	Anaerobes
Mesophiles	Bacillus spp	Coliforms Salmonella spp★ Staphylococcus spp★	Clostridium sporogenes Cl. botulinum★ Cl. welchii (perfringens)★
Cold-tolerant	Micrococcus spp	Lactobacillus spp Streptococcus spp	
Psychrophiles	Pseudomonas spp Acinetobacter spp	Brocothrix thermosphacta	Cl. putrefaciens

★ Commonly associated with food-poisoning

respiratory paralysis. Human botulism is rare in the UK, but incidents in France, Spain and N. America indicate that Cl.botulinum remains a hazard to the food processor. Botulism is most frequently associated with home-processed foods (fruit, vegetables, cured meats, mushrooms), usually due to under-processing, but several recent incidents have involved commercial products. With better bacteriological techniques, Cl.botulinum can readily be demonstrated to be widespread in the environment, and it would seem wise to assume that it is an ever-present contaminant of foods, and must either be killed or controlled.

Clostridium perfringens (welchii) is a common cause of food-poisoning rarely fatal. It grows well in warm meats, and the most common sequence of events is: a light cooking often in a large kitchen, slow cooling allowing the spores which have survived the cooking to grow, overnight storage followed by rewarming before serving but insufficient heating to kill the bacteria present. Large numbers of viable cells must be eaten, hence thorough heating before serving would eliminate the risk, as would storage of the heated meat under good refrigeration. There are several thousand cases of food-poisoning from Cl.perfringens each year, commonly in canteens and institutions. The illness passes rapidly (12–24 hours) after a short incubation period (8–20 hours) diarrhoea and prostration being the main symptoms.

TABLE 2

Minimum Water Activity (a_w) for Growth

Minimum a_w	Organism
0·95	Gram negative rods (e.g. Pseudomonas spp)
0·92	Clostridium spp
0·91	Lactobacillus spp Bacillus spp
0·88	Yeasts
0·86–0·84	Staphylococcus and Micrococcus spp
0·80	Moulds

TABLE 3

Mean Bacterial Counts (log_{10}) per sq cm on Lamb Carcases at Six Commercial Abattoirs

		Abattoir				
	A	B	C	D	E	F
Visit 1	5·1	4·9	4·8	4·7	4·5	4·4
2	4·7	4·7	4·5	4·4	4·4	4·2
3	4·6	4·5	4·4	4·2	4·2	3·8
4	4·5		4·4	4·2	3·7	3·7
5			4·1			3·5
6			4·1			

(Each tabulated value is the mean of counts on at least 20 carcases)

TABLE 4

Minimum Growth Temperatures of Food-Poisoning Bacteria

	°C	°F
Staphylococcus aureus	6·7	44
Salmonella spp	6·7	44
Clostridium botulinum		
type A	10	50
type B	10	50
type B (non-proteolytic)	3·3	38
type E	3·3	38
type F	3·3	38
Clostridium welchii (perfringens)	10	50
enteropathogenic Escherichia coli	10	50

Meat Products

One of the oldest forms of processed food is sausage. Meat, blood and other ingredients, filled into gut and broadly described as sausage, were common items of diet, prior to Roman times.

The word sausage is derived from the Latin *salsus* meaning salted and consequently preserved. In addition to preservation by salt, many sausages were smoked and subject to prolonged drying, thus adding to their keeping qualities and enhancing their flavour.

Generally all sausage is comminuted meat, although small quantities of similar products have been produced from fish, particularly in Japan and China. In Germany where the art of sausage making originated, some 1458 different types have been recorded.

In the United Kingdom relatively large amounts of cereal are legally permitted whilst in other countries its use may be prohibited. In the United States, up to 3% of cereal, starch or dried milk is allowed, under appropriate declaration.

The universal demand for sausage is probably due to the fact that in practically no other food are palatability, utility and good food value combined to such a high degree.

In 1978 in the UK, total sausage usage was estimated to be around 367,100 tonnes of which some 88,575 tonnes were used in the catering trade. About 95% of the domestic use was "fresh" sausage, the balance consisting of frankfurter and similar products, of which a good proportion is imported, frequently canned.

Types of Sausage

DRY SAUSAGE: This is usually consumed without cooking, although in some types relatively high smoking temperatures are employed. In America this type of sausage may be defined as "new sausage" at between 2 and 4 weeks from preparation, "medium dry" from 1 to 2 months, and "dry" at 3 months onwards. Dry sausage is generally prepared from cured meats filled into casings and dried under controlled conditions, whilst some types are smoked. As compared with traditional methods, improved sophisticated techniques have resulted in a marked reduction in the maturing period.

The following represent but a few typical examples: Salami (numerous types), Cervelat (numerous types), Landjaeger, Chorizos, d'Arles, Mortadella, Pepperoni.

COOKED SAUSAGES: These are pre-cooked, ready for consumption although in some cases they may be subjected to some heating prior to serving. They include:

Frankfurters	Liver sausage	Garlic
Wieners	Polonies	Cotechino
Blood sausage (various types)	Savaloys	
Tongue and blood	Hog puddings	
Beef luncheon	White puddings	
Pork luncheon	Bierwurst	

536

FRESH SAUSAGE: This type is generally fried, grilled, oven cooked or in some cases cooked in water, prior to consumption. As previously indicated pork and beef sausages represent a major part of the UK market.

Pork	Tomato pork
Pork chipolata	Bockwurst (beef and pork)
Beef	Bratwurst (pork and veal)
Beef chipolata	Thuringer (pork, veal and in some,
Cocktail	beef)
Cumberland (coarse cut)	Swedish potato

Whilst cereals and spices employed in sausage production are reasonably consistent in composition, the meat/fat content is a different proposition. For example lean pork shoulder trimmings might contain about 19% fat and 17% protein, compared with 41% fat and 11% protein in fat pork shoulder trimmings.

Sausage Ingredients

In the case of fresh pork sausages, the legal requirements stipulate that: the minimum meat content is 65% and of this not less than 80% must consist of pork. The remaining 20% (of the minimum 65%) can be made up with other meat(s), i.e. beef, veal, mutton. In addition there must be not less than 50% of the minimum meat content in the form of lean meat.

Beef sausages must contain a minimum of 50% meat of which at least half must be beef and again the visible fat must not exceed the lean.

The legislation lays down *minimum* standards and any deviation below these minima constitutes an offence.

Owing to margins which can exist in *chemical* analyses, in commercial practice recipes should be based upon a minimum of 67% in pork sausages and 52% in beef.

Thus the selection of meat for sausage manufacture is largely influenced by:

537

1. Cost as prepared, including labour cost and loss of bone and gristle content.
2. Proportion of lean to fat.
3. Condition, to ensure appropriate commercial life.
4. "Binding" characteristics.

MEAT

As an animal acquires physical maturity, the bone ceases to grow and the internal organs do not increase markedly in weight. Consequently any increase in weight will be in the meat. Thus, heavy mature animals will usually be more profitable for sausage manufacture, provided that the proportion of fat is not in excess of requirements.

The bone content of cow beef can vary between about 17–22%, hogs (180 lb) 12–13% and heavy hogs and sows 9–11%. Bobby calves may lose about 33% of their initial carcase weight, in bone and trim.

Boneless Beef

Boneless beef for processing purposes is usually purchased on the basis of its "V/F" (visible fat) content. There is a ready demand for forequarter meats with a 10% V/F. In some cases specification on a "chemical fat" content, may be stipulated. It is not possible to give a specific figure for the relationship between visible fatty tissue and chemical fat (ether extracted) for the following reasons:

1. Ether extract from external fat can vary from about 52% in fat class 1 beef carcases, to 86% in fat class 5.
2. Ether extract from the "seam fat" is usually lower than that from external fatty tissue.
3. The proportion of fat *within* the lean meat varies considerably.

pH Measurements

The value of pH measurements to the meat processing industry, covers a wide field, including assessment of the condition of the meat, potential keeping qualities and the water-holding capacity. In canned products the pH has an influence on the time/temperature of the heat treatment.

In Germany, a colorimetric method is widely used, and 24 hours after slaughter, meat with a pH of 6 is considered to be of "good

538

durability", 6·4 is classified as of insufficient durability, whilst meat with a pH of 6·8 is unfit for food.

In Denmark pH readings of over 6·5 are regarded as evidence of poor keeping quality and such meat must not be sold through ordinary channels, but may be used in sterilised canned products under the control of the meat inspection service. (For further reference to spoilage and pH see pp. 516-17.)

The binding characteristics of a sausage emulsion are good at pH values between 6·0 and 6·3. The formation of jelly pockets and/or the release of fat is often associated with low pH values. In salami and similar types of sausage pH values change throughout the production process. For example it is possible to begin with a pH of 6 during the mixing, dropping to 5·6 during maturation and after drying a final pH value of 4·5 to 5·1 gives the product a slightly acid flavour and good keeping qualities.

Whilst bone, fat and gristle can be controlled by trimming and selection, the "water binding" characteristics of the lean meat is an inherent feature of the flesh. Skeletal muscle from mature bovines is accepted as having good binding qualities, followed by veal, lean pork and pork head meat. The longer the period elapsing from slaughter to use, the poorer the binding effect.

Pork Rind Emulsion

Pre-mix emulsions are frequently used in the composition of sausage and other meat products. They may consist of a cooked rind emulsion or a rind emulsion. Their use presents the problem of what amount can be used to meet the legal requirements, as shown on chemical analysis.

In the regulations "meat" is defined as follows: "Meat means the flesh including the skin, rind, gristle and sinew in amounts naturally associated with the flesh used". Obviously the ratio of rind in pig's head meat would be greater than that from a complete hog carcase. It would appear that in the suggested 67% meat content of pork sausage, a rind constituent of 6% would be acceptable.

Water

Water is an essential ingredient and together with fat, provides juiciness and acceptable texture. A sausage prepared exclusively from meat and cereal would crumble and be unattractive when consumed. With the object of offsetting the temperature rise during manufacture, ice water or cracked ice is frequently used.

Cereals

A wide range of cereal products has been used in sausage. These include various flours and starches, bread and bread crumbs. Today, rusk, prepared from a biscuit dough, is commonly used. After controlled drying the product is ground into various grades, based upon particle size. It is practically sterile and even after soaking retains its physical characteristics providing suitable eating texture. Under test conditions rusk may take up about 3 times its own weight of water, but in practice an absorption of about $1\frac{1}{2}$ lb water per lb of dry rusk, is generally favoured.

Where the "dry mix" system is used i.e. the rusk is not presoaked, but added dry in the bowl cutter, the best results are obtained with a fine grist. This method obviates the labour time involved in soaking and the use of large containers for holding the soaked rusk for a period of about thirty minutes to one hour.

Seasoning

Salt forms the major ingredient of most fresh sausage seasonings. Dendretic salt (I.C.I.) is a type of fine salt, which has a soft floury texture. It has good blending qualities with other finely divided spices, which retards the "dressing out" of salt, when the mixture is vibrated.

Basic seasonings consist of a mixture of salt, pepper, spices and in some cases herbs may be included. An extensive range of "factory-blended" seasonings are available to cover all types of sausage and meat products, in sealed containers, in units for specified batches. These may incorporate preservative, colouring agents and emulsifying substances. Most firms will prepare such seasoning to a customer's specific formula. Thus it is possible for the smaller sausage producer to obtain a high degree of standardisation in his products.

540

With ground spices, the flavour takes some time to diffuse through the mass. The optimum effect in fresh sausage being attained in 24–48 hours from preparation.

The flavouring value of spices and herbs depends largely on its oily constituents. A solvent process is frequently employed, which in addition to extracting the volatile factions, also removes the non-volatile constituents. Thus by total extraction of the spices, the true value and aroma of natural spices is obtained. Obviously these substances are highly concentrated and the minute quantities necessary would involve extreme care in their use; consequently, these substances are incorporated in a spreader base, which would be soluble in, or completely blendable with the food mix. The increased bulk can be more evenly distributed. The principal bases employed may consist of salt, dextrose, or finely ground rusk. In addition to the range of "straight" spices prepared in this form, there are many blends available especially developed for sausage and other meat products.

There is an increased demand for extracted spices, as they are practically sterile and fibre free, and their flavouring effect can be more rigidly controlled than is the case with ground spices and herbs.

Spice combinations commonly employed for fresh pork sausage may include: white pepper, nutmeg, mace, ginger, coriander, cardamom, sage and basil; for beef sausage: black pepper, white pepper, nutmeg, mace, cayenne and ginger.

SAUSAGE ADDITIVES

It is recognised that the use of polyphosphates retards the loss of moisture, meat juices and certain soluble proteins when sausages are cooked. There is a large number of blends of phosphate on the market and frequently they may be incorporated in the spice mix. The quantity employed may run from between 0·1% and 0·3%. When added early in the chopping process it tends to retain the meat juices, whilst if added late it appears to have more influence on reducing fat loss during cooking.

541

M.S.G. (Monosodium Glutamate)

Whilst it does not possess any flavour itself, it has the ability of enhancing the natural flavour of many products. In fresh sausage, its use may vary from about 0·2% to 0·1% of the final product. Used to excess it can produce a taste "carryover" or retention which may not be acceptable to some palates.

Sugars are commonly employed in the US, 6–7 oz of dextrose per 100 lb of pork sausage meat being used. In addition to its contribution to flavour, it has the effect of reducing apparent saltiness and will assist in surface browning on frying. This browning is also obtained by the use of milk powders which contain a fairly high lactose content.

Protein Supplements

In chemical analyses, the figure used for the conversion of protein to assumed meat content is a standard figure for pork and another for beef, and as such does not take into consideration protein variations which may occur in the meat. Consequently some operators consider that the use of a high-protein supplement will tend to stabilise the protein content, and is a form of insurance.

Such products may improve the texture and assist in the retention of fat and water. Those of animal origin include blood plasma, in liquid form, or preferably as a spray dried powder, skimmed milk powder and casein, the chief protein of milk. Caseinate will normally contain around 90% protein. Of the high-protein additives of plant origin, soya bean flour is most widely used. As a straight soya flour it will contain about 40% protein, whilst in soya protein isolates, the figure is around 90% protein. It must be stressed that protein additives in the legal context are a *supplement* and not a *substitute* for meat.

Any egg which is present in any meat product, not exceeding one fifth by weight, may be reckoned as part of the meat content, provided that it bears as part of or in immediate proximity to its name, a description indicating the presence of egg.

Sulphur dioxide is the preservative permitted in sausage and sausage meat, with a maximum limit of 450 p.p.m. Owing to the small amount involved, better distribution is obtained by adding it to the seasoning mixture, or as a solution in water. If 450 p.p.m. is added, a proportion is lost during the chopping stage, so that the finished sausage will be below this figure. Apart from the preservative effect it improves colour retention.

The majority of natural casings are derived from the intestines of pigs, sheep and cattle, the exception being the weasand, from the oesophagus of the ox, and the bladder. The latter may be used for mortadella.

Of recent years the use of reconstituted collagen casings has rapidly increased. Such casings have the advantage of providing a standard product of uniform tenderness and satisfactory cooking characteristics. These casings are used dry, so that the need for soaking, flushing and spooling is eliminated. As no drying is required, sausages can be packed immediately after linking.

Small diameter casings are used mainly for fresh pork and beef sausages, frankfurters, hog puddings and in some cases Scotch white puddings etc.

Various grading systems may be employed and the following is an example of fairly close grading.

Hog Casings

Grade	Approx. Diameter	Approx. Capacity per Bundle*
Narrow	32 mm/−	70– 75 lb
Narrow/Medium	32–35 mm	75– 85 lb
Medium	35–38 mm	80– 90 lb
Medium/Wide	38–42 mm	85– 95 lb
Wide	42–44 mm	90–100 lb
Extra wide	44 mm/+	100–110 lb

Selected hog casings are measured into bundles of 91 metres (100 yards).

* Based upon fresh pork sausages, filling may vary according to locality

Hot Bungs

Part of the large intestines were at one time used for liver sausage; in the UK they have been largely replaced by synthetic casings. In the US and many continental countries they are split, dried and sewn to form "sewed bung ends" as containers for cervelat, thuringer and similar types.

Fat end, from the anal end of the gut is used for traditional-type liver sausage and some types of salami.

Sheep Casings

Typical Grade	Approx. Diameter	Approx. Capacity Per Bundle
Narrow	16–18 mm	25–35 lb
Medium	18–20 mm	30–40 lb
Medium/Wide	20–22 mm	35–45 lb
Wide	22–24 mm	40–50 lb
Specially Wide	24 + mm	45 + lb

Selected sheep casings are measured into bundles of 91 metres (100 yards)

Sheep casings from the small intestines are used for most types of British fresh sausage and frankfurters, wieners and bockwurst.

Liquid Pack Casings

Natural casings are available in "ready to fill" form. Each strand of casing is shirred onto a tube so that they can be directly transferred to the stuffing nozzle. These are packed together with a liquid into a sealed polythene bag.

Collagen Casings

These are available in consistent wall thickness and constant diameter, the calibre ranging from 17 mm to 32 mm. The casing is cut into lengths of 9 m (30 ft) or 11 m (36 ft) and then shirred into 15 cm (6 in.) lengths.

Beef Casings

With the advent of synthetic materials, particularly P.V.C. and cellulose, their use in the UK has drastically declined.

544

The method of preparation will be influenced by the plant available and to some extent by the local demand. The coarse cut Cumberland sausage is readily prepared on a mincer and whilst the finer cut sausage can be manufactured on a mincer, the use of a bowl cutter facilitates production.

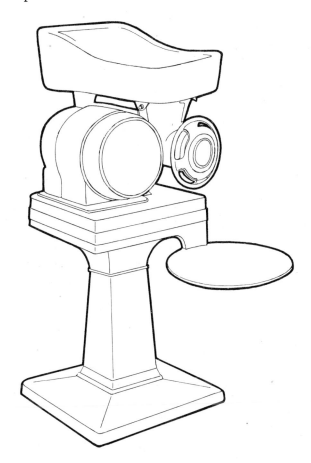

Fig. 1.   Mincer, bench or pedestal machine with quick feed chopper head.

The following will indicate some methods commonly employed.

1. Mincer only. Lean meat and fat, minced, using plates of appropriate size, for fine or coarse mix. Spices and additives sprinkled over lean meat, soaked rusk added and the mass lightly mixed by

545

hand and finally minced fat added, and the whole thoroughly blended. Combination machines incorporating a mincer and mixer are available which obviate the need for hand mixing.

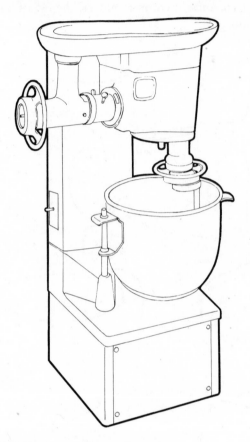

Fig. 2.   Rotabowl for sausage making, 25 lb capacity.

2. Where a bowl cutter is available the following methods may be used.

   (a) Meat prepared, lean meat placed in bowl cutter followed by spices etc. soaked rusk and fat.

   (b) Meat and fat minced, soaked binder placed in bowl cutter followed by seasoning, lean meat and fat.

   (c) "Dry mix", lean meat placed in bowl cutter, spices etc. added. Ice water slowly added, followed by dry fine grist rusk. Chopping continued and cubed fat added last. In some cases, half of the ice water is added early to the lean meat, the balance being added after the dry rusk.

(d) A vertical cutter/mixer has a wide range of applications including pastry mixtures and vegetables. The bowl does not revolve and the knives are attached to a vertical high speed drive shaft and revolve in a horizontal plane creating a vortex effect in the mix. This ensures clean cutting.

Bowl Cutters

A wide range of bowl cutters is available, of British and Continental origin. They can vary from a bowl capacity of a few pounds of sausage meat up to the large sophisticated machines capable of dealing with a 1000 lb batch. A number of the larger modern bowl cutters are equipped to provide variable speeds, mechanical self-emptying, and a thermometer to indicate temperature rise. Some incorporate automatic control which can be pre-set ensuring consistent chopping times for the various ingredients.

Chopping machines operating under a vacuum are common in the US and to a lesser extent in the UK for certain types of mix.

Filling Machines

Whilst sausage filling attachments for mincing machines are available the filling process is slow. Consequently, a separate filling machine is preferable. Basically they consist of two types, i.e. manually or power operated. The former may be of either horizontal or vertical type. The vertical form is essential for filling cut wet mixtures such as blood sausage.

An oil pressure electric system is commonly used on power operated machines and the speed can be readily adjusted according to the product.

Many large plants employ filling machines which provide continuous flow operation. The mix is fed in via a hopper and pumped under vacuum into the filler and casing. Such machines usually incorporate automatic twist-linking of the sausages.

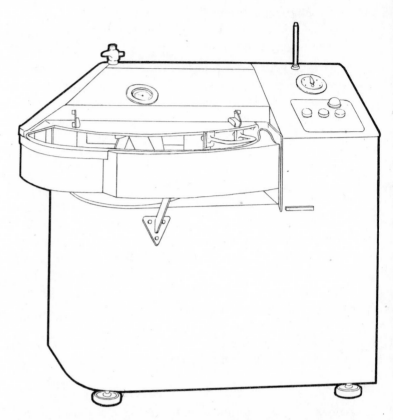

Fig. 3. *A modern filling machine for small and medium scale production.*
Special features include, quick release filling nozzle, which does not require locknut.
Barrel lined with polished stainless steel and meat piston of high impact nylon. Two
models are available.

	A	B
Capacity	40 *lbs.* (20 *ltrs.*)	60 *lbs.* (30 *ltrs.*)
Motor	1.5 *kW* (2 *HP*)	2.2 *kW* (3 *HP*)
Output per. min., max.	38 *ltrs.*	48 *ltrs.*

Both models are available in three finishes.

(Courtesy J. C. Wetter & Co.)

SAUSAGE RECIPES

The following basic recipes are intended primarily for the butcher rather than large scale production.

In general, it is better for the butcher to buy seasonings ready mixed.

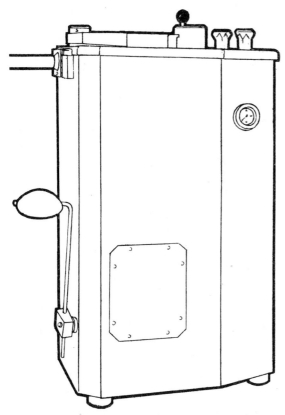

Fig. 4. Cutter-Mixer from 33 to 80 litre capacity for production of all types of cooked and raw sausage.

Most supply houses incorporate such ingredients as preservatives, phosphates and colour retaining agents with the spices, in unit packs. It is recommended that the meat content should slightly exceed the required legal minimum as an insurance against possible errors in blending.

Pork sausage 1
45% lean pork
22% back fat
12% dry rusk
18% cold water
± spices, additives etc.

549

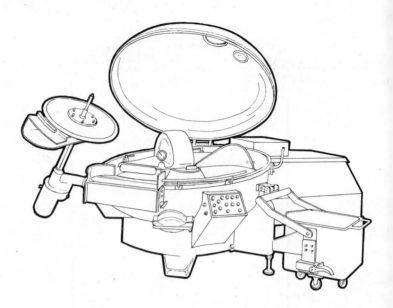

Fig. 5. *Vacuum Cutmix—hydraulic loading, over-the-side unloading*
200, 300, 500 *litre capacity.*

Pork sausage 2
22% lean pork
14% bellymeat (50% fat)
6% pig head meat (frozen)
5% rind emulsion
20% back fat
12% dry rusk
18% water/ice
± 3% spices and additives

Pork sausage 3
Cumberland (coarse cut)
50% lean pork
25% belly meat
8% coarse rusk
14% water/ice
± 3% spices and additives

550

Beef sausage 1
35% lean beef
17% fat beef
18% dry rusk
27% water/ice
± 3% spices and additives

Beef sausage 2
50% plate/flank (approx. 35% fat content)
3% rind emulsion
18% dry rusk
26% water/ice
± 3% spices and additives

Beefburger
50% lean beef
30% fat (preferably beef suet)
3% rusk
12% water/ice
2% spices etc.

Beefburger

In the US a "hamburger" is an all beef product and the inclusion of pork is legally considered as an adulteration. In the UK the name of the principal meat ingredient should be inserted before the word "hamburger" *or* in place of the syllable "ham". Supply houses prepare ready mixed spices for burgers. These may contain colour retention agents, phosphates and if required, onion flavour. "Hamburgers or similar products" may contain sulphur dioxide preservative, not exceeding 450 parts per million. (Statutory Instruments: 1979 no. 752.)

The following method of preparation is suggested for burgers. Precool the meat and fat and break down through the coarse plate of the mincer. Transfer to mixer and whilst mixing add protein additive, rusk and spices, followed by water/ice. The last should be added

551

slowly, until the correct amount is taken up. Run the mass through a fine plate of a mincer and form into patties, usually $2\frac{1}{2}$ oz per piece. For small scale production these can be stamped out on a single burger, manually operated, machine. For large outputs sophisticated automatic units are available.

COSTING

The preparation of sausage is frequently carried out in existing retail premises, by the normal staff. Thus it may be difficult to establish a fair allocation for the cost, but it is desirable that some allowance should be made.

A costing sheet for the ingredients, based upon a given recipe could take the following form:

	Composition %	at per lb	Cost per 100 lb £ p
Meat with fat	65		
Filler—dry	12		
Spices, etc.	3		
Added water	20		
Casings per 100 lb sausage			
Cost of ingredients per 100 lb			

To the above should be added the cost of labour and the overheads. The latter would include such items as allocation of part rent, electricity, interest on capital investment in, and depreciation of plant. The simplest method is to collate these overheads and divide this figure by the output of sausages, over a convenient period, based upon, say per 100 lb of product. The labour cost should be arrived at by dividing the output into the wages directly attributable to such production.

Thus to arrive at the total cost the process would be:

$$\frac{\text{Ingredients} + \text{Labour} + \text{Overheads}}{\text{Total Output (lb)}} \times \frac{100}{1}$$

This would give the actual cost of the product per 100 lb. The following expresses the general method:

Cost of material per 100 lb of product, £ p

Add cost of production, i.e. $\dfrac{\text{Labour}}{\text{Total output (lb)}} \times \dfrac{100}{1} =$

Add $\dfrac{\text{Overheads}}{\text{Total output (lb)}} \times \dfrac{100}{1}$

$$\text{Total cost per 100 lb} \quad £ \underline{}$$

To this total must be added the desired profit.

Frankfurters

The Germans maintain that they should only be given this name if they come from Frankfurt or its immediate surroundings and consist of only lean pork and backfat. In the UK the legal standard for frankfurters requires a minimum meat content of 75% and up to 50% of the meat content will be accepted as fat. However, the frankfurter type of sausage has become international and might consist of the following ingredients:

Lean pork	25%	Spices per 100 lb	
Pork backfat	20%	Curing salt $\pm$ 2 lb (0·6% nitrite)	
Lean beef	15%	Ground white pepper	3 oz
Rind and fat		Ground mace	$1\frac{1}{2}$ oz
emulsion	15%	Ground coriander	1 oz
Ice/water	$\pm 25\%$	Garlic (optional—to taste)	

(Polyphosphate, 0·3% and a small amount of caseinate may be incorporated.)

Where a very fine texture is required a high speed emulsifier may be employed. Using a bowl cutter the following procedure is suggested.

Chop the rind and fat emulsion to a fine consistency, then add the beef followed by pork. Curing salt is added and if used, the polyphosphate and then the ice/water slowly incoporated, so that it is taken up by the mix. Where caseinate is used it should be added

553

at this stage, and then the spices. The pork backfat is added last and the chopping continued to give a smooth texture. Fill into 19 or 21 mm casings and twist into links. Subsequent processing may take either of the following methods.

1. Smoking to obtain the desired colour, *immediately* followed by cooking for 7 to 10 minutes in a vat at 77°C (170°F). The smoking and cooking being considered as part of a continuous heat process. Cooling may be by immersion in cold water or spraying.

2. Smoking/cooking in one operation with hot smoke. Some such units incorporate provision for subsequent cooling.

Black Pudding

The traditional method of preparing black pudding involves the pre-cooking of pearl barley, and in many cases the procurement of fresh pig's blood is difficult.

It is now possible to obtain a dry black pudding mix. This consists of dried red blood cells, plus pearl barley, oatmeal, fine rusk and spices, so that it is only necessary to add the diced pork fat and water. A typical procedure (courtesy Amasal) is as follows:

Dry black pudding mix	26%
Diced pork fat	11% (preferably flair fat)
Water	57%

Add warm water to dry mix slowly, whilst stirring. Allow to stand for about 30 minutes. Add diced fat and thoroughly mix and fill out into beef runners or synthetic casings. Cook in water at 80°C (176°F) for 40–50 minutes according to size.

Salami

There are countless varieties of salami and under modern sophisticated techniques the maturing time, several months in some cases, has been drastically reduced. This has been achieved by the smoking and drying cycles being carried out in temperature and humidity controlled cabinets.

The following is typical of a traditional Milano-type salami:

Fresh lean beef 30%
Lean pork 50%
Fat pork 20%

Cure per 100 lb mix
 (40 oz curing salt (0·6% nitrite))
8 oz sugar
4 oz ground white pepper
3 oz whole white peppercorns
1 oz fresh garlic

Beef broken down through a coarse plate of mincer, with garlic and then through a fine plate. Pork through a coarse plate. All the minced meat, cure and seasoning, thoroughly mixed and packed down in 6 inch layers at 4°C (40°F) until the mix has assumed a good red colour.

The meat may be remixed prior to filling into hot bungs, avoiding air locks. The salamis are hung to dry for about 36 hours and bound with a hemp twine about ½ inch apart. They are then held in a good circulation of air at a temperature of 45–55°F for about 9–10 weeks. (Frequently about half a pint of fresh red wine is added to the mix.)

Cervelat

Beef   36%
Lean pork   30%
Pork backfat 34% (firm)

Spices per 100 lb
 ± 40 oz curing salt (0·6% nitrite)
13 oz brown sugar
 5 oz ground white pepper
75 cc cognac

Meat pre-cured, pork and backfat diced into cubes, beef broken down on a fine plate of mincer (5 mm). Meat well mixed, spices etc. added and the mass passed through 5 mm plate. Mix packed down tightly in shallow trays, avoiding air pockets, and held overnight at 40°F to cure. Fill out tightly in synthetic or beef middles and hang for three days at a steady temperature not exceeding 16°C (60°F). Wash off in tepid water and immediately transfer to smoke chamber. A warm smoke is preferred of around 49°C (120°F), for a period of up to 12 hours. This is followed by cooling under a spray of water, followed by maturing for several days.

BRINES AND CURING

The main objectives with pickling/curing of meats are those of preservation, flavour and attractive colour.

The basic substances consist of salt, saltpetre (nitrate), nitrite and in some cases sugars, spices and polyphosphates may be employed.

Brine curing involves the immersion of pieces of meat for a pre-determined period in a brine solution. To supplement and accelerate the diffusion process, the meat may be injected with brine by using a perforated needle attached to a brine pump, or by introduction via the arterial system using a special needle. Many bacon factories employ automatic injection of pickle by multiple needles, adjusted to ensure the injection of a pre-determined quantity of pickle.

With the brine curing of portions of meat a large number of variables are involved. These include the size of the joint (thickness rather than weight), fat distribution, salt concentration and temperature. Some of the soluble protein is extracted by the brine, which affects the "binding" quality of the meat, an important factor in sausage manufacture. In addition the take up of the brine/salt is difficult to control accurately.

A general purpose brine for the retail business consists of:

10 gallons of water	6 oz saltpetre
20 lb salt	1 lb brown sugar

This will give a brineometer reading of about 60. This figure indicates the percentage of saturation of the brine solution. They are frequently calibrated at 16°C (60°F) and at lower temperatures the readings would be very slightly increased. Where practicable, maintaining a steady temperature of around 4°C (40°F) will eliminate one of the variables. Spices such as peppercorns, juniper berries, pimento and bay leaves are frequently employed in spiced brines.

As meat is taken from the brine, the brine becomes weaker and salt should be added to maintain the correct strength. As a rough guide, $\frac{1}{2}$ oz of salt per gallon of water, per degree rise required, can be added. Before immersion in brine meat should be soaked for a

556

short time in clean running water to reduce the number of surface organisms.

A new brine requires time to mature to ensure a satisfactory colour in the meat. This maturing develops in the following manner. The saltpetre (nitrate) under the action of de-nitrifying bacteria converts it to nitrite.

The nitrite and salt in solution penetrate the membranes of the meat by diffusion. The nitrite with the myoglobin in the meat produces a pink/red colour, which is retained on cooking.

Nitrates, Nitrites and Curing Salts

The Preservatives in Food Regulations: 1979 no. 752 came into force on the 31 July 1979. With reference to "Meat, cured (including bacon or ham)", sodium nitrate and sodium nitrite are "permitted preservatives" subject to the following: Sodium nitrate not exceeding 500 milligrams per kilogram (p.p.m.) and Sodium nitrite not exceeding 200 milligrams per kilogram (p.p.m.).

On the continent extensive use is made of salt containing nitrite. The quantities permitted vary slightly in different countries. In Western Germany there is a limit of $0\cdot5\%$ sodium nitrite in curing salts for meat products. In the Netherlands, sodium nitrite can be added *only* in the curing salts.

Curing Meat For Sausage

More accurate control of the curing ingredients is achieved by dry curing meats intended for sausage manufacture. In addition the meat juices are largely retained. Three methods are commonly employed.

1. Pieces of meat are thoroughly mixed with the dry curing mixture, packed down tightly in shallow trays, until the desired colour is obtained. Some operators prefer to break down the meat on a 1 inch plate of the mixer prior to curing to ensure uniform penetration of the cure and reduce the period necessary. In all dry curing methods the surface should be protected from the atmosphere by covering with grease-proofed paper.
2. The American "emulsion" method consists of breaking down

557

the meat on a $\frac{1}{4}$ inch plate followed by chopping in a bowl cutter with the quantity of cure appropriate to the mix. The ice or icewater is gradually added and the mass is then packed into pans, 6 inches deep and held at 3°C (38°F) to 4°C (40°F) for 12 or 24 hours.

3. In some cases the fresh meat is chopped with the curing ingredients, spices etc. and the sausages filled out and then hung for a period before smoking and/or cooking, to permit the meat to cure.

MEAT TUMBLING

This process is a modern development of the traditional method of rubbing pork meat with a curing mixture until it "sweats". It was probably not appreciated that the "sweat" consisted of salt soluble proteins, exuding to the surface of the meat. A tumbling machine ensures a consistent method of massaging the meat and is widely used in the pickling of pork prior to cooking. The meat is rendered soft and consequently more easy to mould, and on cooking, the coagulation of the protein will hold pieces of meat together.

Various procedures are in general use, including:

1. Meat and brine tumbled together until the brine is taken up. This system is used for small portions.
2. Injection of some brine, followed by maturing, prior to tumbling with additional brine.
3. Tumbling following brine injection, until the brine expressed is reabsorbed.

On account of the quantity of pickle and meat juices retained in the cooked product, the process has its attractions. Using a polyphosphate in the pickle, representative samples may contain about 20% more water than that associated with cooked Wiltshire-cured gammons. Obviously this has an influence on the dilution of the characteristic ham flavour and the moisture content may adversely affect the keeping quality.

COOKED MEATS

In all methods of cooking meat, the fat, on reaching the body

558

temperature of the animal will become liquid. This is in general about 37°C (99°F) and at this temperature very little is lost from the fatty connective tissue. As the temperature increases the connective tissue coagulates and increases the pressure on the fat cells, causing some of them to rupture with a consequent loss. If the fatty tissue is not damaged, the loss may be negligible up to about 80°C (176°F). With higher temperatures as in baking, considerable loss will occur.

With lean tissue a main factor is the coagulation of the protein, which starts at about 50°C (112°F), the maximum effect lying between 65°C (150°F) and 70°C (158°F), and with this a certain loss of water content and soluble protein will occur. In the case of ham/gammon cooking, an internal temperature of about 67°C (152°F) is considered to give a satisfactory product in terms of consumer acceptability, combined with reasonable weight loss.

As previously indicated it is the internal temperature which is the greatest single factor influencing moisture loss on cooking. At a given internal temperature *water* losses are approximately similar, whether they are cooked in water, water vapour or in a vacuum steam oven.

Cooking Gammons and Hams

One method widely used is to start at a relatively high temperature to seal the surface, the heating is then cut off and the cooking completed in a sealed insulated container, giving a gradual fall in the water temperature.

A test on ten gammons cooked in moulds gave the following results:

	lb	oz	% of the initial weight
Initial weight	151	8	
Bones and tendons	13	12	9·1
Cooking loss and evaporation	20	10	13·6
Rind and trim	14	2	9·3
Coldweight (after 24 hours)	103	0	68·0
	151	8	100%

In the above test the cooking period was 5¾ hours, with an initial water temperature of 82°C (180°F) dropping to 77°C (173°F), at the end of the period. The weight loss of an individual gammon can vary considerably. Whilst the average overall loss due to cooking, evaporation, was 13·6% of the initial weight the smallest loss recorded was 10·3% and the greatest 23%. The latter figure is abnormally high, the gammon being excessively fat.

New Processing Method For Hams

The process has been patented in a number of countries and is currently being used in France, Holland and Spain.

Basically the process involves the use of a rigid thermoplastic mould in which the cooking takes place and which replaces the final packaging for sale. The procedure can be summarised as follows.

The portions of cured ham are placed into thermoplastic moulds at a constant weight. This is effected to ensure that the structure of a normal ham is reconstituted. A rigid cover of the same material as the container is placed over the meat and the complete unit is then capped under vacuum. When the unit hermetically sealed is exposed to the air it subjects the ham to heavy pressure and it remains thus throughout the cooking and cooling operations. After cooling the pack can be marketed without any further treatment. The advantages claimed for this process include the production of hams at a constant weight and retention of meat juices and flavour, with the compression giving good slicing characteristics; recontamination after cooking is obviated, which provides good keeping qualities in cool storage; a saving in cost normally incurred by the purchase of metal moulds; and the elimination of pressing, turning out and washing such containers and wrapping hams.

Beef Tongues

Ox tongues may be obtained as "long cut" with the root and some fat attached. "Short cut" with the major part of the root removed or as "blades", consisting of the tongue only. Obviously the yield of the cooked tongue will be greatly influenced by the type of tongue purchased.

Tongues should be first cleansed in cold running water to remove any surface slime, prior to salting. Pumping is preferably carried out by injection of pickle via the lingual arteries or by the use of a perforated needle. This is followed by immersion in cover brine for 5 to 6 days to allow them to mature. (In many wholesale centres ready cured ox tongues are available.)

Trim the root as necessary and cook in water, commencing at boiling point and reducing to 190°F (88°C), cooking for about 4 hours, according to size. Following cooking, the thick mucous cover and the two hyoid bones can be easily removed. When packing in a mould the surface of the tongue(s) should be placed to the outside and the interior filled with the root end. During the handling process a certain amount of surface contamination may occur and immersion in the hot cooking water for a few minutes is desirable. This will provide some degree of surface sterilisation and facilitate packing. Crevices should be filled with a good gelatine glaze prior to pressing and cooling.

Pigs and lambs tongues may in general be treated in a similar manner, although normally these are not arterially pumped and the period in brine and the cooking period is reduced, on account of their size.

In costing cooked products, assuming a gross profit of $33\frac{1}{3}\%$ on sales, the following method can be employed: **Costing**

$$\frac{\text{“P” pence}}{1} \times \frac{\text{original weight}}{\text{weight as sold}} \times \frac{3}{2} = \text{Sale Price in pence per lb}$$

“P” pence is the original cost price. If the required gross profit on sales is 25% the multiplying factor is $\frac{4}{3}$ and for 50% the corresponding factor is $\frac{2}{1}$.

Technical Education and Training in the Meat Industry

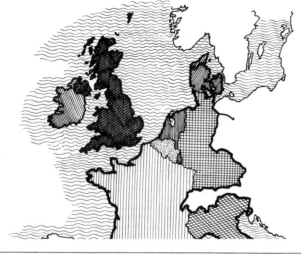

From the earliest records of life, it appears that man has always considered meat, from many species of animals, not only as a source of food, but also as an essential in the diet to ensure growth and maintenance of good health.

As civilisation advanced, the hunting of wild animals for food—as opposed to hunting for "sport"—became impractical and uneconomic as a means of providing the increasing demand for meat and its associated products, particularly in the fast growing areas of concentrated urban population.

Advances in the domestication and evolution of certain animal

HISTORY

563

species to provide meat and by-products naturally went hand in hand with the need for some regular control of production, slaughtering, marketing and retailing of meat.

Education and training within the meat industry as we know it today is the outcome of farseeing, enlightened efforts of butcher craftsmen and traders of the past.

In this country, particularly in the City of London, the earliest records concerning control in and over the meat trade date back almost a thousand years. A study of the historical records of the Worshipful Company of Butchers shows that whilst the actual date of its formation is unknown, it is recorded in Strypes' edition of Stows' *Survey of the Cities of London and Westminster* that "in the year 975 AD, in the Ward of Farringdon, without the City walls, there are situated divers slaughterhouses and a Butchers' Hall where the craftsmen meet". Whilst it is true to say that it is many generations past since the Worshipful Company of Butchers' (London) and other similar companies throughout the country have exercised direct control of the meat industry, it is interesting to recall that originally they controlled livestock and meat prices, day to day trade practices, the flow of entrants to the trade, the indenture of apprentices, and set standards of conduct of all within the trade. A record of the Worshipful Company of Butchers dated 3 August, 1687, states "John Well was summoned to show how he claimed the right to trade as a butcher, and produced a seven years' indenture to his father, but the court did not believe him, and ordered him to be arrested". Today, many of these activities are controlled by government, local and/or central; others have been allowed to lapse altogether.

Guilds of Butchers

There is no doubt that one of the most important functions of the Guilds of Butchers was the indenture of apprentices and the supervision of not only their craft training, but also their physical and moral welfare. Although it should be recognised that apprenticeship in the meat industry is still available under the auspices of the Worshipful Company of Butchers, there has been a decline of the apprenticeship system (and other industries have suffered similarly). Generally nowadays it is under the National Joint Apprenticeship

564

Council for the Retail Meat and Pork Butchery Trades and National Joint Apprenticeship Council for Slaughtermen.

A number of efforts were made to provide training in the various branches of the trade; however it was not until 1923 in the Smithfield Meat Trades Institute, London, that the first successful effort appeared. It soon became apparent that this was no spasmodic effort and on such a modest foundation was laid an educational scheme which was to spread throughout the British Isles and even place its imprint on educational establishments throughout the world. Following the successful establishment of Meat Trades classes in London, the executive of the National Federation of Meat Traders Associations Incorporated prepared a syllabus of instruction and this, with suitable amendment, continued until 1951 when it was mutually agreed by all interested bodies that the Institute of Meat should become the educational examining body.

The Institute of Meat, formed in 1946, is the national meat industry organisation concerned with technical education and training, work which is of vital importance in raising the standard, and with it, the stature of every section of the meat industry. It is the only national meat industry organisation representing everyone in or closely associated with every section of the industry—primary producers, slaughterhouse operators, importers, wholesalers, processors, retailers, veterinary officers, meat inspectors, research workers, as well as those who instruct and teach. In other words, it is the industry's "professional" body where people from every part of the trade meet on common ground with common interests. As with other professional institutes and organisations, there are grades of membership related to qualifications and experience; these are student, associate, affiliate, member and fellow.

Institute of Meat

There is no comparable organisation in Europe or indeed the world. The Institute of Meat recognises no geographical boundaries and members are to be found in all parts of the world and its aims and objectives are educational and professional.

The education and training system of the Institute of Meat takes into account that young people studying today will, on graduation, become full members of the industry during the next few years and will be participating in the industry for some forty years hence. They must therefore be trained in such a way as to enable them to be adaptable to the many changes that are certain to take place.

Adaptability is the key to all training, even in specialist courses, for the man trained in a narrow field specifically for the requirements of today, may be unsuited to the industry of tomorrow. Another important factor in the preparation of syllabuses is their application to the meat industry for, if a student's interest is to be firstly stimulated and then maintained, all subjects taught must and should be as applied subjects to ensure that their application to industry is abundantly clear and its fundamentals well understood.

Main Courses

A three tier system of courses and examinations is operated, each course separate and distinct in itself, yet forming a basis for future higher study.

The present examination structure is:

> Affiliateship
> Associateship
> Membership

and successful students qualify for the appropriate grade of membership of the Institute of Meat.

The syllabus subjects for each of the three courses are based on the modular concept, having common modules or units plus specialist optional modules to meet the requirements of each industrial sector of the meat industry. The student may, therefore, select the group of modules or units best suited for his/her career requirements.

The *Affiliateship* course is designed for young new entrants and aims "To provide the student with an introductory study of the meat industry in terms of its structure, raw materials and processing which will equip the student for skilled craft or higher operative

566

AFFILIATESHIP MODULES

COMPULSORY Section	OPTIONAL Section
1A The Meat Industry.	1D(A) Abbatoir Organ'tn.
1B Calculations. Communications and Documentation.	1D(M) Manufacturing Plant Organisation.
1C Meat Science.	1D(R) Retail Shop Organ'tn.
	1D(S) Supermarket Depot Organisation.
	1D(W) Wholesale Depot Organ'tn.
	1D(CB) Catering Butchery Organisation.
	1D(MC) Meat in Catering Organisation.

Plus:

One module from section 1D and one module from 1E.

PRACTICAL

1E(A)	Abattoir
1E(M)	Manufacturing
1E(R)	Retail
1E(S)	Supermarket
1E(W)	Wholesale
1E(CB)	Catering Butchery
1E(MC)	Meat in Catering

ASSOCIATESHIP MODULES

COMPULSORY Section	OPTIONAL Section
2A The Meat Industry.	2E(A) Abbatoir Super.
2B Business Organ'tn.	2E(M) Manufacturing Plant Super.
2C Meat Technology.	2E(R) Retail Shop Supervision.
2D Anatomy & Physiology of Meat Animals.	2E(S) Supermarket Supervision.
	2E(W) Wholesale Depot Supervision.
	2E(CB) Catering Butchery Supervision.
	2E(MC) Meat in Catering Supervision.
	2F Accounts.
	2G Chemistry & Microbiology.
	2H Meat Hygiene & Inspection.

Plus:

One Module from Section 2E. One module from either 2F–2G–2H and one from 2J.

It is a requirement that Module 2D shall have been taken prior to or in conjunction with Module 2H.

PRACTICAL

2J(A)	Abattoir
2J(M)	Manufacturing Plant
2J(R)	Retail Shop
2J(S)	Supermarket
2J(W)	Wholesale Depot
2J(CB)	Catering Butchery
2J(MC)	Meat in Catering

MEMBERSHIP MODULES

COMPULSORY Section	OPTIONAL Section
3A The Meat Industry.	3D Meat Marketing & Dist.
3B Principles of Management.	3E Administration.
3C Meat Technology.	3F Food Chemistry.
	3G Meat Hygiene.
	3H Meat Processing & Manufacturing.
	3J Meat Factory & Abattoir Op.
	3K Food Engineering.

THE MEMBERSHIP EXAMINATION REQUIRES CANDIDATES TO SIT *THREE CORE SUBJECTS AND TWO SPECIALIST* SUBJECTS.

responsibilities and make him or her aware of adequate safety and hygiene practice."

Specialised studies in the abattoir, wholesale, retail, catering and processing sectors are available. Successful candidates may continue and extend their studies through the Associateship course.

The *Associateship* course aims "To provide the student with knowledge of basic business and management studies and of a more detailed meat technology appropriate to the wider responsibilities of a supervisory nature. It is also intended to lay a foundation for the more advanced studies in management and meat technology provided in the Membership course."

The *Membership* course is designed to:

(a) Enable those associated with the meat industry to gain a senior qualification as a result of an intensive course of study.

(b) Give the student an appreciation of modern principles and techniques used by management with specific relevance to all sectors of the meat industry.

(c) Encourage the student to develop a positive attitude towards management and to help people apply knowledge, skills and understanding gained from a study associated with a business environment.

(d) Develop skills in communicating, both in dealing with people and with the problems linked with a business organisation.

The schedules for the three course levels are shown on p. 567.

Other Courses
Short, specialised and refresher type courses are also offered by the Institute as supplementary to the main educational and training schemes. Some forty-three colleges throughout the UK offer one or more of the Institute of Meat courses on full-time, or more generally, part-time basis.

Successful candidates qualify for an appropriate grade of membership of the Institute of Meat.

Additional to the Institute of Meat educational and training programmes and with the introduction of the Industrial Training Act 1964, the Food, Drink and Tobacco Industry Training Board was formed in 1968. The Board (composed equally of employer, employee and some educational members) in close co-operation with the Institute of Meat, has developed the Meat Modular Training Scheme, a simple "on the job" system of training for manual skills in the meat industry. The system can be adapted to individual company training methods and requirements—including local variation—in butchery, slaughtering and manufacturing.

The Institute of Meat is the Board's appointed body for maintaining a register and the issuing of certificates when particular "modules" of training are successfully completed.

The Board requires that company managers and instructors attend short courses of briefing and instruction at selected colleges throughout the country before using the scheme.

It is but fair to mention that before the advent of this Food, Drink and Tobacco Industry Training Board, individual firms operated training schemes for their young employees, some successfully. The Industrial Training Act 1946 makes training compulsory for all companies within leviable scope of a particular Training Board and some twenty-seven Boards exist, covering between them the vast majority of industry in England, Wales and Scotland.

From the foregoing it can readily be seen that education and training has progressed with the growth and development of the industry from earliest times and, in keeping with the increasingly greater technology now found, education and training schemes in many cases have reached an advanced level of sophistication. In this enlightened age, all engaged within whatsoever sector they operate, from the lowest to the highest, may become qualified to a degree limited only by themselves.

In the final analysis, education and professionalism in their broadest sense should be the ultimate aims of all engaged in this great industry and these aims are exemplified in the Institute of Meat motto "Per

Doctrinum ad Dignitatum"—Through knowledge to Status, and that surely is the desire of all.

Reference:

The Institute of Meat, Boundary House, 91–93 Charterhouse Street, London E.C.1.

The Food, Drink and Tobacco Industry Training Board, Barton House, Barton Street, Gloucester GL1 1QQ.

National Joint Apprenticeship Council for the Retail Meat and Pork Butchers Trades, Boundary House, 91–93 Charterhouse Street, London E.C.1.

The Worshipful Company of Butchers, The Butchers' Hall, 87 Bartholomew Close, London E.C.1.

National Federation of Meat Traders Association Limited, 29 Linkfield Lane, Redhill, Surrey.

Information about the Apprenticeship Scheme for Slaughtermen can be obtained from—

J.I.C. Employers' Secretary:
Mr. R. J. Lickorish, Federation of Fresh Meat Wholesalers, Hart's Corner, Smithfield, London EC1A 9AA.

J.I.C. Employees' Secretary:
Mr. W. H. P. Whatley, Union of Shop, Distributive & Allied Workers, 188 Wilmslow Road, Manchester M14 6LJ.

The Institute of Meat runs Courses at the following forty-three centres:

The College of Business Studies, Brunswick Street, Belfast, N. Ireland.

Birkenhead Technical College, Borough Road, Birkenhead L42 9QD.

Birmingham College of Food and Domestic Arts, Summer Row, Birmingham 3.

Blackpool College of Technology and Art, Palatine Road, Blackpool, Lancs.

Bolton Technical College, Manchester Road, Bolton BL2 1ER.

Brunel Technical College, Ashley Down, Bristol BS7 9BU.

570

Burnley Municipal College, Ormerod Road, Burnley, Lancs.

South Glamorgan Institute of Higher Education, Colchester Avenue, Cardiff CF3 7XR.

North Gloucestershire College of Technology, The Park, Cheltenham, Glos. GL50 2RR.

Chesterfield College of Technology, Infirmary Road, Chesterfield S41 7NG.

Llandrillo Technical College, Llandudno Road, Rhos-on-Sea, Colwyn Bay, Clwyd, North Wales.

North East Wales Institute of Higher Education, Kelsterton College, Connah's Quay, Clwyd CH5 4BR.

Derby College of Further Education, Harrow Street, Wilmerton, Derby.

College of Technology, Cleveland Avenue, Darlington, Co. Durham.

Derby College of Art and Technology, Kedleston Road, Derby DE3 1GB.

Guildford County College of Technology, Stoke Park, Guildford, Surrey.

North Herts. College, Cambridge Road, Hitchin, Herts. SG4 0JD.

Huddersfield Technical College, New North Road, Huddersfield HD1 5NN.

Hull College of Technology, Queens Gardens, Kingston-upon-Hull HU1 3DG.

Thomas Danby College, 5 Roundhay Road, Sheepscar, Leeds, Yorkshire.

South Fields College of Further Education, Aylestone Road, Leicester.

Colquitt Technical College, Dept. of Food Technology, Colquitt Street, Liverpool L1 4DB.

College for the Distributive Trades, Briset House, Briset Street, London EC1M 5SL.

Hollings Faculty, Manchester Polytechnic, Old Hall Lane, Manchester M14 6HR.

Isle of Wight College of Arts and Technology, Dept. of Food, Fashion and Health, Medina Way, Newport, I.O.W.

Norwich City College, Ipswich Road, Norwich, Norfolk NOR 67D.

Clarendon College of Further Education, Pelham Avenue, Nottingham NG5 1AL.

Oxford College of Further Education, Oxpens Road, Oxford OX1 1SA.

Peterborough Technical College, Park Crescent, Peterborough, Northants.

Plymouth College of Further Education, Kings Road, Devonport, Plymouth, Devon PL1 5QG.

Reading College of Technology, Kings Road, Reading, Berks. RG1 4HJ.

Salford College of Technology, Frederick Road, Salford, Lancs. M6 6PU.

Granville College of Further Education, Granville Road, Sheffield S2 2RL.

Southampton Technical College, St. Mary Street, Southampton SO9 4WX.

North Staffordshire Polytechnic, College Road, Stoke-on-Trent ST4 2DE.

Cauldon College, The Concourse, Stoke Road, Stoke-on-Trent, Staffs.

College of Further Education, Tycoch, Swansea SA2 9EB.

South Devon Technical College, Newton Road, Torquay, Devon TQ2 5BY.

Trowbridge Technical College, College Road, Trowbridge, Wilts.

Wulfrum College of Further Education, Paget Road, Wolverhampton, West Midlands.

MEAT TRADE SCHOOLS (other than UK)

BELGIUM

C.E.R.I.A., Avenue Emile Gryzon 1, Brussels B.1070.

Institut Libre des Metiers de la Viande, Rue Sergent de Bruyne 13a, Brussels.

FRANCE

Ecole Superieure des Metiers de la Viande, 37 Boulevard Soult, Paris XII eme.

HOLLAND

Institute Slagersvakonderwijs, Pompelaan 8, Utrecht.

SWITZERLAND

Schweizerische Fachschule fur das Metgereigewerbe, Hotel Belvedere, Spiez.

AUSTRALIA

Wm. Angliss Meat Trade School, Melbourne.

East Sydney Technical College, Forbes Street, Darlinghurst, Sydney, N.S.W.

Queensland Agricultural College, School of Food Studies, Lawes, Q4345, Queensland.

SOUTH AFRICA

School of Meat Cutting, Private Bag 23, Johannesburg.

DENMARK

Meat Trades School, Stagteristozen, Roskilde, Denmark.

IRELAND

School of Commerce and Retail Distribution, 18 Parnell Square, Dublin 1.

GERMANY

C/o Innung fur das Schlachter und Gross Schlachter Handwerk, Marktstrasse 57, 1, Hamburg 6.

Employment and the Law

PER·DOCTRINAM·AD·DIGNITATEM

The term "Contracts of Employment" emphasises that the relationship of employer and employee is a voluntary one governed by the law of contract.

The 1972 Contracts of Employment Act is an Act which every employer and employee should be aware of and in its amended form it covers both the employers' and employees' rights of minimum periods of notice to terminate employment. In addition, it lays a duty on employers to give their employees—of any age—written particulars of their main terms of employment.

In connection with the rights to minimum periods of notice of **Minimum Notice**

575

termination of employment the Contracts of Employment Act states that an employer is required to give an employee:

a. At least 1 week's notice if the employee has been employed continuously for 4 weeks or more.
b. At least 2 weeks' notice if the employee has been employed continuously for 2 years or more.
c. At least 3 weeks' notice if the employee has been employed continuously for 3 years or more.

This ruling applies so that there is one additional week's notice for each further complete year of continuous employment up to twelve weeks' notice if the employee has been employed continuously by the same employer for twelve years or more. Thus, ten weeks' notice for ten or more years' service; eleven weeks' notice for eleven or more years' service.

It is important to notice that these are *minimum* periods of notice and many firms do offer longer periods of notice. Therefore, if a contract gives as much notice as this or longer than the minimum in the Act, then the Act will not alter this. Many monthly paid employees have laid down in their contract that after the first 3 months of probationary employment either party must give the other 1 month's notice in the event of termination of employment. But, if a contract offers less than prescribed by the Act, then the Act will change this, laying down the legal minimum requirements.

The Act also states that an employee is required to give at least one week's notice if he has been with his employer continuously for 4 weeks or more. This does not increase with longer service.

Employees

The minimum periods of notice cause some employers a certain amount of confusion because of the meaning of the word "employee". There are several kinds of employees that are not covered by the provisions of the Act and as far as the meat trader is concerned there are basically two exceptions to the provisions under the Act. Firstly, the Act does not cover part-time employees; that is, those who normally do less than 16 hours a week for their employer.

576

It should be noted that a full-time employee can also be someone who does not less than eight hours a week for the same employer for five years or more. The minimum periods of notice do not apply to certain employees if they are on fixed term contracts. If it is agreed at the outset that a contract of employment is to last for a fixed period, e.g. a year, then the rights to notice will not generally apply. Both the employer and the employee will know from the start when the contract of employment is to terminate.

Continuous Employment

Confusion also arises over the word "continuously" in connection with length of employment. What is meant by "continuous employment"? Does it include strikes and sickness?

Any week counts in which the employee is employed for 16 hours or more by the employer. Any week also counts if in any part of it an employee is covered by a contract with his employer which normally involves employment for 16 hours or more. This means that so long as the contract continues, a week will count even if the employee does not actually work as much as 16 hours in it. Therefore, spells of absence through sickness of whatever length, temporary lay-offs, and holiday breaks all count as part of the period of continuous employment. An occasion might arise when an employer wonders whether or not he has a unique case in what constitutes "continuous employment" in the case of a particular employee. He will find that his local Department of Employment should be able to assist him. Certain regional offices of the Department have Senior Manpower Advisers who are available to answer questions on the Act. Authoritative interpretations of the Act can only be interpreted by the Courts.

Termination of Contract

It is important to note that the provisions do not prevent either the employer or employee from waiving his rights to notice or from accepting payment in lieu of notice. For example, in a company where sales representatives are employed, if a representative hands in his resignation and he is on a month's notice the company often asks him to leave immediately and simply gives him a month's pay. This is done possibly for two reasons. Firstly, the company thinks that the representative's heart is no longer in the job and if he stayed to serve out his notice he may waste the month away driving a company car round the countryside. Secondly, he may "poach" information ready for use in his new employment. Also, if the employee is anxious to leave to start a new post the employer may also waive the right to notice if say he, the employer, has someone ready and waiting to fill the vacant position. Where the employment is terminated because of redundancy and a payment is to be made to the employee under the Redundancy Payment Act 1965, advance notice must be sent to the local Department of Employment by the employer. The problem of redundancy will be dealt with later in the chapter.

Summary Dismissal

There are certain circumstances in which either the employer or employee can terminate the contract without notice if the behaviour of the other party justifies it. For example, the employer could dismiss the employee summarily and thus terminate the contract without notice for embezzlement or where the employee simply refuses to do the work for which he has been employed. The justification for this depends on the particular circumstances and this can be determined only by the Courts.

The question also arises regarding the guarantee of minimum pay during notice. There are guarantees to meet special circumstances. The guarantees do not affect an employee who is working throughout his normal working hours during the period of notice. An employee who does this simply receives what he earns. However, an employee's pay is safeguarded by the Act during the minimum period of notice, if:—

1. he does not work during some or all of his normal working hours because his employer provides no work for him although he is ready and willing to work, or
2. he is incapable of work through sickness or injury, or
3. he is on his holiday, or
4. he has no "normal working hours" like, for example, a sales representative.

578

An employer or employee who considers that he has sustained a loss because he has not been given the correct period of notice (or where an employee is concerned, the minimum pay), can bring an action for damages in the Courts, if he cannot obtain satisfaction by other means. Taking somebody to Court under such circumstances is usually regarded as a last resort.

Written Statement

The Contracts of Employment Act 1972 also deals with the *written* terms of employment an employer should give each employee. Confusion often arises over this part of the Act as many employers tend to think that by the very nature of the title of the Act there must be some contract, such as a fixed term contract drawn up between the two parties. This is not the case. This part of the Act requires that every employer must give every employee a *written* statement containing terms of his employment. These are the terms that are agreed between the employer and employee (and in many cases taken from a collective agreement from the industry) and must be sent to the employee not later than 13 weeks from the date employment began. Thus the statement may show its first sentence as "This statement sets out particulars of the terms and conditions on which I (name of employer) am employing you (name of employee) at (place or work). The employment began on (date)". Statements do not have to be given to part-time employees (those normally on less than 16 hours a week). Statements do not have to be given to a husband or wife of the employer.

Rates of Pay, Pensions, Sick Pay

The written statement should contain:

1. The scale or rate of remuneration, or the method of calculating remuneration e.g. £2 per hour; £80 per week. It should include any terms on such matters as overtime pay and piece rates.

2. The intervals at which payment will be made, e.g. weekly paid; annual salary of £4000 divided into twelve equal parts payable at the end of each month (in arrears).

3. Terms and conditions relating to hours of work, e.g. Basic pay

is to cover a 40-hour week exclusive of 1 hour for lunch each full working day.

4. Any terms and conditions relating to:

a. Pensions and pension schemes.
b. Incapacity for work due to sickness or injury. For example a firm might offer the following:

Employee's length of service	Full pay	Half pay
Less than 6 months	1 week	2 weeks
6 months–2 years	2 weeks	4 weeks
2–3 years etc.	3 weeks	6 weeks

Holiday Pay

c. Entitlement to holidays, including public holidays and holiday pay (sufficient to enable the employee's entitlement including any entitlement to accrued holidays on the termination of employment to be precisely calculated).

e.g. new employees holiday entitlement during the year in which they join a firm:

Before 30th June	2 weeks
1st July–30th September . . .	1 week
1st October–30th December . . .	Nil

But what happens, say, if an employee is entitled to three weeks' holiday, has taken six days and leaves at the end of August. How much holiday pay do you give? The way this is arrived at is by senior management decision but the following is an example:—

Holiday entitlement: 18 working days per year. When terminating employment $1\frac{1}{2}$ days pay for every month worked will be paid in lieu of holiday not taken. Therefore, if an employee has taken 6 days he would be entitled to $12 \times 1\frac{1}{2}$ days pay if he left at 31 December, but in this case he left at the end of August, thus, he is three months short, which is $3 \times 1\frac{1}{2}$ days pay $= 4\frac{1}{2}$ days' pay. If $4\frac{1}{2}$ is then deducted from 12 we find that he is entitled to $7\frac{1}{2}$ days' pay.

d. The length of notice of termination which the employee is obliged to give and entitled to receive.
e. Job title. The employer must give the employee the title of the job which he is employed to do.
f. Disciplinary rules. The employer must specify any disciplinary rules (other than those relating to health and safety at work which are covered by the Health and Safety at Work Act 1974) which apply to the employee or refer to a document reasonably accessible to the employee.
g. The employer must specify by description or name the person to whom the employee can apply and the manner in which such applications should be made:
  i. if he is dissatisfied with any disciplinary decision relating to him.
  ii. for the purpose of seeking redress of any grievance relating to his employment.
h. State whether a contracting out certificate under the Social Security Pensions Act 1975 is in force for the employment in respect of which the written statement is being issued.

There is no prescribed form for giving this information. However, employers and employees could obtain copies of the "Contracts of Employment Act 1972—A Guide to the Act" published free of charge by the Department of Employment. Model written statements are shown in this guide.

Most written terms of employment go well beyond the requirements and give such information as company expenses, use of company vehicles on business, retirement, acceptance of gifts and much more information linked with company conditions.

All the written information has to be kept up to date. If there is a change in the terms, the employer has to inform employees about it not more than one month after its introduction by means of a further written statement. It is worthy of note that the written statement itself does not have to set out all the required information. The written statement may refer the employee to a document or documents of which he has reasonable access to read during the course of his employment. For example, the statement may say,

"For the conditions and benefits on sick pay schemes refer to the company booklet entitled '*Sick Pay—Conditions and Benefits*' available from the Personnel Department".

Before moving on it should be borne in mind that the Contracts of Employment Act 1972 has been embodied in the Employment Protection (Consolidation) Act 1978 which includes as well the Redundancy Payments Act 1965; the Trade Union and Labour Relations Act 1974 and 1976; and the Employment Protection Act, 1975.

Industrial Tribunals

An employee who is dissatisfied because no written particulars have been given to him may require the matter to be referred to an industrial tribunal. This is instigated in the first instance through the Department of Employment. The tribunals can only determine what particulars of the main terms of employment ought to have been given in a written statement (including a further statement). It should be borne in mind, however, that the jurisdiction of the industrial tribunals over the written requirements of the Act began only on 6 December, 1965 and they have no jurisdiction to deal with complaints of failure to give any written statement of the main terms of employment in cases where the employment concerned began before 7 September, 1965.

REDUNDANCY PAYMENTS ACTS 1965 and 1969

The implifications of the Contracts of Employment Act must take into account the considerations of the redundancy payments scheme. The Redundancy Payments Act 1965 and 1969 require employers in certain circumstances to pay redundancy payments to employees who are dismissed. The amount of payment is related to pay, length of service and age, and includes all employees (as far as the meat trader is concerned) except:

a. Those under 18.
b. Those with less than 104 weeks continuous service.
c. Employees who work less than 16 hours a week.
d. Employees whose reckonable service ends on or after the age of 65 for men and 60 for women.

582

It is often noticed that employers ask those men over 65 to leave first where circumstances require it to avoid making redundancy payments. Or they ask those who have been there less than 104 weeks to leave, again to avoid payments.

The Department of Employment can tell an employer on application by reference to a chart how many weeks pay entitlement a redundant employee is to be given. For example, a person of twenty who has been employed in the same firm for two years is entitled to one week's pay. A man finishing his 64th year who has been employed for 20 years (that is the maximum number of years) is entitled to 30 weeks' pay. The maximum that any employer could pay will depend on the maximum weekly pay allowed.

The rate of payment *for each complete year of service* is as follows:

1. From age 41: $1\frac{1}{2}$ weeks' pay.
2. From age 22 (excluding years at 1): 1 weeks' pay.
3. From age 18 (excluding years at 1 and 2): $\frac{1}{2}$ week's pay.

A week's pay *maximum* is £100 at present. If an employee is past his 64th birthday when his employment ends (59th for women) the payment is reduced by 1/12th for each complete month between that birthday and the Saturday of the week in which the employment ends.

How Redundancy Pay Is Calculated

The Act establishes a Redundancy Fund financed by contributions collected from the employers' flat-rate National Insurance contributions. Employers may claim a rebate of 41% of the cost from the fund. However, the rebate must be claimed within 6 months of the date on which the redundancy payment was made.

Redundancy Fund

An employer who is intending to dismiss employees because of redundancy must, where possible, give advance notice to the nearest employment exchange, otherwise the amount of rebate may be affected. If the employer is insolvent and the Department of Employment are satisfied that the employer's claim is correct, they will make payment direct to the employee from the Redundancy Fund and seek recovery from the employer as an unsecured creditor. If the Department are not satisfied that the claim is correct, application to a tribunal may be advised.

No income tax is normally payable by a person who receives a redundancy payment. Contributions to the Redundancy Fund made by an employer are allowable for tax purposes as a business expense. Redundancy payments do not normally affect the entitlement to unemployment benefit.

SEX DISCRIMINATION

The Sex Discrimination Act came into force on 29 December, 1975. Under this Act it is now unlawful to treat anyone less favourably on the grounds of sex difference. A person of one sex must be treated in the same way as one of the opposite sex when the same circumstances apply.

The Sex Discrimination Act specifically states that sex discrimination is not allowed in employment, education, advertising and in the provision of housing, goods, facilities and services. In employment it is also unlawful to discriminate because a person is married.

There are a number of exceptions to the rule on sex discrimination. These apply where the employer has no more than five employees, or if employment is in a private household, or where pension schemes or the age of retirement is concerned. Another exception to the rule is where there is a GOQ—a genuine occupational qualification. For example; in acting or modelling, the sex of the person concerned is rather important! An employer may specify the sex where reasons of privacy or decency are prevalent as, for example, the job of toilet attendant. It should be emphasised that this Act is *not* for the protection of women only, it applies to both men and women.

The Act distinguishes between two kinds of sex discrimination. Firstly, there is *direct* discrimination which involves treating a woman less favourably than a man because she is a woman. Indirect discrimination means that conditions are applied which favour one sex more than the other but which cannot be justified. For example, if

584

an employer in recruiting clerks insists on candidates being 6 feet tall, a case *may* be made out that he is unlawfully discriminating.

As regards employment, an employer in the recruitment or treatment of a person must not discriminate against an employee because of his or her sex. For example, an employer cannot apply a condition or requirement which has the effect of discriminating against married people (or vice versa) because considerably fewer married people can comply with it. Similarly an employer cannot discriminate against a woman with regard to her having access to the opportunities for promotion or training, simply because she is a woman. The employer may not make arrangements for deciding who should be offered a job, for example, by way of the instructions given to a personnel officer or to an employment agency.

As for advertising, words which are indicative of a particular sex will be discriminatory unless they contain a comment that both men and women may apply for the job. Such words as "Waiter" or "Postman" will not be allowed unless the advertisement specifically indicates that both men and women are eligible. Since the Act came into force some amusing advertisements have come into being. For example, "Bricklayer Required. Must be prepared to work topless in the summer months". Even words like "engineer" imply a man and the word "secretary" implies a woman. Difficulties may be experienced here. If an advertisement is illegal—and publishers as well as advertisers take care to see that advertisements do conform to the law—then the Equal Opportunities Commission can bring proceedings. Only the EOC (Equal Opportunities Commission) can bring proceedings in matters to do with advertising, an individual cannot do this. The employer can be fined up to £400. The EOC has been created to ensure the effective performance of the Sex Discrimination Act *and* the Equal Pay Act.

If an employer contravenes the Act the EOC may serve a Non-Discrimination Notice which will require the employer to make changes and to provide any specified information. The EOC must first warn an employer at least 28 days before a Non-Discrimination Notice is served.

Advice on sex discrimination can be obtained from the Equal Opportunities Commission, Overseas House, Quay Street, Manchester M3 3HN. Advice can also be obtained from the local employment office, Job Centre or the Citizens Advice Bureau. Employers may find that for the employment provisions of the Act, the Advisory Conciliation, and Arbitration Service may be of some assistance.

EQUAL PAY

The Equal Pay Act was passed in 1970 but it did not come into effect until 29 December, 1975. It allowed this transitional period because the introduction of equal pay can affect many aspects of a company's activities. It should be emphasised that authoritative interpretations of the provisions of the Act can be given only by courts of law, and in particular by the industrial tribunals and the Industrial Arbitration Board to which, in due course, disputes arising under the Act may be referred. However, the Employment Protection Act 1975 has set up the Central Arbitration Committee (CAC) which will take over the functions of the Industrial Arbitration Board in connection with, for example, equal pay.

The Equal Pay Act establishes the right of men and women to equal treatment as regards terms and conditions of employment. This is achieved in two main ways:

1. By establishing the right of the individual woman to equal treatment when she is employed on work of the *same or broadly similar nature* to that of men.
2. Where women are in a job which, though different from those of men, has been given an equal value to men's jobs under a job evaluation exercise.

This Act applies to men and women. So where women are doing broadly similar work to men and are more favourably treated by, for example, being given longer tea breaks, then this difference should have been removed by 29 December, 1975.

The Act covers all full-time and part-time employees, both male and female, in all manufacturing, services, and all kinds of commercial

586

activity. There are exceptions. The general provisions of the Act do not apply to the Armed Forces and the Police. Other exceptions include pensions and where there are statutory controls over women's hours of work.

It is to be stressed that lower rates may be paid to, say, young workers, provided that age for age, they do not distinguish between the sexes.

The comparisons which a woman may draw with men or with men's jobs are limited to men employed by her employer or an associated employer. If an employer has a female van driver and she is the only driver he employs then there can be no comparison with other jobs in the firm. The female van driver has no case if she complains that a competitor in the town pays his van drivers (be they male or female) a higher rate. The emphasis is on jobs within the company. Where the employer has other establishments in the country women may claim equal treatment with men at these other establishments provided that they observe common terms and conditions of employment either for all employees or for particular classes of employees. So where men and women are doing broadly similar work but the men work either a shift system or do night work then extra payments can be justified for the men so long as it is not based on the sex of the worker.

I have mentioned that men and women are entitled to equal treatment on work which, though different from men's work has been given equal value under a job evaluation scheme. Job evaluation is defined as being a study undertaken with a view to evaluating, in terms of demand made on a worker under various headings (e.g. effort, skill, responsibility, danger), the jobs to be done by all or any of the employees in an undertaking or group of undertakings. The Act does *not* specify that job evaluation studies have to be carried out.

Detailed advice on the Equal Pay Act can be obtained from the Equal Opportunities Commission, the Department of Employment, or the Conciliation and Arbitration Service.

EMPLOYMENT PROTECTION

Employment Protection includes Amendments to Contracts of Employment Act.

The Employment Protection Bill became an Act in November 1975 and its provisions come into force over several months.

This Act covers all full-time and a number of part-time employees who will now be entitled to the same treatment as full-time employees.

Employees who are dismissed are entitled to a written statement of the reasons for dismissal provided that they ask their employer for one. An employee whose complaint of unfair dismissal is upheld by an industrial tribunal is entitled to ask the tribunal for an order for reinstatement or re-engagement. If the tribunal thinks this is practicable it will make such an order. If a financial award is made instead, it will consist of a basic award, usually equivalent to the employee's entitlement to a redundancy payment, as well as a compensatory award to reflect the loss suffered by the employee as a result of dismissal.

Employers are required under the Act to disclose information to the representatives of recognised independent trade unions which it would be good industrial relations practice to disclose. There are, however, limits to this. For example, if the information would cause or is likely to cause substantial injury to the employer's business, or if the information had been given in confidence; if the information is about an individual or had been communicated in confidence, then the employer does not have to disclose the information.

The Employment Protection Act also established new rights for employees and these concern guarantee payments and maternity benefits. Employees who lose pay because of short-time working or lay-offs will be entitled to guarantee payments of a normal day's pay up to a maximum of £6.60 per day. Pay will be guaranteed for 5 days in any calendar quarter. An employer who lays off a worker or puts him on short-time because he cannot provide work must continue to make these payments for the period concerned subject to the limitation of five days in any calendar quarter. He is obliged to make

588

this payment only if a full-day's work is lost; a half-day does not count for payment.

As for maternity rights, this is the first time pregnant employees are protected by law. A pregnant employee may not be dismissed because of pregnancy. If she is dismissed because of this or for any reason connected with it, the dismissal will be unfair and she will be entitled to a reinstatement order, or compensation, from an industrial tribunal. If she is unable to work because of pregnancy she must be offered a suitable alternative job if one is available. If she cannot be found another job she is entitled to maternity pay and reinstatement.

An employee who is away from work because she is having a baby is entitled to pay for the first 6 weeks of her absence. The employer contributes nine-tenths of her normal pay less the amount of the Social Security maternity allowance (whether or not she receives this allowance). Employers will be able to claim from a central fund a refund of any maternity pay they have paid. All employers will contribute to the fund by means of a small addition to the employers social security contributions.

An employee who has been off work to have a baby is entitled to return to her old job at any time up to 29 weeks after her baby is born. Her employer must give her back her old job or if this is not possible then a suitable alternative job must be given to her. There are certain conditions though in which this takes place. The employee must have worked for the same employer for at least 26 weeks by the time she leaves. She must also work until 11 weeks before she expects to give birth. If she wants to return to work, she must tell her employer before she stops working that she will be coming back. An employer who takes on a temporary for an employee who is having a baby must tell the temporary employee at the start that the job is only temporary. When the original employee returns, the employer may dismiss the temporary worker without risking a complaint of unfair dismissal provided the method of dismissal is reasonable.

It should be borne in mind that the law can change and the

Department of Employment currently provides free guides on employment legislation which are quite easy for the "non-legal mind" to understand.

Finally, I would like to draw the reader's attention to the 1979 Agreement of the Joint Industrial Council for the Retail Meat Trade. This is an agreement made up from employers' representatives: Co-operative Employers' Association; National Federation of Meat Traders; Multiple Food Retailers Employers' Association; and the employees' representatives: Union of Shop, Distributive and Allied Workers; Transport and General Workers' Union.

The agreement lays down, among other things, minimum rates of pay, conditions of employment and annual holidays.

For example, the current agreement provides that in addition to Bank and Public Holidays, annual holidays with full pay shall be granted in each year (being the period from 1 April to 31 March), the entitlement shall be three weeks where an employee has at least 12 months continuous service as at 31 March preceding the holiday year. Where the annual holiday is two weeks or more, two weeks shall be granted during the period 1 April–31 October and shall be consecutive.

Regulations Affecting the Meat Trade

PER·DOCTRINAM·AD·DIGNITATEM

There are two problems when writing about the law on any subject —one is that the law changes and therefore any information given may be out of date and the other is that it is inclined to be tedious and boring.

In an attempt to overcome the first of these I have given at the end a short account of proposed legislation and details of some law that may change in the near future. In the case of the second item, I have tried to relate the law, as effectively as I can, to the operation and maintenance of an average meat business using experience gained from inspecting retail butchers' premises and their vehicles, and those premises both wholesale and retail in the Smithfield Market area.

591

In addition, having lectured in Food Law to meat students for some years, I can, I think, indicate some perennial problems which are bound to continue to present difficulty.

To avoid undue repetition, items of legislation are mentioned in full once, and are then given a "short title". Full details of each item are given at the end of the chapter.

Food Hygiene (General) Regulations 1970

There must be few meat trade employees who do not know of the Food Hygiene (General) Regulations 1970. Their knowledge may be substantial or confined to noting these words under a "Now wash your hands" notice in the water closet compartment. However, the regulations do form the basis for the construction of, and behaviour in, all food premises where food is prepared for sale, is handled, or is sold for human consumption.

Necessarily, the Hygiene Regulations are written in general terms, but if the average retail butcher's shop is used as a norm the upgrading required for larger premises will not be difficult.

At the outset, it must be appreciated that meat is naturally a "dirty product". It contains germs or bacteria, some of which may cause food poisoning. It must, therefore, be handled and prepared properly. It is also essential to realise that meat, like all food, is so different from any other item sold by retail. It is actually eaten.

CONTAMINATION

Meat, then, must be protected from sources of contamination. Some will say that meat is rarely eaten raw and therefore cooking will destroy germs. This is partly true, but one cannot justify meat being contaminated by a wound on the hand of the food handler exuding virulent discharges or pus over meat. Nor can one justify, for example, storing meat adjacent to a water closet compartment. One must consider the cross contamination angle when the bacteria in or on meat may be transferred to other foods upon which they may breed.

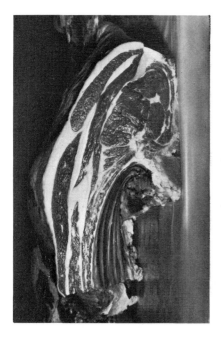

Eye muscle from a Charolais x Friesian, slaughtered at 259 days old at liveweight of 912 lb. Steer fattened at BOCM's Barlby Farm in Yorkshire

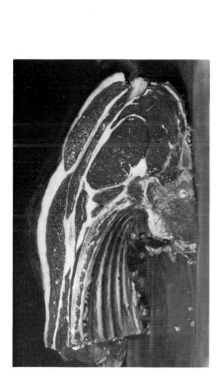

Eye muscle from a Hereford x Friesian, slaughtered at 308 days old at liveweight of 854 lb

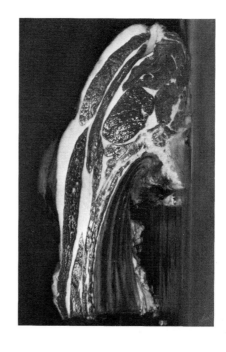

Eye muscle from a Devon x Friesian, slaughtered at 329 days old at liveweight of 833lb

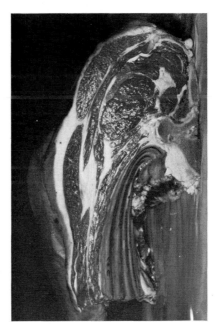

Eye muscle from a Sussex x Friesian, slaughtered at 308 days old at liveweight of 827 lb

Courtesy BOCM

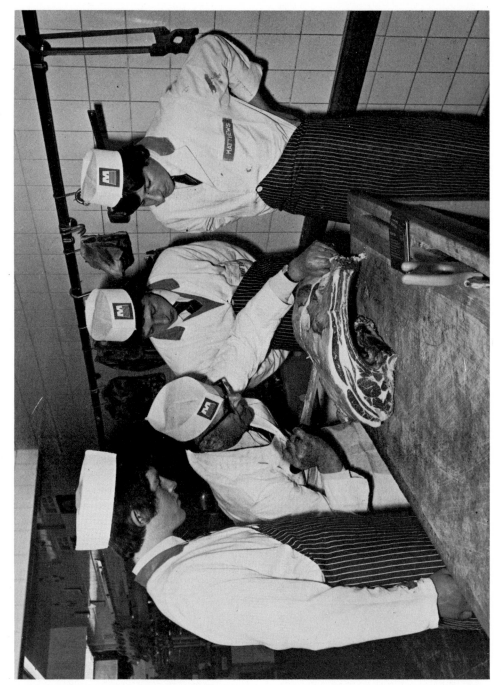

Trainee butchers under instruction

Courtesy Matthews Butchers Ltd.

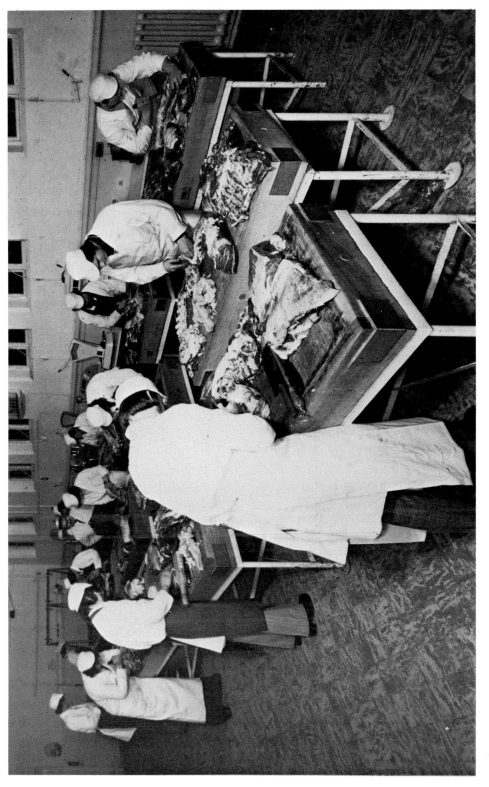

Courtesy The Institute of Meat

Students at work

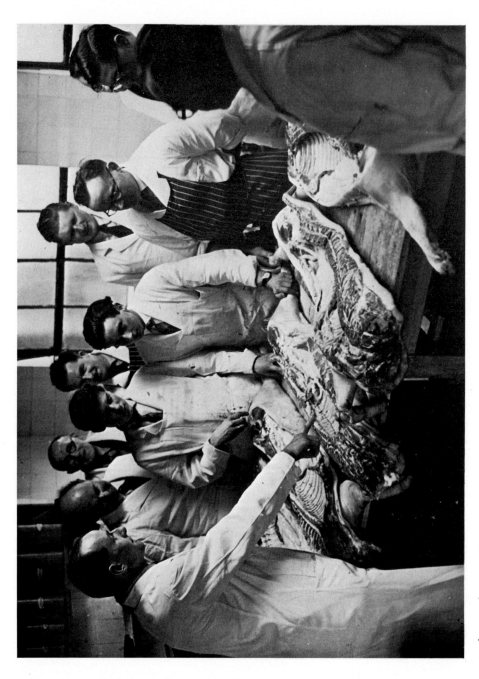

Students under instruction

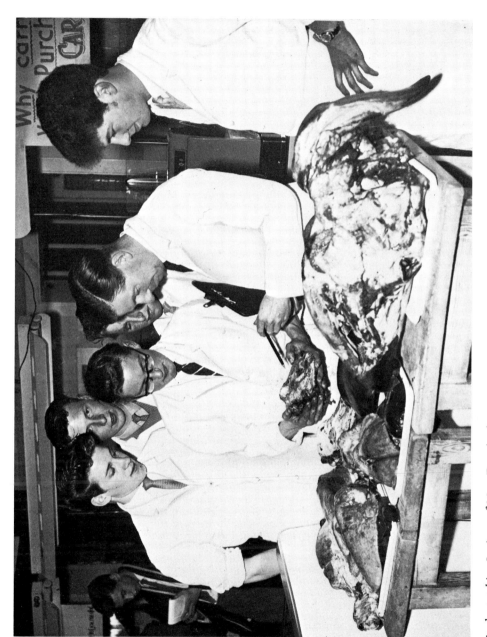

Students taking Institute of Meat Examination

Courtesy The Institute of Meat

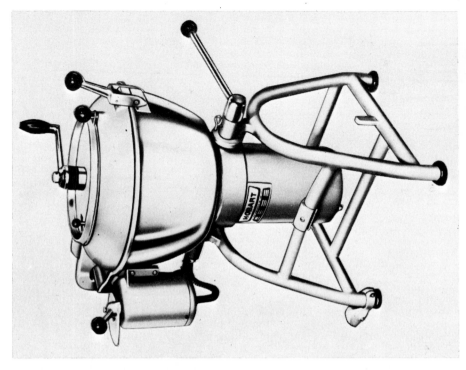

Vertical Cutter Mixer

Courtesy Messrs Hobart Ltd.

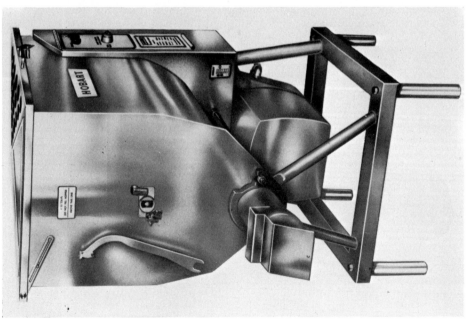

Mixer Grinder

Peerless Rotabowl in use

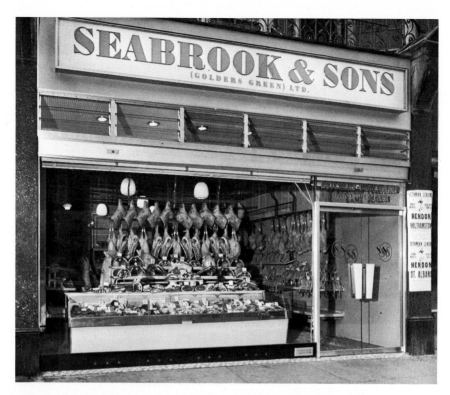

Retail shop window display

Courtesy Messrs Seabrook and Sons

Retail shop lay-out

Courtesy Messrs Dewhurst

The Hygiene Regulations place the onus on the food handler to keep the food, and therefore meat, free from contamination. In particular any food which is unfit for human consumption must be kept separate from fit meat, and all open meat must be effectively screened or protected. Of course, food displayed where customers can cough and sneeze over it must be covered and it should also be protected from street dust, etc.

There are, however, more insidious forms of contamination, e.g. sides of beef hung in a retail shop where customers can brush against them and meat in window displays affected by customers' hair and clothing as they point to their chosen joint. Although the situation envisaged does not often occur the 1970 Hygiene Regulations have prohibited open animal feed from being in a food room containing open food. It must be noted that "animal feed" does not include food which is fit for and normally used for human consumption. We should not assume either that the open topped chilled window display unit gives ultimate protection. It is highly questionable whether these units adequately chill the meat displayed and anyway flies can lay eggs, even at quite low temperatures, and these may well survive and even hatch whilst still in the unit.

As far as the problem of wounds is concerned, the essential point here is that organisms from wounds are, in the main, resistant to heat, being in the staphylcoccal group. Wounds must be covered by a waterproof dressing, and these should be kept in an accessible first aid box. However, it is realised that there are problems in the meat trade particularly in finding an effective waterproof dressing and finger stalls may have to be employed. If no adequate protection can be found it would be necessary for the employee to be temporarily engaged on other work. This is preferable to taking the risk of distributing food poisoning to the customers. (Incidentally no first aid box provided to comply with Office, Shops and Railway Premises Act requirements contains waterproof dressings. These must be added for the benefit of the hygiene regulations.)

To take the illness requirements a stage further, food handlers must notify their employer if they are suffering from or are a carrier of,

INFECTION

typhoid, para-typhoid, any other salmonella infection, dysentery or any staphylcoccal infection likely to cause food poisoning. Once received, the employer should pass the notification on to the proper officer at the local Borough or District Health Department. In the case of the employer himself being affected notification should, of course, be sent direct. It is often said that to report sundry stomach upsets and minor attacks of diarrhoea seems unnecessary, but it must be remembered that all the common types of food poisoning are known as bowel to mouth infections. Any contamination of any food, therefore, by a person who is excreting any of these organisms and who is at all lax in routine matters of personal hygiene may well endanger the health of a member of the public.

Protective Clothing

In the meat trade, the wearing of good protective clothing has been the custom for many years. It must, however, be replaced when it is dirty. The Hygiene Regulations require that the person handling the food must keep all parts of his overclothing as clean as may be reasonably practicable where contact with food is possible. Of course, in the meat trade blood will quickly soil clothing and it is not the intention of the regulations to make the butcher change his overall every few minutes. However, one must not go to the other extreme and use the same overall continuously or use it as a hand-wiping towel. Dirty overalls easily become a breeding ground for germs and this source must be eliminated. The need to maintain clean clothing extends to such things as leather belts and the string ties of aprons. Those involved in the carrying of unwrapped meat are required to protect their head and neck from contact with the meat. Again, there may seem little significance in this, but the appearance of a stray hair in any food is in itself repulsive apart from any contamination problem which may arise.

It is necessary for protective clothing to be properly stored when not in use and the Hygiene Regulations require that proper accommodation should be provided. In addition, outdoor clothing must not come into contact with, and therefore possibly contaminate, the work clothing or even the food itself. No outdoor clothing can be stored in a food room unless it is in a proper locker or cupboard. The well-known prohibition on smoking needs little explanation,

594

but it is worth making the point that smoking is not allowed in a room where there is open food even if no handling of the food is actually taking place at the time. This is to minimise the risk of contamination. A person who is smoking in a food room may well be tempted to handle food without extinguishing his cigarette. Also smoking is disallowed not because of any contamination of the food by ash which, it is agreed, may be a valid reason for complaint, but because it is possible for food to be contaminated by organisms present in the mouth which may be transmitted by the hands to the food.

PREMISES

To turn now to the construction of premises.

The most basic requirement is that the premises must not be in-sanitary. Unfortunately, this word is not defined, but it would generally be interpreted as premises which are so dirty, so poorly constructed and managed that there is a positive risk to health. All food rooms must be kept clean and in good repair. Further, all parts of the food rooms must be constructed of materials which can be effectively cleaned.

Walls

Generally, walls must be finished with an impervious surface and the regulations do not state anything more specific than that. However, the finish must be such that it can be maintained in an impervious state. A gloss-painted plastered wall may be perfectly adequate in some circumstances, but to have that finish at the back of a butcher's chopping block, for example, where damage is inevitable, would be unacceptable. The regulations also require the premises to be kept free, as far as possible, from any risk of infestation of rats, mice or insects. To this end, walls should not be battened out and finished with laminated plastic-surfaced hardboard. This finish looks attractive but between the boarding and the wall there is often a gap of at least three-quarters of an inch which may extend to three or four inches depending upon the state of the wall. These gaps encourage infestations by mice and insects and, in one case in my experience, rats. It is my opinion that such wall construction does not comply with the Hygiene Regulations.

595

Floors

The some time excessive use of water presents particular problems when considering floor construction. Firstly, it is necessary to effectively seal the joints between the wall and the floor at skirting level. The use of a proper skirting tile achieves this and also provides a coved finish which makes cleaning easier. Unfortunately, it is still common practice to litter the floor of butchers' premises with sawdust but I am convinced that careful management and the selection of proper floor tiles could make this practice obsolete. Every effort should be made to minimise the occurrence of accidents but not even this can justify the liberal scatterings of sawdust that occurs in most butchers' shops. Apart from the fact that sawdust encourages food handlers to dispose of fat and other trimmings by throwing them on the floor, the actual dust which rises as people walk across the floor can contaminate the meat. If it must be used then it must be confined to areas of major need, that is around cutting areas for example, and then in minimum quantities. When selecting a floor finish for a butcher's shop or meat preparation area, advice should be sought from one or two of the major tile manufacturers, most of whom operate a free consultancy service, in order that the best finish is provided for the job in hand.

Ceilings

In the construction of ceilings, one is faced with another perennial problem. With mechanised building methods a system of construction has developed which encourages the construction of suspended ceilings. This allows the shell of the building to be constructed in an adequate, but nevertheless rough, manner, knowing that it will all be covered up eventually. It also means that drainage pipes, heating ducts, electric conduits and the like can be hung from the underside of the ceiling slabs. A suspended ceiling then hides all these fittings. However, it is my experience that any enclosed space to which access is not normally necessary or available is likely to be a source of trouble. Whilst the Hygiene Regulations require that the ceiling should be kept clean a suspended ceiling can, on many occasions, do nothing but encourage rodent and insect infestations. It is wise, therefore, to consider the advantages and disadvantages of all types of construction, and after being influenced by the process being carried on, a decision made which shows the major advantage.

596

Mention has already been made of the laziness which accrues when sawdust is used. This general slovenly behaviour presents itself again when no effort is made to tidy the premises or keep them free from accumulations of refuse and rubbish. Sawdust makes premises look untidy and hence may attract other untidiness. It is not uncommon to find rear yard areas containing rubbish of every description. Whilst I am not advocating the disposal of any material which may have further use, it must be remembered that any accumulation which is left undisturbed for a matter of months will inevitably attract rodent infestations which may eventually have serious effects upon the food business. All unnecessary rubbish must therefore be removed.

Waste Disposal

All food premises where open food is handled require a wash hand basin to be fitted and provided with an adequate supply of hot and cold or warm water. The Hygiene Regulations require that its position shall be conveniently accessible to the food handlers. This will depend entirely upon the layout of the premises and the food business being carried on. The essential criteria is that all food handlers must be given the opportunity to wash their hands as necessary. The facilities must be kept clean and provided with the essential soap, towel and nailbrush. It is vital that handwashing is confined to the wash hand basin; to use a sink where either equipment has just been washed up, or in which food is being cleaned is to give an opportunity for germs to be transferred, so commencing a possible food poisoning chain. As has been mentioned earlier, there is a very real need for food handlers to realise the necessity to wash their hands after using the sanitary accommodation. Notice to this effect must be placed in water closet compartments, and the management have an obligation to ensure that their staff are properly educated in this way. Most health departments make small adhesive notices available on request.

Hand Washing

The provision of sinks is based on the same standard—there should be sufficient for the size of the premises, the type of process being carried on and the amount of equipment to be cleaned. They can be of any size and type, but obviously, those which suit the particular situation

Sinks

597

should be chosen. Again, these fittings must be properly supplied with hot and cold or warm water and, like wash hand basins, be properly fixed to the drainage system to comply with the Building Regulations or the GLC Drainage Byelaws.

Lighting

The fact that the Hygiene Regulations are general has encouraged the legal draughtsmen to require that the lighting of food premises should be "suitable and sufficient". The Illuminating Engineering Society recommend 37·2 lumens per square metre (square foot) for areas of food cutting and preparation. This standard is good, but it must not be forgotten that the value given is for the general level of illumination whereas, by bad positioning of light fittings in relation to the working position this standard may not be achieved. Consideration must be given to lighting the actual area of work without involving discomfort glare. After all it is quite usual for an office desk to be provided with a reading lamp, but a very low lighting level is often tolerated at work points where sharp knives and saws are employed.

The Offices, Shops and Railway Premises Act 1963 does give the Minister power to make regulations specifying lighting levels for work positions. However, to date, no regulations have been made and it is understood that none are anticipated.

Equipment

The one matter that still has to be mentioned is the cleanliness of equipment. In the meat trades, there is still an attraction for using wood cutting boards, chopping blocks and benches, and to the experienced butcher there is probably nothing to supersede them. If they still have to be used, and I believe that proper chopping blocks will be with us for some years yet, then attention must be given to using them properly and to keeping them clean. Where possible, manufactured cutting boards should be used but I acknowledge that the main objection is that they are different from the more readily acceptable wooden surface. However, having overcome the unfamiliarity I am sure that they are more hygienic, certainly easier to clean, and they do not absorb moisture.

Flies and other flying insects often cause problems and some form of control is usually necessary. There is no doubt that the most satisfactory method is to use the fitments which attract the insects to an ultra violet light source which is placed behind an electrically charged grill. The body of the fly is retained in the catch tray and thus cannot contaminate any food product. Of course, like all items of equipment maintenance is essential and one must not assume that the quality of light being emitted from the tube is sufficient to attract flies. These tubes must be renewed as advised by the manufacturer. In addition, the attraction area of one of these fitments is limited. One must not expect one device in a huge factory area to be completely effective. It is appreciated that the cost of these fitments may be prohibitive, but some agencies do make them available on hire.

Before leaving the question of hygiene in meat preparation, one must mention the contamination of working surfaces by such 'in-shop processes' as the drawing and cleaning of chickens. There is still considerable disquiet about the apparent continuing incidence of salmonella in battery reared chickens. Until the problem is solved those engaged in the handling of these birds must be aware that this infection emanates from the gut and its contents, and the action of drawing the birds infects the hands, the block or cutting board and any knives or other equipment used. It is therefore vital that there must be no dual use of equipment and cutting surfaces. Poultry handling must be a separate activity and food handlers must wash their hands after handling chickens, etc. As we have seen earlier in this chapter, food handlers must not be content just to wipe their contaminated hands on an overall or wiping cloth for these will soon become very active reservoirs of infection which cannot be tolerated in food premises.

Poultry Handling

The same sort of cross contamination arises from the handling of raw and cooked meats. Cooked meats must be handled separately using separate equipment and if possible, by separate members of staff. Without this separation it would seem that there is a clear contravention of the Hygiene Regulations.

The appearance of domestic animals in food premises also cannot be

599

tolerated. If mice infestations occur the source should be established and a crash baiting programme instigated rather than a cat being depended upon to clear the problem. I have also sincere doubts about the wisdom of continuous baiting; far better to keep the premises in good order or to deal with mice as they occur.

Food and Drugs (Control of Food Premises) Act 1976

This Act follows the power given to London Authorities and to some other Councils by virtue of Private Acts, to apply to the Court for Closure of Food Premises where there exists a danger to health. The Act applies to England and Wales.

It can only be used in conjunction with alleged offences under the Food Hygiene Regulations and then only after the defendant has been convicted by the Court. In addition the alleged offences must include "the carrying on of a food business at any insanitary premises or at any premises the condition, situation or construction of which is such that food is exposed to risk of contamination". Then, if the Court is satisfied that food which is or is likely to continue to be prepared, stored, sold or offered or exposed for sale at those premises *and* that by reason of their situation, construction or insanitary or defective condition or by the condition of the equipment or by infestations of vermin or accumulations of refuse, the carrying on of a food business would be a danger to health they may, subject to an application from the Local Authority, issue a closure order. However, this can only be done if the Local Authority has notified the defendant in writing not less than fourteen days before the hearing that this application for closure will be made on conviction of the alleged offences. Representation from both the Local Authority and the Owner will, of course, be heard by the Court. This seems complicated, verbose and limited in use. The fact that the fourteen days' notice has to be given may even delay the hearing and so lengthen the time before the case can be heard.

However, the Act provides for a speedier interim procedure where it is alleged that offences under the Hygiene Regulations still exist but by their nature or seriousness the handling of food there would

600

present "an imminent risk of danger to health". In these circumstances the period of notice required to be given is three clear days and provided that the Court is satisfied that this has been given and that the information is or has been laid by the local authority it may make "an emergency order" to close the premises until a full hearing can take place. There are compensation clauses to cover the eventuality of the Court at the subsequent hearing failing to uphold the emergency order.

In the case of an emergency order being granted a copy of the Order has, inter alia, to be fixed to the premises.

In both cases the work required to remove the dangers to health has to be specified by the local authority and when the work has been carried out the owner of the business may apply for the Order to be rescinded. Normally the authority's Chief Environmental Health Officer will be appointed to inspect the premises and issue a revocation of the Order if he is satisfied that the danger to health has been removed.

Food and Drugs Act 1955

The Food Hygiene Regulations are made under the provisions of Section 13 of the Food and Drugs Act 1955 but the Act itself puts certain obligations directly upon those engaged in food businesses. The most well used section is Section 2. Basically, this protects the purchaser from being prejudiced when sold food which is not of the nature, substance or quality demanded. In Hoyle v. Hitchman (1879) 4 QBD Lush J. defined prejudice as "that which is suffered by anyone who pays for one thing, and gets another of inferior quality".

It is not easy to define "Nature", "substance" or "quality" and on examination of reported cases, there is no unanimity of enforcement.

Those engaged in the selling of food must ensure that products are delivered as requested and that the food is sold in a satisfactory and

Customer Protection

proper condition. Occasionally, in the meat trade, a customer is sold, say, a shoulder of lamb in which is found an abscess or a liver with a parasitic cyst in the substance. In these cases the vendor would probably have a defence at law by saying that he could not have been aware of its presence when he sold the product. In addition, he could claim warranty under Section 113 of the Act as explained later in the chapter.

Mention must be made of the power of seizure of food and the subsequent action. However, most Food and Drugs Authorities rarely have to use this power, relying on their system of voluntary surrender to prevent unfit food being sold.

Food Examination

Section 9 of the Food and Drugs Act states that an Authorised Officer may examine food intended for sale or sold for human consumption and that he may seize it if it appears to him to be unfit for human consumption. He may then remove it and take it before a Justice of the Peace for condemnation.

It is an obligation of the Authorised Officer to officially notify the person from whom the food was seized of the date and time when it will be taken before a Justice of the Peace because he has the right to be present at that time. In my experience the vendor is reluctant to appear but it is a protection given to him by the law. Further, he is allowed to take with him anybody who he thinks may help his cause, for example, his solicitor. If a Justice of the Peace refuses to condemn the food and it has deteriorated in the time between seizure and the appearance before him, the local authority may be liable to pay compensation. Once the food has been condemned legal proceedings will follow at which the vendor will, in all probability, be legally represented. (For interpretation of the term Authorised Officer reference should be made to the Authorised Officers (Meat Inspection) Regulations 1974 (as amended).)

Now to that part of the Act that requires positive action by those involved in meat preparation.

Some of the processes in the meat trade need to be carried out under controlled conditions, therefore the manufacture and preparation of sausages and any potting, pressing, preserving or pickling of food intended for sale should only be done on registered premises. It is obvious that the range of activities covered, and for which registration is necessary, is enormous and it extends well beyond the meat trade, but, remembering that preservation includes cooking, one would include the boiling of hams as well as the brining and salting of meat and the manufacture of products like brawn etc. Incidentally, there is an exemption for catering premises so that the cooking of meat for a meal does not require the premises to be registered.

Usually the local authority will provide a form of application and will issue a formal Certificate of Registration. If, however, refusal of registration is contemplated, the local authority is bound to serve on the applicant a "time and place notice" which will detail the date, the time and place when the application will be considered by the local authority. Again the applicant has the right to attend and may bring with him his solicitor, other witnesses and any other person who he thinks will be able to assist his application. If after hearing the applicant, the council decides not to issue the registration or to cancel the present one, the applicant has fourteen days in which to request the local authority to provide him with the reasons for their refusal, and they must accede to this request within forty-eight hours. Any appeal against the council's decision under this section is to a magistrate's court. Note that it is the premises that are registered in the name of the present occupier. Any incoming occupier is obliged to notify the local authority.

REGISTRATION OF PREMISES

Local authorities which are also Food and Drugs Authorities or who engage in this work on an agency basis, are entitled to purchase or take samples of food for analysis by the appointed public analyst for the area.

The general system is that the Sampling Officer, and this may be a Public Health Inspector, an Environmental Health Officer or Weights and Measures Inspector, will purchase a product, formally declare himself to the person who served him and then proceed to

Food Analysis

divide the sample into three parts, ensuring that each part is representative of the whole. Finally he will label and seal each part.

During this process, either the owner or the manager of the shop has the right to observe the procedure. This is not necessarily to check on the Sampling Officer but more to acquaint himself with the whole sampling process. All parties concerned in a formal sample must have an equal opportunity of knowing that it has been carried out and the Sampling Officer must inform the person whose name and address appears on the container, within three days, to this effect.

This formal sampling process is the only one from which legal proceedings may arise. Of the three parts one is left with the vendor so that he may have it independently analysed if he so desires. The public analyst is sent one part and the third part is retained by the Food and Drugs Authority for eventual production in court if proceedings are instituted.

The wisdom of the analysis of the vendor's part is exemplified in the notion that the analysis of the part sent to the public analyst may then be challenged if it is in the vendor's interest. If this happens the court must direct that the third part which has been in the possession of the Food and Drugs Authority must be sent to the government chemist for an independent analysis.

When the formal sample is carried out the Sampling Officer will doubtless ask certain questions. These must be answered, subject always to the protection given to the vendor by the law that to answer them he may incriminate himself. Other than this, failure to answer could result in a charge of obstruction under Section 105 of the Act.

If legal proceedings have to be implemented, the information has to be laid before the court within two months of the sample being taken.

It is fair to state that most Food and Drugs Authorities practice what is known as "informal sampling". This is done primarily to save time and to spread the products tested over a wider range. It is not possible for legal proceedings to directly arise from such sampling. It is quite informal and the vendor may not even know that such sampling has taken place.

It must also be stated that when a sampled product is sold in the same state as it was purchased, for example canned or packeted goods, the Food and Drugs Authority may by-pass the vendor in any legal proceedings, and serve the summons directly on the person more responsible, namely the manufacturer. However, the authority has to be satisfied that the circumstances are correct and that a warranty from the manufacturer to the vendor is in existence.

If, for example, a company is prosecuted for adding preservative to minced meat, the company may prefer to rely on the warranty provisions of the Act.

This was the basis of the recent Trade Descriptions Act case, Tesco v. Nattrass (1971 2 All E.R. 127; 1971 2 W.L.R. 1166), where a similar section to that in the Food and Drugs Act is in existence. The outcome was finally decided in the House of Lords.

The Food and Drugs Act gives an Authorised Officer power to enter food premises at all reasonable times. Normal business hours, of course, come within reasonable hours but it could also apply to a period when a shop display is being prepared before the shop is open for business and when meat preparation is taking place behind locked doors, for example when orders are being prepared. In fact, the power of entry provision applies when the food business is active.

ADDITIVES

Butchers still sell a number of raw meat products. These include minced meat and steak and kidney. In these cases, as the meat is still raw and unprocessed (mincing and cutting cannot be regarded as a "process" as defined in Regulation 2 of the Colouring Matter in Food Regulations 1973) it is unlawful to add any colour, permitted or otherwise, to the meat. This prohibition extends to game and poultry again in the raw and unprocessed state.

Colour

Not only does the addition of colour amount to an offence for which there is a penalty, but there is also a daily penalty if the offence continues after conviction. Also the law makes raw meat to which colour has been added unfit for human consumption within the meaning of Sections 8 & 9 of the Act.

The acceptance that meat should only be sold in its original state and should not be disguised extends to the fact that any raw and unprocessed meat intended for sale for human consumption must not have added to it substances known as Ascorbic Acid, Erythorbic Acid, Nicotinic Acid and Nicotinamide or any salt or other derivative of these substances. This is covered by the Meat (Treatment) Regulations 1964.

Reference has been made to the manufacture and preparation of sausages and to various processes associated with meat by-products which amount to registerable processes under Section 16 of the Food and Drugs Act. In all these cases attention must be paid to the Preservatives in Food Regulations 1974.

Preservative

Sausages and sausage meat can contain sulphur dioxide as a preservative, but only subject to a maximum of 450 milligrams per kilogram. In most cases the preservative is already included in the rusk etc. in the prescribed quantity. Any increase in the proportion of this ingredient will, therefore, induce additional preservatives which may result in the maximum level being exceeded in the final product. For the purpose of these regulations, sausage and sausage meat means a mixture of raw meat, cereal and condiments, and includes hamburgers, lamburgers, beef burgers and similar products.

These regulations also impose limits for the sodium and potassium nitrate and nitrite levels for cured meat (including bacon or ham). These are 500 and 200 milligrams per kilogram respectively.

As raw meat is not listed in the Schedule to the Regulations it becomes an offence to add preservative to raw meat including minced meat.

Finally, the new Miscellaneous Additives in Food Regulations 1974 list 130 additives as being permitted in food but which are subject to control for the first time.

The Sausage and Other Meat Products Regulations 1967 detail inter alia the composition of sausages which include by definition chipolatas, frankfurters, polony, saveloy and salami, and the labelling, description and advertisement thereof.

To limit discussion to those types of sausage more directly associated with the normal butchery trades, the description "pork sausage" or "pork sausage meat" can only be used if there is a meat content of 65% of which 50% is lean meat, and of which 80% is pork. For beef and other sausages, except frankfurters, vienna sausage and salami, the meat content is reduced to 50%, of which half must be lean meat and half must be of the named meat.

Having briefly described this rather complicated law, the manufacturer would be best advised to obtain the services of a private analyst who would formally assess the product, interpret the regulations and advise accordingly.

It follows from this that one must consider the labelling of products generally and sausages in particular.

The relative law here is the overriding Section 6 of the Food and Drugs Act. No food shall be falsely described or labelled in such a way that it is likely to mislead a prospective purchaser as to its nature, substance or quality.

Beyond this, the Labelling of Food Regulations 1970 (amended 1972) give much more specific instructions. To take first the display of raw meat in a butcher's shop.

Raw meat description

Displayed with the meat should be a ticket which gives an accurate description of the item referred to. Inevitably, there are bound to be differences in description depending upon geographical areas but

607

one must accept these acknowledging that the essential attribute of a description must be that it is sufficiently specific to indicate to the intending purchaser the true nature of the food. This is required in addition to any indication of origin required by the Trade Descriptions Act 1972. This labelling of non-prepacked food extends to sausages etc. and other meat products displayed for sale in that condition. But products consisting of two or more ingredients need to have displayed, in addition to the appropriate designation (the official term given to the name describing the true nature of the food), a declaration if permitted preservatives, antioxidents or colouring matters are present in the food. This information must be clearly visible to the intending purchaser.

Pre-packed food

If food is pre-packed more stringent conditions apply. Food sold in this condition generally has to bear an appropriate designation, a list of ingredients (the greatest being stated first) and the name and address of the packer or labeller.

The Labelling of Food Regulations lay down minimum type sizes for the appropriate designation and the list of ingredients, and also quite detailed instructions as to the conspicuousness of this information and its position on the label. The only exception is that food pre-packed on the premises from which it is sold need not comply with this part of the regulations.

There are laws governing the wholesaling of food and, in general, these require the product to be labelled as for retail sale. Alternatively, it can be sold in an unlabelled condition and then within fourteen days the purchaser has to be furnished with such details as will allow him to label the goods to comply with the regulations.

Tenderised and other meat products

Finally on labelling, meat which has come from animals which have been treated with proteolytic enzymes must have the word "tenderised" included in the appropriate designation.

There are three other very complex sets of regulations which govern the composition and labelling of meat products. These are the

Canned Meat Product Regulations 1967 (amended 1968, the Meat Pie and Sausage Roll Regulations 1967 and the Fish and Meat Spreadable Products Regulations 1968).

As these regulations are very detailed any attempt to indicate their contents would tend to over-simplify the issue, and possibly lead the reader into making false assumptions. Sufficient to say, therefore, that the law does prescribe legal standards for the products mentioned in these regulations and in order to evaluate how they can affect any particular product, a very detailed study would have to be made.

The Materials and Articles in Contact with Food Regulations 1978 are designed to produce a framework to ensure that materials and articles that come into contact with food are manufactured so that they do not transfer their constituents to food in quantities which could endanger human health or bring about a deterioration in the organoleptic characteristics of food or unacceptable changes in its nature, substance or quality.

DELIVERY VEHICLES

The Food Hygiene (Markets, Stalls and Delivery Vehicles) Regulations 1966 (as amended) put mobile shops, street market stalls and meat delivery vehicles on the same basic hygienic footing as fixed premises.

The stall or vehicle must be constructed so that the food is not exposed to a risk of contamination, and in addition, of course, the stall or vehicle must again not be insanitary.

As regards the cleanliness of equipment, protection from contamination, personal cleanliness, the wearing of protective over-clothing by the food handlers, the carriage and wrapping of food and persons suffering from certain infections, the provisions are much the same as the Food Hygiene (General) Regulations. All stalls and vehicles must have the name and address of the person carrying on the food business legibly and conspicuously displayed and an address at which the stall or vehicle is normally kept or garaged (if different).

| **Cleanliness and washing facilities** | The attempt to put stalls and delivery vehicles on a par with fixed premises extends to the necessity to provide handwashing facilities, washing-up facilities and first-aid materials. There are specially manufactured units available to suit these requirements providing for the hot water to be stored in an insulated container or in the case of vehicles, heated by the engine. In the case of a stall operated by a butcher or poulterer, the local authority may grant an exemption from the need to provide a sink at the stall, provided that it is satisfied that satisfactory facilities are available in a nearby food premises. No such exemption, however, would be available for a mobile butcher's shop. |

Vehicles used entirely for the delivery of meat are exempted by Regulation 24 (as amended) from the need to provide these facilities on the vehicle, subject to certain conditions prevailing. These are that the vehicle operates between places or premises at each of which handwashing facilities, a sink and first-aid facilities are provided, that these facilities are conveniently, readily and freely available, and that the meat on the vehicle is handled only for the purposes of loading and unloading. This latter exemption will mean that most meat delivery vehicles will not have to provide facilities actually on the vehicle although any operator which does do this would obviously be at an advantage.

Stalls, mobile shops and delivery vehicles have to be provided with adequate lighting. In the case of vehicles, this may mean that additional lights have to be provided in the body of the vehicle.

All stalls should be provided with adequate receptacles for waste and other refuse and care should be taken to ensure that undue amounts of refuse do not accumulate at the stall.

In the case of a street market stall selling raw and cooked meats, it will be necessary to enclose the stall top, back and sides.

Construction of vehicles

Finally, these regulations give specific details of construction for vehicles used for the transport of meat. These are in addition to those requirements in the main body of the Regulations given above.

Excepting vehicles used for the sole transport of unskinned rabbits or hares or unplucked game and poultry, it will be necessary for the vehicle to be completely enclosed, to be provided with an impervious floor (or fitted with movable duckboards), and provided with covered receptacles to separately contain all offal. It will be necessary for all receptacles, duckboards, rails, hooks etc. to be kept thoroughly clean and properly maintained. Also the bins provided for the reception of offals must be used and kept clean. The need to use these separate containers does not extent to vehicles used solely for the transport of uncleaned tripes, stomachs, intestines or feet and unskinned or unscalded heads. Also frozen offal which remains frozen and the giblets which have not been removed from poultry are exempted from the need to be carried in separate offal bins.

It must be noted that the provisions of the Food and Drugs (Control of Food Premises) Act, 1976 referred to earlier apply equally to "stalls" as defined in the Markets, Stalls Regulations.

Date marking

In 1972 the well publicised question of the open date marking of pre-packed food was the subject of the Food Standards Committee's Report. It has recommended that pre-packed short life foods, that is up to three months' shelf life, should bear a "sell by date", that foods subjected to a special process, for example vacuum-packed bacon, should bear an "open by date" and that long life food should be marked with a manufacturing or packaging date. There are certain exceptions for foods such as wine and sugar which do not deteriorate with age provided that they are properly stored.

Following on from the FSC Report on the Date Marking of Food, the Steering Group on Food Freshness produced its final report in 1976. This report deals with various aspects of food freshness and with matters that have arisen from the earlier Report. The discussions now includes packaging, stock control and distribution, retailer education and informative labelling together with that vital link in the food freshness chain—temperature control.

No draft regulations have yet been published. It does seem, however, that a good number of manufacturers of perishable foods are already offering "sell-by" information in a comprehensible form although prepacked sausages, meat pies etc. still use the Meat Manufacturers Association's coded "last day for delivery" scheme. With the greater awareness on the part of the public it may be that this voluntary arrangement will suffice for some years to come.

Acts of Parliament:
Statutory Instruments and other Items of Reference

The Food and Drugs Act 1955 4 Eliz 2 Ch. 16 HMSO

Food and Drugs (Control of Food Premises) Act 1976 Eliz 2 Ch. 37 HMSO

Food Hygiene (General) Regulations 1970 S.I. 1970 No. 1172 HMSO

Food Hygiene (Market, Stalls and Delivery Vehicles) Regulations 1966 S.I. 1966 No. 791 HMSO

Food Hygiene (Markets, Stalls and Delivery Vehicles) (Amendment) Regulations 1966 S.I. 1966 No. 1487 HMSO

Colouring Matter in Food Regulations 1973 S.I. 1973 No. 1340 HMSO

Colouring Matter in Food (Amendment) Regulations 1974 S.I. 1974 No. 1119 HMSO

Colouring Matter in Food (Amendment) Regulations 1975 S.I. 1975 No. 1488 HMSO

Colouring Matter in Food (Amendment) Regulations 1976 S.I. 1976 No. 2086 HMSO

Preservatives in Food Regulations 1979 S.I. 1979 No. 752 HMSO

Meat Pie and Sausage Roll Regulations 1967 S.I. 1967 No. 860 HMSO

Canned Meat Product Regulations 1967 S.I. 1967 No. 861 HMSO

Canned Meat Product (Amendment) Regulations 1968 S.I. 1968 No. 2046 HMSO

Sausage and Other Meat Products Regulations 1967 S.I. 1967 No. 862 HMSO

Sausage and Other Meat Products (Amendment) Regulations 1968 S.I. 1968 No. 2047 HMSO

Fish and Meat Spreadable Products Regulations 1968 S.I. 1968 No. 430 HMSO

Meat (Treatment) Regulations 1964 S.I. 1964 No. 19 HMSO

Authorised Officers (Meat Inspection) Regulations 1978 S.I. 1978 No. 884 HMSO

Miscellaneous Additives in Food Regulations 1974 S.I. 1974 No. 1121 HMSO

Miscellaneous Additives in Food (Amendment) Regulations 1975 S.I. 1975 No. 1485 HMSO

Manchester Corporation (General Powers) Act 1971. 1971 Chap lxvii

Greater London Council (General Powers) Act 1973. 1973 Chap xxx

Food Hygiene Code of Practice No. 5 "Poultry Dressing and Packing" HMSO

Food Hygiene Code of Practice No. 8 "Hygiene in the Meat Trades" HMSO

Food Standards Committee Report on the Date Marking of Food and Steering Committee Review on Food Freshness 1972. HMSO

Health and Safety at Work etc. Act, 1974 Eliz 2 Ch. 37 HMSO

Health and Safety (Enforcing Authority) Regulations 1977 S.I. 1977 No. 746 HMSO

Food Standards Committee Report on Novel Protein Foods 1974. HMSO

Bell and O'Keefe's Sale of Food and Drugs (14th edition) Butterworth & Co. (Publishers) Ltd, and Shaw & Sons Ltd

Food Labelling. The Food Manufacturers Federation (Inc) 1–2 Castle Lane, London SW1E 6DN

Is it Legal? Institute of Practitioners in Advertising and the Advertising Association
Trade Descriptions 68–72 by Bowes Egan, The Commercial Publishing Co.
The Cleaning, Hygiene and Maintenance Handbook. T. McLaughlin B.Sc.,
    ARIC Business Books Ltd, London
A Guide to Improving Food Hygiene. Graham Aston and John Tiffney, North-
    wood Publications Limited
Food Hygiene and Food Hazards. A. B. Christie and Mary Christie, Faber
Leaflet—10 Point Code for Food Trade Workers—The Health Education Council
Film—Key to Cleanliness—J. Lyons & Co. Ltd
Film—The Germ War—The Gas Council
Film—Alice in Label Land—Central Office of Information, for the Ministry of
    Agriculture, Fisheries and Food
Film—The Germ that came to Dinner—Central Film Library

INDEX

615